ADVANCES IN
HEAT TRANSFER

Volume 29

Advances in
HEAT
TRANSFER

Volume Editor / Serial Associate Editor

George A. Greene
Department of Advanced Technology
Brookhaven National Laboratory
Upton, New York

Serial Editors

James P. Hartnett
Energy Resources Center
University of Illinois at Chicago
Chicago, Illinois

Thomas F. Irvine, Jr.
Department of Mechanical Engineering
State University of New York at Stony Brook
Stony Brook, New York

Serial Associate Editor

Young I. Cho
Department of Mechanical
   Engineering
Drexel University
Philadelphia, Pennsylvania

Volume 29

ACADEMIC PRESS
San Diego   Boston   New York
London   Sydney   Tokyo   Toronto

This book is printed on acid-free paper.

Academic Press
525 B Street, Suite 1900, San Diego, California 92101-4495, USA
http://www.apnet.com

Academic Press Limited
24-28 Oval Road, London NW1 7DX, UK
http://www.hbuk.co.uk/ap/

International Standard Serial Number: 0065-2717
International Standard Book Number: 0-12-020029-5

PRINTED IN THE UNITED STATES OF AMERICA
  97 98 99 00 01 02 BC 9 8 7 6 5 4 3 2 1

CONTENTS

Heat Transfer from Heat-Generating Pools and Particulate Beds

V. K. DHIR

Hydrogen Combustion and Its Application to Nuclear Reactor Safety

JOHN H. S. LEE AND MARSHALL BERMAN

Heat Transfer and Fluid Dynamic Aspects of Explosive Melt–Water Interactions

D. F. FLETCHER AND T. G. THEOFANOUS

Heat Transfer During Direct Containment Heating

Martin M. Pilch, Michael D. Allen, and David C. Williams

CONTRIBUTORS

Numbers in parentheses indicate the pages on which the authors' contributions begin.

MICHAEL D. ALLEN (215), Sandia National Laboratories, Albuquerque, New Mexico 87185.

MARSHALL BERMAN (59), Sandia National Laboratories, Albuquerque, New Mexico 87185.

V. K. DHIR (1), Mechanical and Aerospace Engineering Department, University of California, Los Angeles, California 90095.

D. F. FLETCHER (129), Department of Chemical Engineering, University of Sydney, New South Wales 2006, Australia.

JOHN H. S. LEE (59), McGill University, Montreal, Quebec H3A 2K6, Canada.

MARTIN M. PILCH (215), Sandia National Laboratories, Albuquerque, New Mexico 87185.

T. G. THEOFANOUS (129), Departments of Chemical and Mechanical Engineering, Center for Risk Studies and Safety, University of California, Santa Barbara, California 93106.

DAVID C. WILLIAMS (215), Sandia National Laboratories, Albuquerque, New Mexico 87185.

PREFACE

Nuclear reactor technology has been a field of considerable interest and research in the United States and abroad during the previous three decades due to an increasing reliance upon nuclear power to meet the escalating global demand for electricity within both the developed and under-developed countries of the world. As would be the case for any high-risk industry, the commerical nuclear power industry, as well as governmental regulators have placed a great deal of emphasis on reactor accident analysis and safety research. Since the nuclear reactor accidents at the Three Mile Island Nuclear Power Station in the United States in 1979 and at the Chernobyl Nuclear Power Station in the Ukraine in 1986, a great deal of effort has gone into fundamental research in severe nuclear reactor accident phenomenology in order to improve the understanding of the physics and chemistry that would occur in the event of another severe nuclear reactor accident.

The purpose of this volume is to present the fundamentals and applications of heat transfer in nuclear reactor safety in four important and prominent areas of severe accident phenomenology: heat transfer from heat-generating pools and particulate beds, hydrogen combustion, explosive melt-water interactions, and heat transfer during direct containment heating. The authors have been selected on the basis of their contributions to the literature in the subject area and their standing in the nuclear safety community as preeminent experts in their fields. The chapters represent the most thorough, in-depth, and comprehensive presentations on severe accident phenomena published to date. This topical volume is the twenty-ninth in the *Advances in Heat Transfer* series and is expected to serve as a desktop reference book for scientists and engineers in the nuclear industry, both in research as well as in operations. The volume can also serve as a primary text for graduate-level courses in nuclear safety and for guidance in thesis research in these areas.

The editors wish to thank Helen K. Todosow for compiling the guide for document retrieval and to congratulate the authors for a fine job in preparing the manuscripts.

G. A. Greene

Heat Transfer from Heat-Generating Pools and Particulate Beds

V. K. DHIR

Mechanical and Aerospace Engineering Department, University of California, Los Angeles, California

I. Introduction

Natural convection in volumetrically heated layers is of interest in many fields of science and engineering, including chemical, geophysical, and nuclear. However, during the last two decades, it is nuclear reactor safety that has motivated research in this area of heat transfer. The research on dryout heat fluxes in volumetrically heated porous layers has also received impetus from the need to analyze severe core damage accidents in liquid-metal-cooled reactors and light-water reactors.

Prior to the accident at Three Mile Island, Unit 2 (TMI2), most of the research on natural convection in volumetrically heated pools and on dryout heat flux in particulate layers was keyed to resolution of issues related to postaccident heat removal (PAHR) in liquid-metal-cooled fast breeder reactors. For these loop type reactors two kinds of hypothetical accidents were considered [1]. In the loss-of-flow accident (LOF), it was assumed that the pumps driving the sodium flow through the core lost power and the protective systems failed to scram the reactor. In the transient-over-power (TOP) accidents, a positive reactivity was assumed to be inserted in the absence of reactor scram. A rapid increase in power due to reactivity insertion can lead to substantial melting of fuel prior to failure of the cladding. The molten fuel-cladding materials interact with liquid sodium in the coolant passages. The rapid transfer of thermal energy to the coolant and conversion of some of the thermal energy into mechanical energy leads to ejection of the core material mostly in the axial direction. The interaction of core debris with subcooled sodium in the upper and lower plena leads to further fragmentation of the debris into smaller particles. In the course of time, debris particles settle on horizontal surfaces in the upper core support structure and on the bottom of the reactor vessel. In a loss-of-flow accident, reduced flow of sodium will cause boiling in the coolant channel and subsequent melting of cladding and fuel, as well as relocation of the molten material in the axial direction. The mechanical force created by sodium vapor and gravity play a role in relocating the fuel in the upward and downward directions, respectively. The quenching of debris, with subcooled sodium in the upper and lower plena, can lead to further fragmentation of the debris. As in the TOP accident, the debris settles on horizontal surfaces in the upper core structure and on the bottom of the reactor vessel. Removal of fuel from the active core region causes the reactor to shut down. However, the core debris will continue to generate heat as fission products decay. If a coolable state of the core debris cannot be maintained, the debris will dry out, remelt, and pose a threat to the integrity of the support structures. Failure of the support structures can lead to release of radioactivity from the reactor vessel. If a large amount of debris collects at the bottom of the vessel, the dryout and remelting of the debris can cause failure of the reactor vessel and the guard vessel along with release of sodium and core debris into the reactor cavity.

Thus, a knowledge of maximum possible heat removal rate from a particulate bed without dryout and heat transfer rate to the walls bounding a pool of molten liquid is essential for assessment of the hazards of severe core damage accidents and for development of strategies to mitigate their consequences. Installation of core-catchers, both inside and outside of the vessel, are plausible mitigation strategies that have been considered.

In light-water reactors, continued undercooling of the core can result in loss of integrity of the fuel pins and of the structural materials in the core. During core degradation, several processes, such as Zircaloy steam reaction and melting and freezing of the material, take place. Any attempt to inject water during core degradation can lead to quenching and further fragmentation of core material. In such a case, no well-defined coolant paths will exist and the core will behave like rubble bed. When an impervious blockage does not form at the inlet of the core, maintaining a coolable configuration of the rubble bed depends on water flow rate, which can be sustained through the bed under the available driving head. If an impervious blockage forms, the coolability of the debris bed will depend on the heat removal rate that is possible with liquid flowing downwards and vapor moving upwards. If the heat flux due to decay heat is below the maximum possible heat removal rate without dryout for this countercurrent flow configuration, the debris bed will be in a coolable state. Otherwise, a portion of the debris will dry out and the particles in the dried-out region may remelt. In certain circumstances, as was the case in the TMI2 accident, a crucible containing molten core material may form in the reactor core. Failure of the crucible will lead to relocation of the molten material in the lower head of the vessel. Fragmentation and resolidification of molten core debris will occur if water is present in the vessel lower head when the debris relocates. With time, more debris may accumulate in the lower head of the vessel. Quenching and subsequent continued cooling of the debris bed will be possible if a sufficient supply of water is made available and decay heat from the debris can be removed without dryout. Absence of water in the lower plenum, failure to quench the debris, and dryout of the partially quenched debris can lead to creep rupture failure of the vessel lower head and, in turn, release of core material into the cavity. If water is present in the cavity, fragmentation of the molten core material and quenching of the debris will occur. It will be possible to maintain a coolable state of the debris if the decay heat can be removed without dryout. In case the debris dries out or complete quenching of the debris does not occur, the debris may continue to penetrate into the concrete basement. Thus, a knowledge of dryout heat flux in debris beds is important to determine in-vessel and ex-vessel progression of a severe core damage accident in light-water reactors.

Flooding of the reactor cavity with water is currently being considered as a severe accident management strategy for both the advanced and the current light-water reactors. Implementation of this strategy requires that the vessel's lower head and a substantial portion of the cylindrical section of the vessel be submerged in water prior to relocation of the core material into the lower head. No injection of water into the reactor vessel has been considered, and most of the core material is assumed to accumulate in the vessel's lower head in the form of a molten pool. The molten pool is separated from the vessel wall by a crust of frozen core material. The viability of this accident management strategy depends on an ability to dissipate all of the decay heat in the core material to the water outside without any substantial melting of the vessel wall. In order to determine the stable configuration of the pool and the crust, a knowledge of the natural convection heat transfer coefficient on the boundaries of the volumetrically heated pool contained in a spherical segment is needed.

II. Molten Pool Heat Transfer

From the preceding discussion it can be inferred that during severe accidents in nuclear reactors the geometry of pools of molten fuel and structural material can be considered to be bounded by several basic geometries: rectangular cavities, cylindrical cavities, and spherical cavities. The progression of a severe accident and the physical state of the pool depends on the heat transfer at the pool boundaries. Heat transfer from the pool to the boundaries is mostly by natural convection. In the following, the experimental and numerical studies reported in the literature for both single- and two-phase convection and volumetric heat generation in the liquid are reviewed. The studies are subdivided according to the geometry of the cavities.

A. RECTANGULAR CAVITIES

The earliest experimental study of natural convection in rectangular cavities containing volumetrically heated liquids is that of Kulacki and Goldstein [2]. In their experiments, the lower and upper surfaces of the cavity were maintained at the same temperature, whereas the side walls were insulated. The cavity had a square cross-section with a width of 25.4 cm. The layer depth was varied parametrically from 1.27 to 6.35 cm, and dilute aqueous silver nitrate was used as the heat generating liquid. It was found that above a Rayleigh number of 35,840, a larger amount of energy generated in the fluid was removed at the upper plate and maximum temperature occurred above the midplane. The Nusselt number based on the total depth, L, of the layer, the heat flux, q_w, at the wall, the maximum fluid temperature, T_{max} in the layer, the wall temperature, T_w, and fluid thermal conductivity, k, was defined as

$$\mathrm{Nu} = \frac{q_w L}{(T_{max} - T_w)k} = \frac{L\,|dT/dz|}{(T_{max} - T_w)}.$$ (1)

The temperature gradient, dT/dz, in Eq. (1) was obtained from interferograms. The data for the Nusselt number were correlated in terms of Rayleigh number, which was defined as

$$\mathrm{Ra} = \frac{g\beta \dot{Q}_v L^5}{k\nu\alpha}.$$ (2)

In defining the Rayleigh number, Kulacki and Goldstein used half of the layer thickness as the characteristic length. However, here the total liquid layer depth is used. The correlations for heat transfer at the top and bottom plate are

$$\mathrm{Nu} = c\mathrm{Ra}^m.$$ (3)

For layer depth, L, to width, W, ratios varying from 0.05 to 0.25, Rayleigh number varying from 3.71×10^4 to 2.42×10^7, and Prandtl number varying from 5.76 to 6.35, the constants c and m for the top plate were found to be 0.329 and 0.236, respectively, whereas for the bottom plate the values were 1.43 and 0.094, respectively. The experimental uncertainty in Ra was 2–3%, whereas for Nu it was between 3.2 and 6.5%. The fraction of energy transported to the upper plate was found to increase with Rayleigh number. The critical Rayleigh number was found to be 39,680 ± 1920. From the fringe patterns it was noted that temperature profiles in the layer became asymmetric as the Rayleigh number was increased beyond the critical Rayleigh number. The degree of asymmetry increased with Rayleigh number. Figure 1 shows the fringe pattern for Ra = 1.69×10^5. For a Rayleigh number less than 6×10^5, the fluid motion was found to be quasi-periodic. Individual isothermal regions were separated either by cold flows from the upper boundary or upflows of warm liquid entrained from the central region of the liquid layer. At Ra > 6×10^5 turbulent mixing became important and periodicity in the temperature began to diminish. The turbulent convection was characterized by a well-mixed isothermal core and downward flow of cold thermals from the upper boundary. The release of the cold thermals from the upper boundary was found to be a random process. Figure 2 shows the release of thermals from the boundary layer on the upper surface. From the observation of mean temperature field, the flow field was characterized as laminar for Ra < 6×10^5. In this regime, conduction still played an important role in the transport of energy. For $3 \times 10^5 \leq$ Ra $\leq 6 \times 10^5$ laminar convection was fully developed. Transition to turbulent convection occurred at Ra $\simeq 6 \times 10^6$.

Mayinger *et al.* [3] reported results of heat transfer in a geometrical configuration similar to that used by Kulacki and Goldstein. In their experiments water was used as the test liquid (Pr $\simeq 7$) and both upper and lower surfaces were maintained at the same temperature. The heat transfer data were correlated as given by Eq. (3). For layer depth-to-width ratios

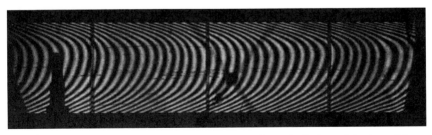

FIG. 1. Interferogram of natural convection in a volumetrically heated horizontal layer cooled at top and bottom (Ra = 1.69×10^5, $L = 1.91$ cm) (Kulacki and Goldstein [2], reprinted with permission of Cambridge University Press).

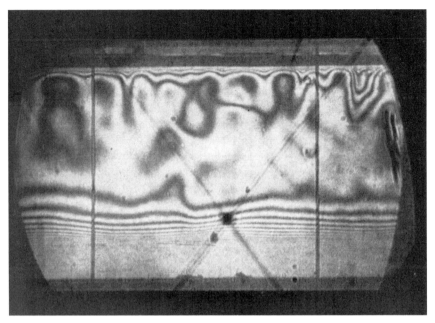

FIG. 2. Interferogram of turbulent convection in a volumetrically heated horizontal layer cooled at top and bottom (Ra = 8.13 × 10⁷; L = 5.08 cm) (Kulacki and Goldstein [2], reprinted with permission of Cambridge University Press).

of less than 0.5 and for $4 \times 10^4 < \text{Ra} < 5 \times 10^{10}$, the constants c and m for heat removal from the upper surface were 0.345 and 0.223, respectively, while for the lower surface the constants had a value of 1.389 and 0.095, respectively. These values are very close to those obtained by Kulacki and Goldstein.

Jahn and Reineke [4], on the other hand, found that for $10^6 < \text{Ra} < 10^9$ and for Pr = 7 the constants c and m have values of 0.78 and 0.2, respectively, for the upper surface and 2.14 and 0.1, respectively, for the lower surface. Consistent with the observation of Kulacki and Goldstein, Jahn and Reineke noted that even at low Rayleigh numbers, the fluid motion was continuously changing. A stable layer of cold liquid formed at the lower boundary, whereas near the upper boundary convective motion broke down into a series of nonuniform eddies which changed in size, position, and number with time.

Steinbrenner and Reineke [5] reported heat transfer data at the top and bottom plates for a layer depth-to-width ratio of 1 and for $10^{13} \leq \text{Ra} < 10^{14}$ and concluded that their data were consistent with the correlation of Jahn and Reineke.

Numerical evaluation of heat transfer to the top and bottom walls when surfaces of both are held at the same temperature while side walls are insulated has been performed by Mayinger *et al.* [3]. The correlation for Nusselt number based on average heat transfer coefficient for a layer-aspect (depth-to-width) ratio of 0.5 was obtained from numerical calculations as

$$\text{Nu}_1 = 0.292\text{Ra}^{0.23}\text{Pr}^{0.085} \quad \text{for } 10^5 \leq \text{Ra} \leq 2 \times 10^{10}$$

and (4)

$$\text{Nu}_2 = 1.235\text{Ra}^{0.1} \qquad 0.04 \leq \text{Pr} \leq 10.$$

For the lower surface, the dependence of Nusselt number on Rayleigh number obtained from numerical analysis is about the same as that obtained from the experiments. However, Nusselt number for the top surface shows an additional dependence on Prandtl number beyond that noted from the experiments in which Prandtl number was varied over a very narrow range. From the numerical calculations, it was found that the magnitude of the Prandtl number could significantly influence the large-scale convective motion. However, it had only a weak effect on Nusselt number.

Emara and Kulacki [6] also carried out numerical calculations for the temperature and velocity field in a volumetrically heated liquid layer bounded from top and bottom with two isothermal rigid walls. They noted that at low Rayleigh numbers the upflows and downflows occupy an equal portion of the span of the horizontal liquid layer. With an increase in Rayleigh number, the downflowing cold liquid was confined to a narrow region between upflowing hot liquid thermals. For laminar convection with $\text{Ra} < 10^8$, they found constants c and m in Eq. (3) to have values of 0.328 and 0.23, respectively, for the upper plate. The constants c and m for heat transfer to the lower plate had values of 1.042 and 0.119, respectively. Consistent with the results of Mayinger *et al.* [3], it was found that flow patterns were generally influenced by the magnitude of the Prandtl number. However, for $1 \leq \text{Pr} \leq 20$, no additional dependence of Nusselt number on Prandtl number was noted.

As a first step in understanding the behavior of fluid layers with unequal temperatures at the top and bottom, experimental and numerical studies have been reported for natural convection in rectangular cells when the lower surface and the bounding walls are insulated while the upper surface is maintained at a constant temperature. Fiedler and Willie [7] were the first to experimentally and analytically study natural convection in such a condition. In their experiments, the width of the test section was relatively small and the aspect ratio of the liquid layer was varied from 0.096 to 1.65. As such, their data are suspected of suffering from wall effects.

Kulacki and Nagle [8] used a much wider test section and varied the aspect ratio from 0.05 to 0.25. They correlated the data for heat transfer to the upper plate as

$$Nu_1 = 0.305Ra^{0.239} \quad \text{for } 1.5 \times 10^5 \leq Ra \leq 2.5 \times 10^9,$$

$$\text{and } 6.21 \leq Pr \leq 6.64. \quad (5)$$

The combined uncertainty in their steady state results was 5–6% for Ra and 4–5% for Nu. Kulacki and Nagle noted that the temperature in the liquid layer fluctuated with time. These fluctuations were most intense in the region spanning 7 to 13% of the layer depth from the upper surface, and the region was outside the edge of the thermal layer on the upper surface.

Mixing of cold liquid thermals in the upper and central regions of the layer tended to dampen fluctuations in the lower 60% of the layer. Kulacki and Nagle [8] also attempted to predict the rate of heat transfer to the upper surface by using correlations developed by Kulacki and Goldstein for heat transfer to upper and lower surfaces when both surfaces were held at the same temperature. These analyses were based on the fact that in a liquid layer bounded on both sides with surfaces held at equal temperatures a certain fraction of energy generated in the layer was removed at the upper surface, whereas the remainder was dissipated at the lower surface. The ratio of the distance of the plane of zero heat flux from the upper surface to the layer depth, L_1/L, can be written as

$$\frac{L_1}{L} = \frac{Nu_1}{Nu_1 + Nu_2}, \quad (6)$$

where Nu_1 and Nu_2 are defined according to Eq. (1) with total layer depth as the characteristic length and are obtained from Eq. (3) by appropriately choosing constants c and m for the upper and lower surfaces maintained at equal temperatures. Kulacki and Nagle used L_1 evaluated from Eq. (6) in conjunction with their expression for Nusselt number for the upper surface when the lower surface was insulated and compared it to prediction from the correlation of Kulacki and Goldstein. A good agreement was found for $Ra < 10^8$. However, at $Ra \simeq 10^9$ the Nusselt number predicted from Kulacki and Nagle correlation was about 40% higher than that obtained from the correlation of Kulacki and Goldstein. It was argued that this discrepancy was due to the fact that the correlation of Kulacki and Goldstein was valid only up to $Ra \simeq 10^8$.

Kulacki and Emara [9] extended the range of Rayleigh number studied in the experiments of Kulacki and Nagle. For layer aspect ratios varying from 0.025 to 0.5 Kulacki and Emara [9] correlated their data for heat transfer to the upper surface as

$$Nu_1 = 0.338Ra^{0.227} \quad \text{for } 3.78 \times 10^3 \leq Ra \leq 4.34 \times 10^{12}$$

$$\text{and } 2.75 \leq Pr \leq 6.86, \quad (7)$$

The data had an experimental uncertainty of 5–7% for Ra and 4–5% for Nu. From the correlation, the authors deduced that the critical Rayleigh number for onset of convection (Nu = 2) was 2509. This value is about 6.7% lower than that obtained from linearized instability theory (Kulacki and Goldstein [10]) for a Biot number (ratio of resistance on the fluid side to that of the wall) $\simeq 65$. The chosen value of the Biot number is consistent with the experimental setup. Starting with a Rayleigh number of 8.4×10^5 and up to a Rayleigh number of 3×10^{10}, discrete transitions in heat flux were observed each time the Rayleigh number was increased by about an order of magnitude. However, for Ra $> 10^{10}$ transitions were found to occur more frequently (with every fivefold increase in Rayleigh number). Emara and Kulacki [6] carried out numerical calculations of heat transfer to the upper surface when the lower surface was insulated. They obtained results both for a rigid upper surface and for a free upper surface. For a rigid surface, correlation for heat transfer to the upper surface was obtained as

$$Nu_1 = 0.412Ra^{0.210} Pr^{0.041} \quad \text{for } 10^4 \leq Ra \leq 10^9 \text{ and } 0.05 < Pr < 20 \quad (8)$$

However, for the free surface the correlation was

$$Nu_1 = 0.759Ra^{0.178} Pr^{0.059}$$

$$\text{for } 10^5 < Ra < 10^9 \text{ and } 0.05 < Pr < 20 \quad (9)$$

It is seen from Eqs. (8) and (9) that an explicit dependence of Nusselt number on Prandtl number is quite weak. For Pr = 6.5 the predictions obtained by using Eq. (8) are within a few percent of the data. It should be noted that the rate of heat transfer is slightly higher when the upper surface is free in comparison to that for the rigid surface.

In applications dealing with postaccident heat removal in liquid-heat fast-breeder reactors, the temperatures at the upper and lower boundaries of the liquid layer are different and the interest is in determination of the fraction of the thermal energy generated in the layer that is transferred downwards. Suo-Antilla and Catton [11] experimentally investigated natural convection in volumetrically heated layers with unequal temperatures at the top and bottom. From their work in which Ra was varied up to 10^{12}, they noted that the effect of a destablizing temperature difference was to reduce the fraction of energy transferred to the bottom surface. The effect of imposition of a stabilizing temperature gradient was the opposite.

Baker, Faw, and Kulacki [12], using the data and observation of Kulacki and Goldstein [2] for a volumetrically heated layer cooled at the top and the bottom with the bounding surfaces maintained at equal temperatures, have developed a correlation for the fraction of thermal energy transferred downwards when the two bounding surfaces have unequal temperatures. The cornerstone of the Baker *et al.* approach is division of the layer into two sublayers, with all the energy generated in the upper layer transferred to the upper surface and all the energy generated in the lower layer dissipated at the lower bounding surface. The plane dividing the total layer into two sublayers represents the insulated boundary. Using the correlation developed for Nusselt number for the total layer, they showed that when these correlations were written for sublayers conduction was found to be the main mode of heat transfer in the lower layer. This is consistent with the observation of Kulacki and Goldstein [2] and others that little heat transport into the lower layer occurs as a result of penetration of eddies from the upper portion.

For a horizontal layer in which the upper and lower surfaces are maintained at the same temperature, the fraction of energy that is transferred downwards can be written as

$$\eta = \frac{\mathrm{Nu}_2}{\mathrm{Nu}_1 + \mathrm{Nu}_2}, \qquad (10)$$

where

$$\mathrm{Nu}_1 = \frac{Q_v L_1 L}{(T_{\mathrm{max}} - T_1)} = \frac{q_{w1} L}{(T_{\mathrm{max}} - T_1)k} \qquad (11)$$

and

$$\mathrm{Nu}_2 = \frac{Q_v L_2 L}{(T_{\mathrm{max}} - T_2)} = \frac{q_{w2} L}{(T_{\mathrm{max}} - T_2)k}. \qquad (2)$$

Nu_1 and Nu_2 are evaluated from correlations such as those given by Eqs. (3) or (4). If the layer is artificially subdivided into two layers, with one of the bounding surfaces of these layers being insulated as shown in Fig. 3, the sublayer Nusselt numbers can be written as

$$\mathrm{Nu}_1' = \frac{Q_v L_1 L_1}{(T_{\mathrm{max}} - T_1)k}, \qquad (13)$$

$$\mathrm{Nu}_2' = \frac{Q_v L_2 L_2}{(T_{\mathrm{max}} - T_2)k}. \qquad (14)$$

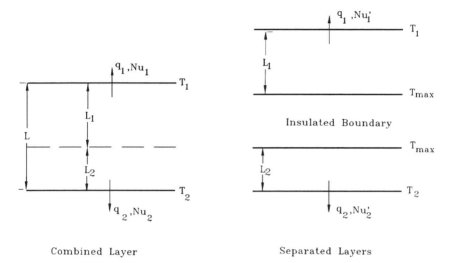

Combined Layer   Separated Layers

FIG. 3. Partitioning of a volumetrically heated horizontal layer into two layers at the plane of zero heat flux.

For $T_1 = T_2$, Eqs. (11) and (12), yield an L_1/L ratio as given by Eq. (6), whereas L_2/L is obtained as

$$\frac{L_2}{L} = \frac{\mathrm{Nu}_2}{\mathrm{Nu}_1 + \mathrm{Nu}_2} \equiv \eta. \qquad (15)$$

Substituting for L_1 and L_2 from Eqs. (6) and (15) into Eqs. (13) and (14), expressions for Nu_1' and Nu_2' are obtained as

$$\mathrm{Nu}_1' = \frac{\mathrm{Nu}_1^2}{\mathrm{Nu}_1 + \mathrm{Nu}_2}; \qquad \mathrm{Nu}_2' = \frac{\mathrm{Nu}_2^2}{\mathrm{Nu}_1 + \mathrm{Nu}_2}. \qquad (16)$$

Substitution of expressions for Nu_1 and Nu_2 from Eq. (3) (with appropriate values of the constants) into Eq. (16) yields for Nu_2' a value of about 2 for all Rayleigh numbers. This is consistent with the observation that conduction is the dominant mode of heat transfer in the lower layer, that is, below the plane of zero average heat flux. However, for the upper layer Nu_1' increases with Rayleigh number, which is indicative of the enhanced effectiveness of heat removal at the upper surface.

If heat transfer correlations developed for volumetrically heated layers with insulated lower surface and cooled upper surface are to be employed to determine Nusselt numbers for the sublayers, an assumption must be made that replacement of the previous zero heat flux boundary with a rigid insulated boundary does not alter the transfer of heat at the cooled surface. Since the middle portion of the layer is nearly isothermal (well mixed) and eddy transport is the main mechanism of heat transfer, the above assumption can be considered to be realistic. Rayleigh numbers of the sublayers are related to the Rayleigh number of the total layer as

$$\text{Ra}'_1 = \text{Ra}\left(\frac{L_1}{L}\right)^5 \quad \text{and} \quad \text{Ra}'_2 = \text{Ra}\left(\frac{L_2}{L}\right)^5 \qquad (17)$$

or

$$\text{Ra}'_1 = \text{Ra}\left(\frac{\text{Nu}_1}{\text{Nu}_1 + \text{Nu}_2}\right)^5 \quad \text{and} \quad \text{Ra}'_2 = \text{Ra}\left(\frac{\text{Nu}_2}{\text{Nu}_1 + \text{Nu}_2}\right)^5. \qquad (18)$$

Thus, the Nusselt number correlations developed for a horizontal layer of depth L_1 with an insulated lower boundary can be written in terms of the Rayleigh numbers based on the total layer height, L.

For a layer with unequal temperatures at the top and bottom, an additional Rayleigh number, based on the temperature difference between top and bottom surfaces, needs to be defined. This Rayleigh number can be positive or negative depending on whether the top surface temperature is higher or lower than the bottom surface temperature. The external Rayleigh number based on the temperature difference between the bounding surfaces is defined as

$$\text{Ra}_E = \frac{g\,\beta L^3(T_1 - T_2)}{\alpha\nu}. \qquad (19)$$

The ratio of external to internal Rayleigh number can be written as

$$\frac{\text{Ra}_E}{\text{Ra}} = \frac{k(T_1 - T_2)}{Q_v L^2} \qquad (20)$$

Dividing the layer into two sublayers of heights L_1 and L_2 at the plane of zero heat flux, the Nusselt numbers for the two layers can be written (see Eqs. (13) and (14)) as

$$\text{Nu}'_1 = \frac{Q_v L_1^2}{k(T_{\max} - T_1)} \quad \text{and} \quad \text{Nu}'_2 = \frac{Q_v L_2^2}{k(T_{\max} - T_2)}. \qquad (21)$$

Since conduction is considered to be the dominant mode of heat transfer in the lower layer,

$$\text{Nu}'_2 \equiv 2. \qquad (22)$$

Combining Eqs. (22), (21), (20), and (15), the ratio of temperature difference across the lower and upper sublayers is obtained as

$$\frac{T_{\max} - T_2}{T_{\max} - T_1} = \frac{Ra\eta^2}{Ra\eta^2 - 2Ra_E}. \tag{23}$$

Nusselt numbers for the total layer can be written in terms of the Nusselt numbers of the two sublayers as

$$Nu_1 = Nu_1' \frac{L}{L_1} = Nu_1' \frac{1}{(1 - \eta)}.$$

Substituting for Nu_1' in terms of Ra_1' from Eq. (7), we obtain

$$Nu_1 = 0.338(Ra_1')^{0.227} \frac{1}{(1 - \eta)}$$

$$= \frac{0.338(Ra)^{0.227}(1 - \eta)^{1.135}}{(1 - \eta)}$$

$$= 0.338(Ra)^{0.227}(1 - \eta)^{0.135} \tag{24}$$

and

$$Nu_2 = Nu_2' \frac{L}{L_2} = Nu_2' \frac{1}{\eta}$$

$$= \frac{2}{\eta}. \tag{25}$$

From Eqs. (11) and (12), the ratio of temperature difference for heat transfer at the lower and the upper surface is obtained as

$$\frac{T_{\max} - T_2}{T_{\max} - T_1} = \frac{Nu_1 L_2}{Nu_2 L_1} = \frac{Nu_1 \eta}{Nu_2(1 - \eta)}. \tag{26}$$

Substituting for Nu_1 and Nu_2, from Eqs. (24) and (25) the above ratio becomes

$$\frac{T_{\max} - T_2}{T_{\max} - T_1} = \frac{0.169\eta^2 Ra^{0.227}}{(1 - \eta)^{0.865}}. \tag{27}$$

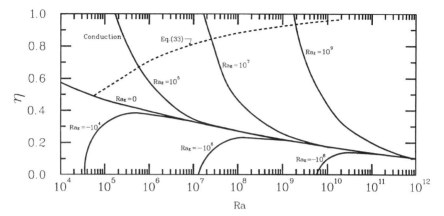

FIG. 4. Dependence on Ra and Ra$_E$ of fraction of heat transferred downward.

Eliminating $(T_{\max} - T_2)/(T_{\max} - T_1)$ between Eqs. (23) and (27), a relation between the fraction of heat transferred downwards and the internal and external Rayleigh numbers is obtained as

$$\frac{(1 - \eta)^{0.865}}{\eta^2 - 2\dfrac{\text{Ra}_E}{\text{Ra}}} = 0.169\text{Ra}^{0.227}. \tag{28}$$

In Fig. 4 values of η are plotted as a function of Ra when negative and positive external Rayleigh numbers are varied parametrically. Negative Rayleigh numbers occur when upper surface temperature is smaller than lower surface temperature. For a given internal Rayleigh number, an increase in positive external Rayleigh number increases downward heat transfer. However, the opposite is true if the negative Rayleigh number is decreased. From Eq. (28), the limit corresponding to $\eta = 0$ is obtained as

$$-\text{Ra}_E = 2.9\text{Ra}^{0.773}. \tag{29}$$

At $-\text{Ra}_E$ equal to or above that given by Eq. (29), all of the energy generated in the layer will be transferred upwards.

If conduction were to be the dominant mode of heat transfer in the whole layer (i.e, in both the upper and lower sublayers), both Nu$'_1$ and Nu$'_2$ will be equal to 2. For this case,

$$\text{Nu}_1 = \frac{2}{1 - \eta} \quad \text{and} \quad \text{Nu}_2 = \frac{2}{\eta}. \tag{30}$$

Substitution of Nu_1 and Nu_2 from Eq. (30) into Eq. (28) yields

$$\frac{T_{max} - T_2}{T_{max} - T_1} = \frac{\eta^2}{(1 - \eta)^2}. \tag{31}$$

Elimination of $(T_{max} - T_2)/(T_{max} - T_1)$ between Eqs. (31) and (23), results

$$\eta = \frac{Ra_E}{Ra} + 0.5. \tag{32}$$

The boundary between conduction and convection is obtained by eliminating Ra_E between Eqs. (28) and (32) and is given by

$$\eta = 1 - 4.79 Ra^{-0.2}. \tag{33}$$

The dotted line in Fig. 4 marks the conduction–convection limit. In the region above the dotted line, conduction is the dominant mode of heat transfer throughout the layer. As such, when all of the heat is transferred downwards ($\eta = 1$), all of the layer will be in conduction.

The results presented in Fig. 4 can be used to calculate the downward heat loss in a given hypothetical accident scenario as long as the thermophysical properties of the molten core material and heat generation rate are known. It should be noted that the correlation used in arriving at the η values plotted in Fig. 4 were obtained for $Ra < 10^9$. For use of the information plotted in Fig. 4 when internal Rayleigh numbers are greater than 10^9, a priori validation of the correlation used in developing Fig. 4 should be carried out.

Mayinger et al. [3] have given correlations for upward, downward, and sideward heat transfer when the upper, lower and side walls of a rectangular cavity are maintained at a constant temperature. For a layer height-to-width ratio less than 0.5, $Pr = 7$, and $10^7 \leq Ra \leq 3 \times 10^{10}$, their correlations are

$$Nu_1 = 0.345 Ra^{0.233}, \tag{34}$$

$$Nu_2 = 1.389 Ra^{0.095}, \tag{35}$$

$$Nu_s = 0.6 Ra^{0.19}. \tag{36}$$

It should be noted that the proportionality constants and exponents in Eqs. (34) and (35) are almost the same as those obtained by Mayinger et al. for a cavity with insulated side walls. They also gave a correlation for the combined heat transfer to upper and lower horizontal surfaces as

$$Nu_1 + Nu_2 = 1.05 Ra^{0.19} \tag{37}$$

Steinbrenner and Reineke [5] have reported a correlation for heat transfer to the side walls when the upper and lower horizontal surfaces are adiabatic. Their correlation for layer height-to-width ratios of less than

0.044, Pr = 7, and $8 \times 10^{12} \leq \text{Ra} \leq 4 \times 10^{13}$ is

$$\text{Nu}_s = 0.85\text{Ra}^{0.19}. \tag{38}$$

It is interesting to note from Eqs. (36) and (38) that the rate of side walls heat transfer is about 30% higher when only side walls rather than all the walls bounding the cavity are cooled.

Greene [13] has conducted experimental and analytical investigations of natural convection heat transfer to a vertical wall bounding a pool of volumetrically heated liquid. He could not obtain a complete similarity solution of the governing equations under the imposed boundary conditions. As such he invoked local similarity to solve for the temperature and velocity fields. From analysis he obtained an expression for the local Nusselt number as

$$(\text{Nu}_z)_s = 0.148(\text{Ra}_z)^{1/4} \cdot \left(\frac{1}{\text{Pr}}\right)^{1/4} \tag{39}$$

and for the Nusselt number based on the average heat transfer coefficient as

$$\text{Nu}_s = 0.130(\text{Ra})^{1/4}\left(\frac{1}{\text{Pr}}\right)^{1/4}. \tag{40}$$

In Eq. (39), the distance from the leading edge of the convective boundary layer is used as the characteristic length, whereas in Eq. (40) the total height of the plate is used as the characteristic length. In contrast to natural convection on a vertical wall in the absence of volumetric heating, where the heat transfer coefficient decreases with distance from the leading edge, Eqs. (39) and (40) show that the heat transfer coefficient increases with distance from the leading edge. This cannot be true near the leading edge, where assumption of local similarity can be grossly in error. The local and average heat transfer coefficient data obtained in the experiments from a 30.5-cm tall vertical plate were correlated in the form of Eqs. (39) and (40) when the empirical constant in Eq. (39) was increased to 0.153 and in Eq. (40) it was decreased to 0.127. The experimental uncertainty in the data for Ra was 10% and that in Nu was 6%. It should be noted that Eq. (39) tended to underpredict the data at low Rayleigh numbers or near the leading edge. As such, Eqs. (39) and (40) were not recommended for $\text{Ra}/\text{Pr} < 10^7$, and the upper limit of applicability of Eqs. (39) and (40) being $\text{Ra}/\text{Pr} < 10^{13}$. Although the data were also found to be consistent with laminar boundary layer analysis, large-amplitude fluctuations in wall temperature were observed. The magnitude of oscillations was found to increase along the boundary layer. However, a critical condition for transition to turbulent flow was not identified. Equations (39) and (40) were also found to be applicable to inclined walls with an angle of inclination with the vertical less than 30° when g in Ra was replaced by $g \cos \theta$.

Greene *et al.* [13] have also reported correlations similar to Eqs. (39) and (40) for heat transfer from internally heated boiling pools. For an average pool void fraction $\bar{\alpha}$, the correlations for Nusselt number based on local and average heat transfer coefficients were given as

$$(\mathrm{Nu}_z)_s = 1.05(\mathrm{Ra}_z \bar{\alpha})^{1/4} \left(\frac{1}{\mathrm{Pr}}\right)^{1/4}, \tag{41}$$

$$\mathrm{Nu}_s = 0.99(\mathrm{Ra}\,\bar{\alpha})^{1/4} \left(\frac{1}{\mathrm{Pr}}\right)^{1/4}. \tag{42}$$

B. CIRCULAR AND SPHERICAL CAVITIES

1. *Circular Cavities*

Mayinger *et al.* [3] were the first to report experimental and numerical results for natural convection heat transfer in volumetrically heated circular segments (troughs). Figure 5a shows the configuration of the cavity. Both the lower curved surface and the upper surface of the trough were maintained at the same temperature. The height of the liquid layer and the radius of the circular segment were varied parametrically. The data for the heat transfer at the upper horizontal surface and at the lower curved surface were deduced from the fringe pattern. Nusselt numbers based on an average heat transfer coefficient at the upper horizontal and lower curved surfaces were correlated as

$$\mathrm{Nu}_1 = 0.36\mathrm{Ra}^{0.23} \quad \text{for } 10^7 < \mathrm{Ra} < 5 \times 10^{10} \tag{43}$$

and

$$\mathrm{Nu}_2 = 0.54\mathrm{Ra}^{0.18} \left(\frac{L}{R}\right)^{0.26} \quad \text{and} \quad \mathrm{Pr} = 7. \tag{44}$$

It should be noted that the radius, R, of the circular cavity was used as the characteristic length in defining the Nusselt and Rayleigh numbers. In Eq. (44) L is the depth of the liquid layer and the correlation Eqs. (43) and (44) were obtained for $0.3 < L/R < 1$. The heat transfer coefficient was found to vary along the curved surface. The lowest heat transfer coefficient occurred at the stagnation point, whereas it had the highest value near the intersection of the horizontal and curved surfaces. However, no correlation was proposed for the angular dependence of the heat transfer coefficient.

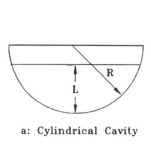

a: Cylindrical Cavity

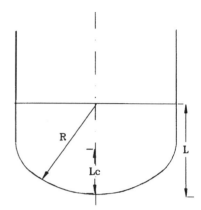

b: Two Dimentional (Slice) Cavity
of a Torospherical Bottom of
VVER−400 Reactor

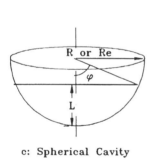

c: Spherical Cavity

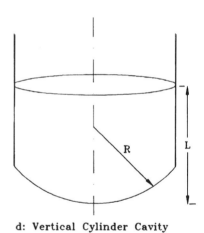

d: Vertical Cylinder Cavity
with a Spherical Bottom

FIG. 5. (a–d) Cylindrical and spherical cavity configurations.

From the numerical calculations using THEKAR code with and without the turbulence model, Mayinger *et al.* [3] obtained expressions for heat transfer to the upper and lower surfaces of a semicircular cavity as

$$\mathrm{Nu}_1 = 0.73 \mathrm{Ra}^{0.18} \quad \text{for } \mathrm{Ra} \leq 4 \times 10^6 \qquad (45)$$

and

$$\mathrm{Nu}_2 = 0.25 \mathrm{Ra}^{0.14} \qquad \qquad \mathrm{Pr} = 7. \qquad (46)$$

For $10^7 \leq Ra \leq 10^{11}$, the numerical predictions were correlated as

$$Nu_1 = 0.36Ra^{0.23} \qquad (47)$$

and

$$Nu_2 = 0.54Ra^{0.18}. \qquad (48)$$

Comparing Eqs. (47) and (48) with Eqs. (43) and (44), it is seen that predictions for a semicircular cavity are in complete agreement with the data. As shown in Fig. 6, the numerical predictions of the local heat transfer coefficient as made by Jahn and Reineke [4] are also in general agreement with the data. The maximum heat transfer coefficient is found to be as high as two times the average heat transfer coefficient.

Mayinger et al. [3] also carried out numerical calculations for flow and temperature field during natural convection in a volumetrically heated liquid pool contained in a vertical cylinder. The upper, lower, and side surfaces of the cylinder were assumed to be rigid and were held at a constant and equal temperature. The cylinder diameter was equal to the cylinder height and calculations were carried out for laminar natural convection. The cylinder radius was used as the characteristic length in defining Nusselt and Rayleigh numbers.

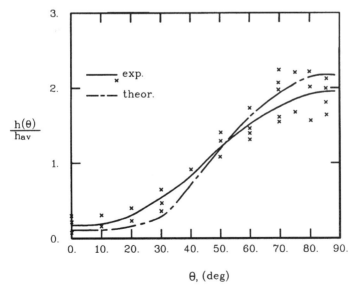

FIG. 6. Dependence of normalized heat transfer coefficient in a cylindrical cavity on angular position.

From the numerical calculations, the Nusselt number for the upper surface was correlated as

$$\text{Nu}_1 = 0.34\text{Ra}^{0.22} \quad \text{for } 10^5 \leq \text{Ra} \leq 1.5 \times 10^7 \quad \text{and Pr} = 7 \quad (49)$$

and

$$\text{Nu}_1 = 2.8\text{Ra}^{0.09} \quad \text{for } 2 \times 10^7 \leq \text{Ra} \leq 10^9 \quad \text{and Pr} = 7. \quad (50)$$

The Nusselt numbers based on average heat transfer coefficients at the lower and side surfaces were correlated as

$$\text{Nu}_2 = 0.935\text{Ra}^{0.1} \quad \text{for } 10^5 \leq \text{Ra} \leq 10^9 \quad \text{and Pr} = 7 \quad (51)$$

and

$$\text{Nu}_s = 0.495\text{Ra}^{0.18} \quad \text{for } 10^5 \leq \text{Ra} \leq 10^9 \quad \text{and Pr} = 7. \quad (52)$$

It is noted from Eqs. (50), (51) and (52) that the highest rate of heat transfer takes place on the side wall whereas the lowest rate of heat transfer occurs on the bottom surface.

Kymäläinen et al. [15] have reported cavity wall heat transfer results when a pool of volumetrically heated liquid was formed in a two-dimensional slice of the torospherical lower head (see Fig. 5b) of a VVER 440 reactor. The width of the test section of the facility (COPO) was 1.77 m and maximum pool depth was 0.8 m. The thickness of the slice was 0.1 m. The electrodes were placed on the flat lateral surfaces of the test section and $ZnSO_4$–H_2O was used as the electrolyte. The lower, side, and upper surfaces of the pool were cooled. The heat flux on the vertical (side) surface was found to be nearly uniform and the Nusselt number for $1 \times 10^{14} < \text{Ra} < 1.6 \times 10^{15}$ was correlated with Eq. (38) of Steinbrenner and Reineke [5]. Similarly, the Nusselt number for the upper surface was found to approximately correlate with Eq. (34). For both the side wall and top surface, total pool depth was used as the characteristic length. The Nusselt number based on average heat transfer coefficient on the lower (curved) surface was found to be in agreement with the correlation in Eq. (44) of Mayinger et al. for a semicircular cavity. Consistent with the observations of Mayinger et al. for a semicircular cavity, the downward heat transfer coefficient was found to depend strongly on the angular position. However, the variation of heat transfer coefficient with angular position showed less nonlinearity in comparison to that for the circular cavity.

2. Spherical Cavities

The first experimental study of natural convection in volumetrically heated pools contained in spherical containers is that of Gabor et al. [16]. In the experiments hemispherical containers made of copper and cooled at the curved surface were used. The curved surface of the containers served both as a nearly constant temperature heat-transfer surface and as one of the electrodes. A copper disc placed in the middle of the pool free surface acted as the second electrode. This arrangement of electrodes led to nonuniformity in the heat generation rate in the pool. In the experiments a very dilute solution of H_2SO_4 was used as the electrolyte and hemispherical containers of three different radii were employed. The height of the pool in the containers was varied parametrically. No attempt was made to measure the local heat transfer coefficient along the curved surface, but the Nusselt number based on average heat transfer coefficient was correlated as

$$\mathrm{Nu} = 0.55 \mathrm{Ra}^{0.15} \left(\frac{L}{R} \right)^{1.1} \quad \text{for } 2 \times 10^{10} \leq \mathrm{Ra} \leq 2 \times 10^{11}. \quad (53)$$

The radius of the cavity was used as the characteristic length in defining the Nusselt and Rayleigh numbers in Eq. (53).

Asfia et al. [17] experimentally studied natural convection in spherical segments in 1996. In their experiments, refrigerant 113 was used as the test liquid and the liquid pool was formed in the spherical head of glass bell jars. The radii of the bell jars and the pool height were varied parametrically. Microwaves were used to volumetrically heat the test liquid and test containers were cooled from the outside with water. Local and averaged heat transfer coefficients were obtained for pools with a free surface, pools bounded with rigid nearly insulated surfaces, and pools with rigid cooled surfaces. Figure 7 shows the variation of the normalized heat transfer coefficient as a function of angular position along the curved surface for pools of different depths. The maximum heat transfer coefficient occurs near the pool free surface, while the minimum value occurs at the lower stagnation point. For a hemispherical cavity the maximum heat transfer coefficient is found to be as high as 2.5 times the time- and area-averaged heat transfer coefficient. The local heat transfer coefficients were obtained by dividing the conductive heat flux across the cavity wall by the difference between the maximum pool temperature and the temperature of the inner wall of the cavity. Asfia et al. noted the pool and wall temperatures to oscillate with time. However, the magnitude of the oscillations was less than 1% of the difference between the maximum pool temperature and the temperature of the outside coolant. As such, the reported heat transfer coefficients are time-averaged heat transfer coefficients. Also, the inner surface of the cavity was not isothermal. The curved wall temperature was higher near the pool surface and was lowest near the stagnation point. The local heat transfer coefficient data, such as those plotted in Fig. 7, were

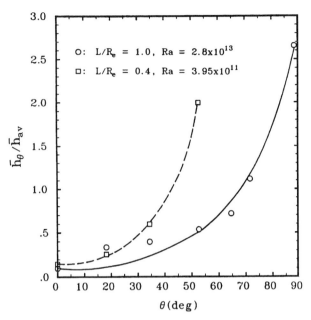

FIG. 7. Dependence on angular position of normalized heat transfer coefficient in a spherical cavity with a free surface pool.

correlated as

$$\frac{\overline{h}(\theta)}{\overline{h}_{av}} = C_1 \sin \Theta - C_2 \cos \Theta \quad \text{for } 0.73 < \frac{\theta}{\phi} \leq 1, \quad (54)$$

where

$$C_1 = -1.20 \cos \phi + 2.6,$$
$$C_2 = -2.65 \cos \phi + 3.6,$$

and

$$\frac{\overline{h}(\theta)}{\overline{h}_{av}} = C_3 \sin^4 \Theta + C_4 \quad \text{for } 0 \leq \frac{\theta}{\phi} \leq 0.73, \quad (55)$$

where

$$C_3 = -0.31 \cos \phi + 1.06,$$
$$C_4 = 0.24 \cos \phi + 0.15.$$

In Eqs. (54) and (55), θ is the angular position measured from the lower stagnation point and ϕ is the pool angle, which depends on the depth of the pool. These parameters are shown schematically for a spherical geometry in Fig. 5c. The parameter Θ is defined as

$$\Theta = \frac{\theta}{\phi} \cdot \frac{\pi}{2}. \tag{56}$$

By interpolating the temperature data from thermocouples placed in the pool and on the curved wall of the spherical segment, Asfia et al. [17] were able to obtain an isotherm pattern in the pool. The isotherm pattern revealed the existence of a nearly well-mixed layer in the upper portion, a nearly stagnant stratified pool in the lower portion, and a thin boundary layer along the curved wall of the cavity.

From experiments in which a rigid wall bounded the pool upper surface, Asfia and Dhir [18] noted that the effect of the changed boundary condition was to push downward the location of the maximum heat transfer coefficient along the curved wall of the cavity. However, the normalized magnitude of the local heat transfer coefficients deeper into the pool and the maximum heat transfer coefficients were not affected by the changed boundary condition. Figure 8 shows the distribution of the heat transfer coefficient for two pool depths and a rigid upper wall. Asfia [19] has also reported heat transfer results of experiments in which freezing of R-113 occurred on the cavity walls. In these experiments water replaced liquid nitrogen as the coolant. Formation of crust along the spherical segment provided an isothermal boundary for the pool. The variation of the heat transfer coefficient with angular position in the experiments in which a crust separated the pool from the inner wall of the bell jar was found to be similar to that for pools with a free surface and no freezing on the inner wall. However, the magnitude of the heat transfer coefficient near the stagnation point was higher in the presence of crust than without the crust. Since the heat transfer coefficients with crust did not differ much from those obtained in the absence of a crust, it is believed that nonisothermal curved surfaces in the experiments without crust did not appreciably distort the measured distribution of the heat transfer coefficients.

The Nusselt number based on the area- and time-averaged heat transfer coefficient for pools with free surface was correlated as

$$\mathrm{Nu} \equiv \frac{\overline{h}_{av} L}{k} = 0.54 \mathrm{Ra}^{0.2} (L/R_e)^{0.25} \quad \text{for } 2 \times 10^{10} \leq \mathrm{Ra} \leq 1.1 \times 10^{14}$$

and

$$\mathrm{Pr} = 8.2. \tag{57}$$

In Eq. (57) R_e is the equivalent radius of the cavity, and the Nusselt and Rayleigh numbers are based on the pool depth, L. Within the uncertainty

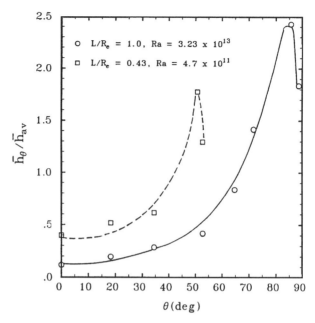

FIG. 8. Dependence on angular position of normalized heat transfer coefficient in a spherical cavity with a rigid surface.

of the average heat transfer coefficient data, which was generally less than $\pm 16\%$, the average heat transfer coefficients obtained with pools bounded by a rigid wall at the top and with a crust separating the pool from the cavity wall were found to correlate with Eq. (57).

Mayinger *et al.* [3] used their numerical code validated with data from a circular cavity (trough) to predict natural convection heat transfer from a volumetrically heated pool to the isothermal wall of a hemispherical cavity. The cavity was bounded on top by a rigid wall. In defining the heat transfer coefficient the difference between maximum pool temperature and the wall temperature was used. From the numerical results the average heat transfer coefficients at the upper and lower (curved) isothermal walls held at the same temperature were correlated as

$$\mathrm{Nu}_1 = 0.4\mathrm{Ra}^{0.2} \quad \text{for } 7 \times 10^6 \leq \mathrm{Ra} \leq 5 \times 10^{14} \quad \text{and } \mathrm{Pr} = 7 \quad (58)$$

$$\mathrm{Nu}_2 = 0.55\mathrm{Ra}^{0.2}. \quad (59)$$

In Eqs. (58) and (59), the pool depth, L (equal to pool radius, R), is used as the characteristic length. The Nusselt numbers predicted from Eq. (59) are plotted in Fig. 9 as a function of Rayleigh number. In this figure, the

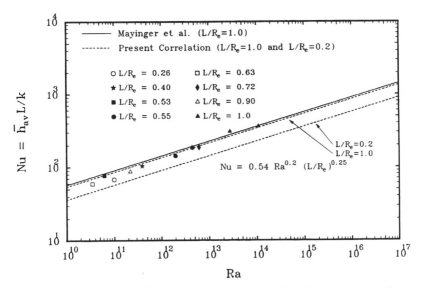

FIG. 9. Correlation of data for average heat transfer coefficient in a spherical cavity.

predictions from Eq. (57) and the data of Asfia *et al.* [17] for different pool depths are also plotted. It is seen that for a hemispherical cavity the average heat transfer coefficients predicted from the numerical calculations of Mayinger *et al.* [3] are very close to the predictions from the correlation of Asfia *et al.* [17].

Kelkar *et al.* [20] have numerically analyzed natural convection in a volumetrically heated pool contained in a hemispherical cavity. The lower curved surface of the cavity and the upper-bounding (flat) surface were assumed isothermal rigid boundaries. Calculations were carried out for $10^3 < Ra < 10^8$ and $Pr = 0.1$, 1, and 10. For $Ra = 10^4$ the convective motion was found to be very weak and the isothermal pattern similar to that for pure conduction. At $Ra = 10^8$ the hot fluid was found to be pushed upward, causing the heat to be removed mostly from the horizontal surface and upper portion of the curved surface. At the curved surface the local heat flux was maximum at an angular position of 75° from the lower stagnation point. The heat flux was zero at the line of contact of the horizontal and curved surfaces and had a small value at the stagnation point. The Nusselt number based on the heat transfer coefficient averaged over angular positions from 0 to 45° was found to decrease with Ra, but the opposite was true for the heat transfer coefficient averaged over angular positions from 45 to 90°. The effect of changing the Prandtl number from 0.1 to 10 had little effect on the Nusselt number based on average Prandtl number. However, the heat transfer coefficient distribution was seen to be influenced by the magnitude of the heat transfer coefficient. The Nusselt numbers calculated for the upper surface were higher than those on the lower curved surface. This is in contrast to Eqs. (58) and (59) of Mayinger et al. [3], which yield an average heat transfer coefficient on the lower surface higher than that on the upper surface. Unfortunately, the authors do not provide the temperature different they used to determine the heat transfer coefficients. For Ra = 10^9 the authors could not obtain steady-state solution of the governing equations.

In a subsequent work, Schmidt et al. [21] assumed that the upper surface of the pool was insulated and they used both no-slip (rigid wall) and slip (free surface) conditions at the upper surface. The effect of the changed boundary conditions on the flow pattern in the pool and on heat transfer to the curved surface was minimal. In contrast to their earlier results, in this work the authors found the Nusselt numbers based on heat transfer coefficients averaged over angular positions from 0 to 45° as well as from 45 to 90° to increase with Ra. They noted that cooling of the upper surface represented an unstable flow situation, as opposed to the case for an insulated top, in which the fluid in the upper portion was stratified. Consequently, Nusselt numbers on a curved surface with cooling at the top were higher than those obtained for an insulated top. This is in contradiction to the experimental results of Asfia and Dhir [18], who found little effect of a cold rigid wall at the top on the average heat transfer coefficient on the curved surface.

Natural convection in a volumetrically heated pool contained in a spherical segment having a total polar angle of 60° and attached to a right circular cylinder (Fig. 9d) has been experimentally studied by Min and Kulacki [22]. In the experiments the pool was bounded at the top by a rigid plate. The rigid plate was cooled and served as one of the electrodes. The cylindrical and spherical surfaces were insulated, while the spherical surface served as the second electrode. A dilute (maximum 0.091% by weight) aqueous copper sulfate solution was used as the electrolyte. The depth of the pool was varied by moving the top bounding plate upward or downward. The difference between the temperatures at the center of the lower curved surface and the upper bounding surface was used to define the heat transfer coefficient. The Nusselt number based on the average heat transfer coefficient at the upper horizontal surface was correlated in the form of Eq. (3). The depth of the layer measured from the lowest point on the spherical surface was used as the characteristic length in defining the Nusselt and Rayleigh numbers. The correlation parameters c and m were obtained for layer depth to diameter ratios of 0.1, 0.15, 0.4, and 0.7. Table I

TABLE I

MIN AND KULACKI'S CORRELATION CONSTANTS FOR USE IN EQ. (3)

L/D	Ra	c	m	Pr
0.108	$9.12 \times 10^7 - 1.27 \times 10^{10}$	2.086	0.127 ± 0.006	2.09–5.94
0.148–0.158	$3.72 \times 10^8 - 3.54 \times 10^{11}$	2.32	0.125 ± 0.006	2.70–7.01
0.403	$3.5 \times 10^{10} - 1.25 \times 10^{13}$	1.092	0.166 ± 0.005	2.27–6.82
0.686–0.703	$5.54 \times 10^{11} - 1.55 \times 10^{14}$	0.28	0.227 ± 0.006	1.90–5.76

lists the values of these parameters along with the range of Rayleigh and Prandtl numbers studied in the experiments. The Prandtl numbers were changed by changing the temperature of the test liquid. The layer depth-to-diameter ratios of 0.1 and 0.15 correspond to layers contained only in the spherical segment. From Table I it is seen that for large L/D ratios the constants c and m tend to approach those obtained by Kulacki and Emara [9] for a horizontal layer with a cooled top and adiabatic bottom.

As has been noted by Min and Kulacki [22], at high Rayleigh numbers (deep pools) the heat transport process is governed by global circulation within the fluid and a predominantly vertical transport by eddies and mixing. The thermal boundary layer on the upper surface is thin and not influenced by the lower surface. For $L/D \leq 0.15$ fluid is contained only in the spherical segment and the results diverge significantly from those for a horizontal layer. For shallower pools (low Ra) a thicker boundary layer is expected to form at the cooled surface, and over a large portion of the pool conduction will be the dominant mode of heat transfer.

C. TRANSIENT CONVECTION

After a pool of molten core material forms in the core or molten core material relocates to the vessel lower head, it will be some time before the convective process in the pool attains a steady state. During the transient period the heat transfer coefficients at the bounding walls will increase and eventually attain steady-state values. The time it takes for the pool to attain steady state is thus an important parameter.

Kulacki and Emara [9] experimentally studied transient natural convection in a volumetrically heated horizontal layer with an adiabatic bottom and cooled at the top. For development of natural convection in the fluid layer, starting with the layer at a uniform temperature, they initiated volumetric heating and thereby imposed a stepped increase in the Rayleigh number. Volumetric heating was also increased in steps after a steady state was reached subsequent to an earlier increase in power. For studies of decay of natural convection, the power was suddenly decreased when steady-state convection initially existed in the layer. During transient development and decay of natural convection, the temperature profiles in the layer was determined as a function of time. For turbulent convection $(\text{Ra} > 10^9)$ an overshoot in the temperature near the upper cold surface was observed during early stages of the transient period. Overshoot or the presence of a knee in the temperature profile was related to local instability resulting from downward transport of cold eddies from the upper surface and strong mixing effects near the edge of the thermal boundary layer. With initiation of volumetric heating, the rate of temperature rise in the conduction-dominated region near the upper surface was different from that in the core, where mixing effects produced a lower mean temperature. Thus, starting with a stable layer near the upper surface, eddy transport was delayed, which in turn led to temperature overshoot. When the core temperature attained a high value, the strength of eddies originating from the upper (cold) surface increased and temperature overshoot was observed to disappear. For a stepped increase in volumetric heating that led to laminar convection $(\text{Ra} < 10^5)$ in steady state, no temperature overshoot was observed. This is due to the fact that during the early stages of flow development heat transfer throughout the layer is mostly due to conduction. The dimensionless time for development of steady state convection was related to the change in Rayleigh number with the change in power; it was correlated as

$$\text{Fo} = 13.42(\Delta\text{Ra}^{-0.213}), \tag{60}$$

where $\text{Fo} = \alpha t_{\text{max}}/L^2$, and ΔRa is based on the total depth of the layer and is given by Eq. (2); t_{max} is the time at which the temperature at a given location in the layer reaches within 2% of its steady-state value. The time for convection to decay when a stepped decrease in power was applied is correlated as

$$\text{Fo} = 13.88(\Delta\text{Ra}^{-0.215}). \tag{61}$$

From Eqs. (60) and (61) little hysteresis is seen to exist between developing and decaying natural convection in the volumetrically heated horizontal liquid layer.

Min and Kulacki [22] experimentally determined the maximum time required for development and decay of natural convection in a volumetrically heated layer bounded from below by a segment of a sphere (Fig. 5d). As stated earlier, in these experiments the lower and side surfaces were insulated whereas the top surface bounding the pool was cooled. The time required for development of convection upon sudden initiation of volumetric heating was correlated as

$$\text{Fo} = 23.34(\Delta\text{Ra})^{-0.223 \pm 0.028} \quad \text{for } z/L = 0,$$
$$\text{Fo} = 58.06(\Delta\text{Ra})^{-0.226 \pm 0.51} \quad \text{for } z/L = 0.5. \tag{62}$$

Similarly, starting with steady-state convection, the time needed for convection to decay when power to the electrodes was suddenly switched off was correlated as

$$\text{Fo} = 23.54(\Delta\text{Ra})^{-0.224 \pm 0.028} \quad \text{for } z/L = 0,$$
$$\text{Fo} = 57.45(\Delta\text{Ra})^{-0.287 \pm 0.052} \quad \text{for } z/L = 0.5. \tag{63}$$

In Eqs. (62) and (63) Fo and Ra are based on the maximum depth of the layer and the correlations cover the ranges $0.108 \leq L/D \leq 0.681$ and $1.27 \times 10^8 \leq \text{Ra} \leq 2.9 \times 10^{13}$. Also, z is the vertical distance from the lowest point on the curved surface. Comparing Eqs. (60) and (61) with Eqs. (62) and (63), it is seen that L/D has no effect on the form of correlations for time of development and decay of convection. It is seen that transient times for a layer bounded from below by a spherical segment are about 70 to 400% higher than those for a horizontal layer. It was argued that the increase in time is due to the inhibiting effect of the curved surface on turbulent transport processes near the cold surface. The authors did not attempt to correlate transient heat transfer coefficients with time.

In the experiments of Asfia [19], microwave heating was initiated starting with a pool of R-113 at a nearly uniform temperature. During the development of convection the temperatures at various locations in the pool were monitored as a function of time. At a given location in the pool, initially the temperature increased monotonically with time. However, at later times the temperatures reached their steady-state values asymptotically. Prior to steady state being achieved the temperatures were observed to oscillate. The magnitude of the temperature oscillations was about 1% of the temperature difference between the pool and the outside cooling water. The period of oscillations varied from 20 to 60 minutes. At upper locations in the pool, temperature overshoot was observed prior to onset of oscillations. The overshoot is believed to result from penetration and mixing of fluid from the core with the upper stratified layer. Asfia also correlated the time needed for the convective process to develop and the temperature near the top of the pool to reach a quasi-static value. For free-surface pools formed in spherical segments cooled at the curved surface, the time for steady state convection was correlated as

$$\text{Fo} = 40(\Delta\text{Ra})^{-0.3}. \tag{64}$$

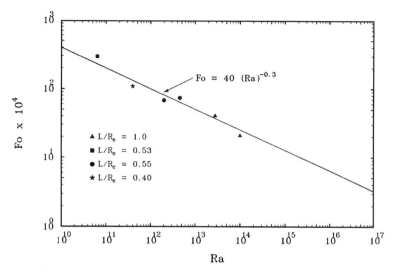

FIG. 10. Dependence on Rayleigh number of time to reach steady-state convection in a spherical cavity.

Figure 10 shows the correlation of the data. In Fig. 10 Ra = ΔRa. Correlation Eq. (64) covered the ranges $0.4 \leq L/R_e \leq 1$ and $5 \times 10^{10} \leq$ Ra $\leq 10^{14}$. No dependence of Fo on L/R_e was observed. It is interesting to note that the proportionality constant in Eq. (64) is about average for $z/L = 0$ and $z/L = 0.5$ in Eq. (62). The exponent of the Rayleigh number also has a value comparable to that obtained by Kulacki and Emara [9] and Min and Kulacki [22]. Asfia [19] also showed that during development of convection the heat transfer coefficient at a given location on the curved wall increased and reached a quasi-static value asymptotically. However, she did not correlate the transient heat transfer coefficient data. Using correlation Eq. (64) and employing the typical properties of core material as given in Table II, Asfia calculated the time to achieve steady-state natural convection in a pool formed in the lower head of a pressurized

TABLE II

TYPICAL THERMOPHYSICAL PROPERTIES OF CORE MATERIAL (2900 K)

Density	8200 kg/m^3
Specific heat	470 J/kg K
Thermal conductivity	3.6 W/mK
Thermal diffusivity	9.3×10^{-7} m^2/sec
Kinematic viscosity	4×10^{-7} m^2/sec
Coefficient of isobaric expansion	1.4×10^{-4} K^{-1}
Volumetric heat generation rate	8×10^5 W/m^3

water reactor to be about $1\frac{1}{2}$ hours. In her calculations Asfia assumed that the pool remained at a constant temperature (melting temperature) for the first half hour as the solids present in the pool melted.

D. CONVECTION IN SUPERPOSED LAYERS

Depending on the manner in which a severe core damage accident progresses, it is possible to form a pool in which stratified layers of different materials exist. For example, for a pool formed in the lower head of the vessel of a light-water reactor, it is quite possible that a layer of steel will overlay a layer containing molten uranium dioxide, Zircaloy, and some steel. As such it is important to know the natural convection heat transfer behavior of the superposed layers with volumetric heating of the lower layer or both layers.

Studies reported in the literature on natural convection in superposed layers are very limited. Schramm and Reineke [23] experimentally and numerically investigated natural convection in two immiscible liquid layers contained in a rectangular cavity. In the experiments the upper layer was silicon oil and the lower layer water. Volumetric heating was present only in the lower layer and the total layer was cooled both at the top and bottom. The ratio of the height of the lighter liquid layer to the total layer height was varied from 0.11 to 0.75. The Rayleigh number based on the total depth and thermophysical properties of the heavier liquid was varied from 2.2×10^6 to 1.1×10^{12}. The isotherm patterns in the upper layer where no volumetric heating was present was similar to Bénard convection, whereas in the lower volumetrically heated layer the isotherms were similar to those obtained in the internally heated layer bounded by rigid cooled walls at the top and bottom.

Fieg [24] has also studied natural convection in superposed horizontal layers bounded at the top and bottom by rigid cooled walls maintained at nearly the same temperature. Heptane and water were used as the lighter and heavier liquids, respectively, and only the lower layer was heated volumetrically. Fieg's conclusions were similar to those of Schramm and Reineke in that the temperature profile in the lower layer was similar to that for the internally heated layer cooled at the top and bottom. The temperature profile in the upper layer was similar to that for Bénard convection. The interface between the two layers behaved like a rigid conducting wall separating the two layers. The interface temperature determined the energy from the volumetrically heated layer that was transferred upwards. It also determined the temperature profile for Bénard convection in the upper layer. The heat transfer data were found to be within 10% of that predicted by using correlations available in the literature for Bénard convection and for an internally heated layer cooled at the top and bottom.

Experimental and numerical studies of natural convection in superposed layers of silicon oil and water and heptane and water have also been reported by Kulacki and Nguyen [25]. In their experiments the heavier layer was volumetrically heated. The layers were bounded by an adiabatic rigid wall at the bottom and a cooled wall at the top. The Nusselt numbers based on the average heat transfer coefficient for silicon–water ($Pr_1/Pr_2 \simeq 12$ and $k_1/k_2 \simeq 0.2$) were correlated as

$$\mathrm{Nu}_2' = 0.186(\mathrm{Ra}_2')^{0.1986} \quad \text{for } \frac{L_1}{L_2} = 0.035$$

$$\text{and } 9.4 \times 10^7 \leq \mathrm{Ra}_2' \leq 4.2 \times 10^{11},$$

$$\mathrm{Nu}_2' = 0.183(\mathrm{Ra}_2')^{0.1988} \quad \text{for } \frac{L_1}{L_2} = 0.111 \tag{65}$$

$$\text{and } 3.4 \times 10^7 \leq \mathrm{Ra}_2' \leq 3.0 \times 10^{11},$$

$$\mathrm{Nu}_2' = 0.115(\mathrm{Ra}_2')^{0.2262} \quad \text{for } \frac{L_1}{L_2} = 0.433$$

$$\text{and } 3.4 \times 10^4 \leq \mathrm{Ra}_2' \leq 9.6 \times 10^{10}.$$

For heptane–water ($Pr_1/Pr_2 = 1$ and $k_1/k_2 \simeq 2$) the data were correlated as

$$\mathrm{Nu}_2' = 0.126(\mathrm{Ra}_2')^{0.255} \quad \text{for } \frac{L_1}{L_2} = 0.04$$

$$\text{and } 1.06 \times 10^7 \leq \mathrm{Ra} \leq 2 \times 10^{11},$$

$$\mathrm{Nu}_2' = 0.112(\mathrm{Ra}_2')^{0.232} \quad \text{for } \frac{L_1}{L_2} = 0.111 \tag{66}$$

$$\text{and } 1.6 \times 10^7 \leq \mathrm{Ra} \leq 1.6 \times 10^{11},$$

$$\mathrm{Nu}_2' = 0.135(\mathrm{Ra}_2')^{0.232} \quad \text{for } \frac{L_1}{L_2} = 0.433$$

$$\text{and } 8.6 \times 10^5 \leq \mathrm{Ra} \leq 4.0 \times 10^{10}.$$

In Eqs. (65) and (66), Nu_2' is defined as

$$\mathrm{Nu}_2' = \frac{q_1(L_1 + L_2)}{k_2 \,\Delta T}. \tag{67}$$

where ΔT is the temperature difference between the insulated lower surface and the cooled upper surface. Ra_2' is defined in the same manner as in Eq. (2), except that the characteristic length is the depth of the lower layer. The uncertainty in the experimentally observed Nusselt and Rayleigh numbers is reported to be less than 5.3%, and predictions from the numerical analysis were found to be in general agreement with the data.

From Eqs. (65) and (66) it is seen that the exponent of Ra'_2 is similar to that obtained by Kulacki and Emara [9] for a single volumetrically heated layer insulated at the bottom and cooled at the top. The empirical constant multiplying Ra'_2, however, is smaller because of the additional resistance imposed by the overlying liquid layer. From the plot of Nu'_2 as a function of L_1/L_2 for fixed Ra_2, Kulacki and Nguyen noted that, starting with $L_1/L_2 = 0$, the Nusselt number decreased, reaching a minimum value, and then increased as L_1/L_2 was increased. The reduction in heat transfer in Nusselt number was attributed to the conductive resistance of the thin upper layer. However, as the thickness of the upper layer increases and turbulent convection develops in the upper layer, the resistance of the layer decreases, which in turn leads to an increase in the Nusselt number. In contrast, Schramm and Reineke [23] noted from their numerical results for superposed layers cooled at both top and bottom that the fraction of heat transported upwards reached a maximum value when the ratio of the upper layer to total layer depth was about 0.25. The reduction in heat transfer for depth ratios less than 0.25 was attributed to conductive resistance imposed by the upper layer, whereas the reduction in heat transfer for depth ratios greater than 0.25 resulted from increased convective resistance of the upper layer with an increase in the depth of the upper layer.

From the reported studies it can be concluded that hydrodynamic coupling between the superposed layers is negligible. As such, correlations reported in the literature for each layer bounded between rigid walls with appropriate boundary conditions can be used to determine heat transport across each layer. The effect of the presence of the upper layer is to impose an additional resistance for transfer of heat from the lower volumetrically heated layer.

E. GENERAL REMARKS

A significant database exists in the literature for natural convection heat transfer in a volumetrically heated pool contained in rectangular, circular, and spherical cavities. The correlations based on the data cover Rayleigh numbers up to 10^{14} for horizontal layers, up to 5×10^{10} for circular cavities, and up to 10^{14} for spherical cavities. For a two-dimensional slice of a torosphere the heat transfer data have been obtained for Ra up to 10^{15}. The Rayleigh number for a molten pool formed in the lower head of a light-water reactor vessel is of the order of 10^{17}. Thus, for applications to a nuclear reactor the correlations must be extrapolated by about three orders of magnitude. Because of the very weak dependence of the heat transfer coefficient on the ratio of pool height to cavity radius, it appears that no significant error will be made if the existing correlations are extended to the reactor case. However, the uncertainty that exists in the thermophysical properties of the core material can lead to a much larger error.

Numerical calculations for natural convection in the three geometries have been performed with and without turbulence models in the governing equations. A good comparison of the numerical predictions with the data has been reported for Ra up to 2×10^{10}, 10^{11} and 10^{14} for rectangular, circular, and spherical cavities, respectively.

Data for the time needed to develop steady-state convection with a stepped increase in volumetric heating and for decay of convection when heating is stopped have been reported. The correlations of these data for a rectangular geometry suggest little hysteresis in the times for development and decay of convection. A correlation for time needed to develop steady-state convection in a volumetrically heated pool contained in a spherical cavity has also been reported. The use of this correlation for a pool formed in the lower head of a reactor vessel suggests that up to 90 minutes will be needed for the convective process to attain steady state.

A limited number of data and correlations have been reported for natural convection in superposed layers of immiscible liquids. In the experiments only the lower layer of heavier liquid was volumetrically heated. It was found that two layers behaved as if separated by a rigid conducting wall. As such, correlations available in the literature that are applicable to each layer can be used. However, no data exist in the literature for where the overlying layer has a low Prandtl number liquid, such as for steel.

III. Dryout Heat Flux

Several studies of the energy that can be removed from a liquid-saturated porous medium without drying out or overheating any region of the medium have been reported in the literature. These studies were motivated by their application to postaccident heat removal in early versions of liquid-metal-cooled reactors, coolability of a degraded core in light-water reactors, nuclear waste disposal, and single- and two-phase convection in geological systems. In the context of dryout heat flux, particle diameter represents an important length scale. For very small particles (particles on the order of tens or hundreds of microns in diameter), capillary and particle drag forces dominate. For particles larger than a few millimeters in diameter, fluid inertia, particle and interfacial drag, and buoyancy are important. In the intermediate range all of the above identified forces need to be considered. Since fluid properties and system variables (e.g., pressure) affect the magnitude of these forces, a critical particle diameter for transition from one regime to another will depend on pressure. When considering dryout heat flux, a distinction between consolidated and unconsolidated media, especially for small particles, must also be made.

The focus of most of the studies reported in the literature has been the onset of dryout conditions rather than the heat transfer process associated with generation of vapor at the heated surface. As such, dryout is considered a hydrodynamic process where, in the absence of an imposed flow, vapor and liquid flow countercurrently in the porous medium. Generally, in carrying out the analysis the following assumptions are made:

1. The pore space is shared by vapor and liquid.
2. Steady-state conditions exist.
3. Fluids are incompressible.
4. The porous medium is isotropic and the bounding walls do not interfere with the flow.
5. Acceleration terms are small.
6. Fluids are at their saturated state when heating is present.

Under these assumptions, the momentum equations for the two phases can be written as[1]

$$\vec{\nabla} P_v = \rho_v \vec{g} + \frac{\vec{F}_{pv}}{\alpha_e \epsilon} + \frac{\vec{F}_i}{\alpha_e \epsilon} \qquad \text{(for the vapor phase)} \qquad (68)$$

and

$$\vec{\nabla} P_l = \rho_l \vec{g} + \frac{\vec{F}_{pl}}{(1 - \alpha_e)\epsilon} - \frac{\vec{F}_i}{(1 - \alpha_2)\epsilon} \qquad \text{(for the liquid phase)}. \qquad (69)$$

The first term on the right-hand side of Eqs. (68) and (69) represents the hydrostatic head, the second term represents the drag of the particles on vapor or liquid, and the last term represents interfacial drag. The forces $\vec{F}_{pv}$, $\vec{F}_{pl}$, and $\vec{F}_i$ are defined per unit of total porous layer volume.

The pressure difference between vapor and liquid is equal to the

---

[1] Here α_e is the effective void fraction.

capillary pressure, that is,

$$P_v - P_l = \bar{P}_c. \tag{70}$$

The equivalent void fraction, α_e, is defined as

$$\alpha_e = \frac{\alpha - \alpha_{min}}{\alpha_{max} - \alpha_{min}}. \tag{71}$$

In Eq. (71), α_{max} and α_{min}, respectively, correspond to the maximum void fraction or residual saturation and minimum void fraction or maximum saturation in the porous medium. The early studies with packed columns of sand revealed that α_{max} could be less than unity and $\alpha_{min} > 0$. In the context of dryout, if no noncondensible gas is trapped in the porous matrix, α_{max} will be equal to unity and $\alpha_{min} = 0$.

A. ONE-DIMENSIONAL MODELS BASED ON EMPIRICAL RELATIONS FOR RELATIVE PERMEABILITIES

Most of the reported studies on dryout heat flux are one-dimensional in nature. For such a case, Eqs. (68) and (69) can be written as

$$\frac{dP_v}{dz} = \rho_v g + \frac{F_{pv}}{\epsilon \alpha_e} + \frac{F_i}{\epsilon \alpha_e}, \tag{68a}$$

$$\frac{dP_l}{dz} = \rho_l g + \frac{F_{pl}}{\epsilon(1 - \alpha_e)} - \frac{F_i}{\epsilon(1 - \alpha_e)}. \tag{68b}$$

The concept of relative permeabilities is used to apply Darcy's equation or the modified Darcy's equation (Forschheimer's equation) to two-phase flows. Expressions for particle drag on gas and liquid are written as

$$F_{pv} = \alpha_e \epsilon \left[\frac{1}{Kk_v} \mu_v j_v + \frac{1}{\eta \eta_v} \rho_v j_v |j_v| \right], \tag{72}$$

$$F_{pl} = (1 - \alpha_e) \epsilon \left[\frac{1}{Kk_l} \mu_l j_l + \frac{1}{\eta \eta_l} \rho_l j_l |j_l| \right], \tag{73}$$

where K is the permeability of the porous media and is defined as

$$K = \frac{D_p^2 \epsilon^2}{C_5(1 - \epsilon)^2}. \tag{74}$$

Values of constant C_5 varying from 150 to 180 have been used in the literature. The parameter η is an inertial counterpart of K and is defined as

$$\eta = \frac{D_p \epsilon^3}{1.75(1 - \epsilon)}.$$ (75)

The parameters, κ_v, κ_l, η_v, and η_l are the relative permeabilities.

Usually the functional dependence of capillary pressure on saturation is obtained from Leverett's [26] data obtained with clean unconsolidated sands under both drainage and imbibition conditions. Hystersis in capillary pressure was observed under the two modes of experiments. For dryout of an initially liquid-saturated porous layer, the drainage condition is more appropriate. The capillary pressure has been correlated as

$$\bar{P}_c = \sqrt{\frac{\epsilon}{K}} \cdot \sigma \cos \theta \cdot J(S \equiv 1 - \alpha_e),$$ (76)

where θ is the liquid–solid contact angle and J is the Leverett function.

Eliminating the pressure difference between vapor and liquid from Eqs. (68a) and (68b) through the use of the capillary pressure, neglecting F_i, and substituting for F_{pv} and F_{pl} from Eqs. (72) and (73), an equation relating the superficial velocities of vapor and liquid and the void fraction is obtained as

$$\sqrt{\frac{\epsilon}{K}} \, \sigma \cos \theta \, \frac{dJ}{d\alpha_e} \cdot \frac{d\alpha_e}{dz} = -(\rho_l - \rho_v)g + \left[\frac{1}{Kk_v} \mu_v j_v + \frac{\rho_v}{\eta\eta_v} j_v |j_v| \right]$$

$$- \left[\frac{1}{Kk_l} \mu_l j_l + \frac{\rho_l}{\eta\eta_l} j_l |j_l| \right].$$ (77)

Equation (77) is very general and is applicable to both co- and counter-current flow of vapor–gas and liquid.

If phase change occurs at the heated surface, steady-state liquid flow towards the surface must equal vapor outflow from the surface. As such, the mass conservation condition can be written as

$$\rho_v j_v = -\rho_l j_l.$$ (78)

For a bottom-heated porous medium with all of the dissipated energy being utilized in evaporation, the energy balance at the surface yields

$$\rho_v j_v = -\rho_l j_l = q/h_{fg}.$$ (79)

If, on the other hand, a porous medium is volumetrically heated, the cross-sectionally averaged heat flux will be a function of distance in the vertical direction. For this case,

$$q = \int_0^z Q_v (1 - \epsilon) \, dz. \tag{80}$$

Substitution of Eq. (79) into (77) and nondimensionalization yields

$$-\frac{1}{\sqrt{K^*}} \cos\theta \, \frac{dJ}{d\alpha_e} \frac{d\alpha_e}{dz'} = -q^* \left[\frac{\nu_l}{\nu_v} \cdot \frac{1}{\kappa_l} + \frac{1}{\kappa_v} \right]$$

$$- q^{*2} \mathrm{Gr} \left[\frac{\rho_v}{\rho_l} \frac{1}{\eta_l} + \frac{1}{\eta_v} \right] + 1, \tag{81}$$

where

$$q^* = \frac{q \nu_v}{K h_{fg} g (\rho_l - \rho_v)}, \tag{82}$$

$$K^* = \frac{K g (\rho_l - \rho_v)}{\epsilon \sigma}, \tag{83}$$

$$\mathrm{Gr} = \frac{g (\rho_l - \rho_v) K^2}{\rho_v \nu_v^2 \eta}, \tag{84}$$

$$z' = \frac{z}{\sqrt{\dfrac{\sigma}{g (\rho_l - \rho_v)}}}. \tag{85}$$

1. Small Particles

For small particles (very small Gr), the second term on the right-hand side of Eq. (81) can be neglected. If the capillary pressure contribution is also assumed to be negligible, as has been done by Hardee and Nilson [27] and Bau and Torrance [28], Eq. (81) reduces to

$$q^* = \left[\frac{\nu_l}{\nu_v} \frac{1}{\kappa_l} + \frac{1}{\kappa_v} \right]^{-1}. \tag{86}$$

Hardee and Nilson [27] and Bau and Torrance [28] also assumed that relative permeabilities κ_v and κ_l varied with α_e as

$$\kappa_v = \alpha_e \quad \text{and} \quad \kappa_l = (1 - \alpha_e). \tag{87}$$

Upon substitution for κ_v and κ_l, Eq. (86) becomes

$$q^* = \frac{\alpha_e(1 - \alpha_e)}{(1 - \alpha_e) + \alpha_e \dfrac{\nu_l}{\nu_v}}. \tag{88}$$

Equation (88) yields two possible roots for α_e. The only physical solution is the one which yields an increase in α_e with q^*. It should be noted that for a bottom-heated porous layer α_e is constant throughout the porous layer whereas for a volumetrically heated layer it will be a function of distance in the vertical direction. Bau and Torrance reported that dependence of α_e on q^* given by Eq. (88) was consistent with the data of Cornwell et al. [29].

Equation (88) involves two unknowns. To obtain the maximum heat flux another relation is needed. This relation is obtained by optimizing the heat flux with α_e. The value of α_e corresponding to the maximum heat flux need not be equal to unity. Since the maximum heat flux condition corresponds to a breakdown in the force balance, a heat flux slightly higher than the maximum will cause the surface to eventually dry out. The maximum heat flux will thus be nearly the same as the dryout heat flux. The dryout heat flux corresponding to Eq. (88) is

$$q_d = \frac{K h_{fg} g (\rho_l - \rho_v)}{\left(\sqrt{\nu_v} + \sqrt{\nu_l}\right)^2}. \tag{89}$$

According to Eq. (89), the dryout heat flux in porous layers of small particles should vary as D_p^2.

Several investigators (e.g., Ogniewicz and Tien [30], Udell [31], and Chuah and Carey [32]) have shown that for very small particles capillary pressure cannot be neglected. Several expressions correlating Leverett's data have been reported in the literature. For example, Lu and Chang [33] used Udell's [31] correlation of Leverett's data as

$$J = a\alpha_e + b\alpha_e^2 + c\alpha_e^3, \tag{90}$$

where, a, b, and c are empirical constants having values of 1.417, 2.12, and 1.263, respectively. Lipinski [34], on the other hand, has used the expression

$$J = \frac{\left(\dfrac{\alpha_e}{1 - \alpha_e}\right)^{m_1}}{\sqrt{5}} \tag{91}$$

where m_1 is an empirical constant having a value of 1.75.

Several expressions other than those used by Hardee and Nilson [27] and Bau and Torrance [28] for relative permeabilities have also been reported in the literature. Lipinski [34] and Lu and Chang [33], among others, have used theoretical expressions derived by Brooks and Corey [35] and Corey [36] as

$$\kappa_v = (1 - \alpha_e)^n \tag{92}$$

and

$$\kappa_l = \alpha_e^2 \left(1 - (1 - \alpha_e)^m\right), \tag{93}$$

where

$$n = 3 + 2/\lambda, \tag{94}$$

$$m = 1 + 2/\lambda. \tag{95}$$

In Eqs. (94) and (95), λ is an index of pore size distribution. Corey [36] suggests a value of 2 for a typical porous medium. For a uniform pore size structure $\lambda \rightarrow \infty$. Values of λ substantially lower than 2 have been suggested for well-structured porous media and porous media having secondary porosity.

When capillary pressure is retained, Eq. (81) for small particles can be written in integral form as

$$-\frac{\cos \theta}{\sqrt{K^*}} \int_{\alpha_{eb}}^{\alpha_{et}} \frac{dJ}{d\alpha_e} d\alpha_e = \int_0^{L'} \left[1 - q^*\left(\frac{\nu_l}{\nu_v} \frac{1}{\kappa_l} + \frac{1}{\kappa_v}\right)\right] dz'. \tag{96}$$

For a porous layer of a given dimensionless height, L', Eq. (96) represents one equation involving three parameters, namely, void fraction α_{eb} at bottom, void fraction α_{et} at top, and heat flux q^*. Thus, one parameter has to be specified to obtain a relationship between the other two. Since capillary pressure does not play a role in deep beds, the dryout flux for volumetrically (heat flux evaluated at the top) or bottom-heated beds should be the same. However, the data obtained with unconsolidated porous layers show that the observed dryout heat fluxes in volumetrically heated beds can be twice as large as in bottom-heated porous layers. This behavior appears to be caused by the formation of vapor channels in the porous layer which have not been included in the model and will be discussed later.

2. Large Particles

With an increase in the diameter of the particles, Gr increases and the buoyancy force dominates the viscous force. In this case the second term on the right-hand side of Eq. (81) cannot be neglected. In fact, for very large particles the second term dominates the first term. For very large particles (large K) the capillary contribution also becomes small. Thus, in this limit Eq. (81) yields

$$q^* = \mathrm{Gr}^{-1/2} \left[\frac{\rho_v}{\rho_l} \frac{1}{\eta_l} + \frac{1}{\eta_v} \right]^{-1/2}. \tag{97}$$

Once a functional dependence of η_v and η_l on α_e is specified, Eq. (97) provides a unique relationship between the heat flux and the void fraction. To obtain the dryout heat flux, a second relation can be obtained by optimizing the heat flux with α_e. In volumetrically heated particulate layers it is the heat flux near the top of the layer. In the literature several empirical forms for η_v and η_l have been given without physical justification. Lu and Chang [33] have assumed that

$$\eta_v = \kappa_v \quad \text{and} \quad \eta_l = \kappa_l, \tag{98}$$

where κ_v and κ_l are given by Eqs. (92) and (93). Lipinski [34], on the other hand, has assumed

$$\eta_v = \alpha_e^5 \quad \text{and} \quad \eta_l = (1 - \alpha_e)^5. \tag{99}$$

From Eq. (97) it can be deduced that

$$q_d \sim D_p^{1/2}. \tag{100}$$

Thus in the limit of very large particles, the dryout heat flux varies as the square root of the particle diameter, whereas for very small particles, as was seen earlier, the dryout heat flux varies as the square of the particle diameter. In the intermediate region both the first and second terms on the right-hand side of Eq. (81) are equally important. Figure 11 shows a plot of maximum heat flux (dryout heat flux) as a function of particle diameter for very deep beds. The predictions are obtained for water at one atmosphere pressure by using the complete right-hand side of Eq. (81). The bed porosity was assumed to be 0.4 and Lu and Chang's relative permeabilities were employed. The predicted results should be valid for both volumetrically and bottom-heated particulate layers. In Fig. 11 data obtained by several investigators with volumetrically and bottom-heated porous layers are also plotted [37]. It is noted that the model tends to underpredict the volumetrically heated data for both small and large particles. In several studies the empirical formulations of relative permeabilities have been adjusted to obtain better agreement between the data and the predictions.

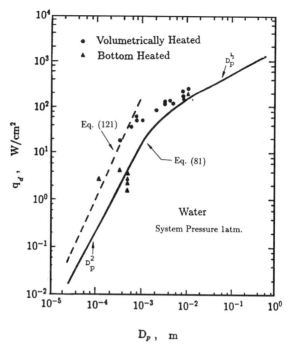

FIG. 11. Comparison of predicted and observed dryout heat flux in a volumetrically heated porous layer.

B. A Mechanistic Model for Interfacial Drag and Relative Permeabilities in Porous Media Composed of Large Particles

The differences between model predictions and data can be attributed to uncertainties in several interacting factors. In unconsolidated porous layers of small particles some of the contributing factors are the formation of vapor channels, uncertainty in the functional dependence of capillary pressure and relative permeabilities on saturation, and the properties of the porous media. In porous layers of large particles, uncertainty in specification of the phasic permeabilities and assumptions such as neglect of interfacial drag and the hydrodynamic interaction between the particular bed and the overlying liquid layer can also play an important role. For particles of all sizes the phase change process at the heated surface can also influence the dynamics of the flow. Although several empirical constants embedded in various constitutive relationships can be adjusted to match the available data, such an approach is destined to fail as the scope and range of the database expands.

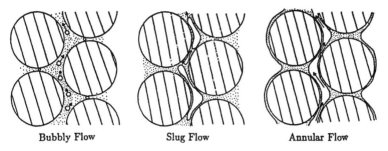

Bubbly Flow   Slug Flow   Annular Flow

FIG. 12. Two-phase flow patterns in a porous layer of large particles.

To eliminate some of the empiricism associated with the modeling effort described earlier for large particles, Tung and Dhir [38] have attempted to obtain mechanistic models for relative permeabilities and interfacial drag. These models rely heavily on visual observations of co- and countercurrent flow of air and water in beds composed of large-diameter ($D_p \geq 3$ mm) glass particles. The visual observations have suggested that flow patterns similar to those found in tubes also exist in porous layers [38]. Figure 12 shows the flow patterns. As the gas flow rate is increased while keeping the liquid flow rate constant, the flow pattern changes from bubbly, to slug, to annular flow. Pure bubbly flow persists until discrete bubbles begin to merge. The onset of bubble merger occurs at

$$\alpha_e = 0.6(1 - \gamma)^2, \tag{101}$$

where

$$\gamma = D_b/D_p$$

and

$$D_b = 1.35 \left[\frac{\sigma}{g(\rho_l - \rho_v)} \right]^{1/2}. \tag{102}$$

Since as $\gamma \to 0$ flow in the pores will approach that in tubes, the upper limit on Eq. (101) is 0.3. Transition from bubbly to slug flow is complete when $\alpha_e = \pi/6$, pure slug flow persists up to $\alpha_e = 0.6$, the slug–annular transition lasts up to $\alpha_e = \pi\sqrt{2}/6$ and pure annular flow continues to persist up to $\alpha_e \cong 1$. In bubbly and slug flows, the relative permeabilities for particle drag on gas were evaluated [39] as

$$\kappa_v = \left(\frac{1 - \epsilon}{1 - \epsilon\alpha_e} \right)^{4/3} \alpha_e^3; \qquad \eta_v = \left(\frac{1 - \epsilon}{1 - \epsilon\alpha_e} \right)^{2/3} \alpha_e^3 \tag{103}$$

and in the annular flow as

$$\kappa_v = \left(\frac{1 - \epsilon}{1 - \epsilon\alpha_e}\right)^{4/3} \alpha_e^2; \qquad \eta_v = \left(\frac{1 - \epsilon}{1 - \epsilon\alpha_e}\right)^{2/3} \alpha_e^2. \qquad (104)$$

In the transition between slug and annular flow, a weighted average was used. It should be noted that according to the above expressions the porosity of the porous layer directly affects the relative permeabilities. In contrast, in the empirical forms used by Lu and Chang [33] and Lipinski [34], porosity affected saturation only through the Leverett function. An expression for liquid–relative permeabilities applicable to all flow regimes was found to be

$$\kappa_l = \eta_l = (1 - \alpha_e)^3. \qquad (105)$$

To evaluate the interfacial drag, a drift velocity approach was used. The drift velocity is defined as the velocity of the gas bubble or slug relative to the two-phase mixture:

$$j_s = j_v \frac{1 - \alpha_e}{\alpha_e} - j_l. \qquad (106)$$

The drag on a bubble was assumed to contain both viscous and inertial components.

For bubbly flow, an expression for the interfacial drag was found to be

$$F_i = C'_v \frac{\mu_l j_s}{D_b^2} + C'_i \frac{[\rho_l(1 - \alpha_e) + \rho_v \alpha_e] j_s^2}{\epsilon D_b}, \qquad (107)$$

where

$$C'_v = 18\alpha_e f, \qquad (108)$$

$$C'_i = 0.34(1 - \alpha_e)^3 \alpha_e f^2. \qquad (109)$$

The above expressions are valid only as long as

$$\alpha_e < \frac{\pi}{3} \frac{(1 - \epsilon)}{\epsilon} \gamma(1 + \gamma)[6\eta_0 - 5(1 - \gamma)] \equiv \alpha_0. \qquad (110)$$

The parameter η_0 is defined as

$$\eta_0 = \left[\frac{\pi\sqrt{2}}{6(1 - \epsilon)}\right]^{1/3}. \qquad (111)$$

The geometric parameter f in Eqs. (108) and (109) is a correction factor

[39] for the slip velocity when the bubbles move along the surface of the particles. At higher void fractions the coefficients C'_v and C'_i are given as

$$C'_v = 18(\alpha_0 f + \alpha_e - \alpha_0),$$ (112)

$$C'_i = 0.34(1 - \alpha_e)^3(\alpha_0 f^2 + \alpha_e - \alpha_0).$$ (113)

By assuming that in slug flow the viscous coefficient C'_v was given by viscous flow around an ellipsoid and an inertial coefficient could be obtained from the correlation of Ishii and Chawla [40], the coefficients C'_v and C'_i were obtained as

$$C'_v = 5.21\alpha_e,$$ (114)

$$C'_i = 0.92\alpha_e(1 - \alpha_e)^3.$$ (115)

In bubbly slug flow, a weighted average between bubbly and slug flows was used. In annular flow, the liquid film follows the particles and an expression for the interfacial drag force per unit of total bed volume was obtained as

$$F_i = \frac{\epsilon \mu_v}{\kappa \kappa_v} j_s + \frac{\alpha_e \epsilon}{1 - \alpha_e} \frac{\rho_v j_s^2}{\eta \eta_v},$$ (116)

where κ_v and η_v are given by Eq. (104).

If capillary pressure is neglected and the pressure gradients in gas and liquid are eliminated between Eqs. (68a) and (68b), a single equation is obtained containing j_v, j_l, and α. In principle such an equation can be used to evaluate α_e for given j_v and j_l. However, this procedure requires an iterative process since the drag forces depend on the flow regime and, in turn, on the void fraction. The procedure is facilitated if j_l and α_e are assumed to be given and the resulting equation is solved for j_v. The equation is quadratic, and the possibility of obtained a parasitic root exists. Tung and Dhir [38] evaluated this equation for several particle sizes and found a very good agreement between predicted and experimentally observed void fractions and pressure gradients under both co- and counter-current flows of gas and liquid over nonheated particles.

Figure 13 shows the various dimensionless forces acting on the two phases for the parameters corresponding to those shown in the figure itself. For the solution shown by the solid line, the particle–liquid drag can be seen to increase monotonically with j_v and has an infinite slope at the flooding limit. The particle–gas drag increases very rapidly at small j_v and α_e due to the α_e^3 term in the relative permeabilities. It reaches a maximum value at $j_v \simeq 0.015$ m/s and then decreases monotonically since the flow area (α_e) increases at a rate faster than j_v^2. In a large range of j_v

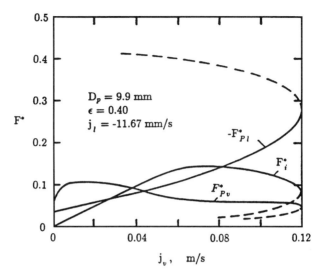

FIG. 13. Magnitude of various forces on approach to flooding.

near the flooding limit, the particle–gas drag, however, remains fairly constant. The interfacial drag initially increases with j_v due to an increase in the interfacial area associated with an increase in the void fraction. At higher void fractions the term $(1 - \alpha_e)^3$ in the drag coefficient results in a decrease in the interfacial drag. Figure 13 shows that for $j_v > 0.06$ m/s the interfacial drag can be more than twice the particle–gas drag. Neglect of this term can therefore result in a serious error in the predictions and can also obscure some of the physically observed phenomena. It should also be mentioned that the void fraction, α_e, used in the above analysis is based on the assumption that $\alpha_{min} = 0$ and $\alpha_{max} = 1$. In reality some inactive voids may exist in the porous layer, causing $\alpha_{min} \neq 0$. If the particles are heated and phase change occurs, α_{max} will always approach unity.

The flow regimes that exist in the porous layer at the onset of counter-current flooding depend on the particle size and superficial velocity of the liquid or gas. Table III lists the range of liquid superficial velocities for a particular flow regime at countercurrent flooding conditions for particles of 19, 9.9, and 5.8 mm in diameter. The predictions are for an air–water system at room temperature and pressure. It is interesting to note that, in contrast to flow in tubes, onset of the countercurrent flooding condition in porous layers can occur in any flow regime. The countercurrent flooding predictions of the above model are found to compare well with the correlation of Marshall and Dhir [41] at a medium range of dimensionless

TABLE III
PREDICTED FLOW REGIMES AT FLOODING

| | $j_l^{*1/2}$ | | |
Flow regime	D_p = 19 mm	D_p = 9.9 mm	D_p = 5.8 mm
Bubbly	0.83–1.07	0.91–1.22	1.13–1.24
Bubbly-slug	0.55–0.83	0.49–0.91	0.44–1.13
Pure slug	0.45–0.55	0.40–0.49	0.33–0.44
Slug-annular	0.21–0.45	0.17–0.40	0.11–0.33
Pure annular	0–0.21	0–0.17	0–0.11

liquid superficial velocities:

$$j_l^{*1/2} + j_v^{*1/2} = 0.875; \qquad 0.15 \le j_l^{*1/2} \le 0.73, \qquad (117)$$

and with that of Schrock *et al.* [42] at low superficial velocities.

$$j_l^{*0.38} + j_v^{*0.38} = 1.075; \qquad 0 \le j_l^{*1/2} \le 0.45, \qquad (118)$$

$$j_l^* = |j_l| \left[6\rho_l(1 - \epsilon)/g(\rho_l - \rho_v)\epsilon^3 D_p \right]^{1/2}. \qquad (119)$$

$$j_v^* = |j_v| \left[6\rho_v(1 - \epsilon)/g(\rho_l - \rho_v)\epsilon^3 D_p \right]^{1/2}. \qquad (120)$$

C. EFFECT OF SYSTEM VARIABLES ON DRYOUT HEAT FLUXES

1. Unconsolidated Porous Media

In most of the applications with respect to nuclear safety, the particles are not constrained and can rearrange themselves depending on the nature of the forces that act on them. Visual observations (see, e.g., Dhir and Catton [43]) of liquid-saturated volumetrically heated beds of small particles show the existence of vapor channels that tend to grow in diameter in the direction of vapor flow. In fact, the first model on dryout heat fluxes proposed by Dhir and Catton relied heavily on the premise that vapor flowing through the channels (well-defined flow paths) encountered minimal resistance while the liquid flowing through the interstitials encountered particle resistance. The drag experienced by the liquid was balanced by buoyancy resulting from vapor flowing in the channels. When particle–vapor drag, interfacial drag, and capillary pressure were neglected and Darcy's equation was used to obtain particle drag on the liquid, an expression for the dryout heat flux was obtained as

$$q_d = C_6 \frac{\epsilon^3 D_p^2}{(1 - \epsilon)^2} \frac{h_{fg} g(\rho_l - \rho_v)}{\nu_l}.$$ (121)

The constant C_6 was determined empirically and its value of 10^{-4} for a bed of volumetrically heated particles was found to be twice that for a bottom-heated porous layer. The difference was rationalized by the fact that in volumetrically heated bed liquid did not encounter resistance in the upper portion of the particle layer, where channels were well developed and particles were alleviated.

It has been argued [34] that in the unchanneled portion of the porous layer (between the base of the particulate layer and the base of channels) countercurrent flow of vapor and liquid takes place in the interstitials of the particles. In this case Eq. (81) is applicable in the lower region of the particulate layer. Hence, the effect of the presence of the channels is to reduce the effective height of the volumetrically heated porous layer and thereby lead to a higher dryout heat flux. In bottom-heated beds the presence of channels does not affect the calculated dryout heat flux because a certain flow distance is required for the channels to form.

If the particulate layers become very shallow, the base of the channels in a volumetrically heated bed will extend to the bottom of the porous layers. In this type of situation the particle layer may behave as a fluid with saturation close to unity everywhere. In porous layers of very small particles, burping behavior (similar to that reported by Schrock et al. [44]) leading to cyclic release of vapor from a porous layer may also be observed. At present, little mechanistic understanding of dryout heat flux in shallow beds exists. Dhir and Catton [43] did develop a semitheoretical model for dryout heat flux in shallow beds. The model employed concepts similar to those for pool boiling on a flat plate and yielded dryout heat fluxes that increased as bed depth was decreased.

2. Overlying Liquid Layer Depth

The depth of the overlying liquid layer is found to have little effect on the dryout heat flux in deep beds of small particles as long as the liquid is at its saturation temperature. However, as the size of the particles becomes large the overlying liquid layer can start to interact hydrodynamically with the porous layer underneath. Marshall and Dhir [41], from their adiabatic experiments with air and water as test fluids, have identified the existence of a transition layer in which adjustment of the void fraction in the porous layer to that in the liquid layer takes place. Consequently, the effective height of the particulate layer in determining the limiting conditions is reduced, producing an effect similar to that of the channels in beds of small particles. For a volumetrically heated layer, this reduction in effective bed height corresponds to (for a fixed flooding limit) an increase in the dryout heat flux. Since the limited data from the hydrodynamic experiments showed that the height of the transition layer was nearly constant, its effect on dryout heat flux diminishes as the bed becomes deep or is bottom heated. Our present understanding of the transition layer is only qualitative. Further work is required to quantify the behavior of the transition layer.

3. *Liquid Subcooling*

In an unconsolidated porous medium the presence of a layer of sub-cooled liquid can limit the penetration of the channels in the subcooled zone. In addition, in both unconsolidated and consolidated media the overlying liquid layer may convectively interact with the porous layer underneath.

Liquid subcooling has little effect on dryout heat flux in bottom-heated layers since condensation of vapor occurs away from the surface and the liquid near the surface is saturated. In volumetrically heated porous layers of large particles, liquid subcooling will enhance dryout heat flux since some energy in the upper region will be utilized in heating the down-flowing liquid to its saturation temperature. In unconsolidated volumetrically heated porous layers of small particles, liquid subcooling can enhance or degrade dryout heat flux depending on the type of interaction that occurs between the vapor channels and the subcooled liquid.

4. *System Pressure*

In some scenarios the system pressure can be much different than one atmosphere; hence, it is natural that the effect of pressure be scaled. Since the two-phase momentum equations and the energy equation contain fluid properties, the effect of pressure on dryout heat flux and on the relationship between superficial velocities and void fraction is already built into these equations. However, because of the scatter in the data and the influence of other extraneous variables, it is not always possible to determine if the empirical constants used in defining relative permeabilities continue to hold at higher pressures and for different fluids. To circumvent this difficulty, Catton and Jakobsson [45] have attempted to correlate their R-113, methanol, acetone, and water data obtained at pressure above and below one atmosphere with reduced pressure. Using corresponding state arguments similar to those advanced by Lienhard and Schrock [46], they have obtained an expression for the scaling heat flux as

$$q_s = P_c \frac{(g\sigma)^{1/4}}{(\rho_l - \rho_v)} \left[\frac{8MP_c}{3\mathscr{R}T_c} \right]^{3/4} . \tag{122}$$

It was noted that for particles larger than 1.6 mm in diameter the data for the three fluids collapsed on a single curve. The acetone data for 3.2-mm-diameter particles, however, showed a large scatter. With an increase in particle diameter, fluid inertia becomes more important than the viscous drag, and it is not surprising that the authors have been able to scale the effect of fluid properties and pressure in large particles. For small particles ($0.59 \leq D_p \leq 0.79$ mm) the scaling appears to fail since the data for the three fluids lie on three distinct curves. This behavior is also not unexpected since for small particles the viscosity of the fluids plays an important role in determining particle drag. Equation (122) for the scaling heat flux was obtained without including molecular viscosity.

5. Forced Flow

Dryout heat fluxes in volumetrically heated layers can be increased with the feeding of liquid from below. At low mass flow rates with an overlying liquid layer, both co- and countercurrent flow conditions may exist in the lower and upper parts, respectively, of the porous layer. However, as the mass flow rate is increased or if a liquid layer is not present over the porous layer, cocurrent flow will exist throughout the particulate bed. Detailed discussion of dryout heat flux and heat transfer in porous layers with feeding of liquid from below is given by Dhir [47].

D. MULTIDIMENSIONAL FLOWS

In many instances, the porous medium may contain regions of distinctly different permeabilities even though each region by itself may be isotropic. In some instances volumetrically heated particles may be surrounded by nonheated particles. Tung and Dhir [37] have used mechanistically derived relative permeabilities, as described earlier to study multi-dimensional flows. They included interfacial drag in the model but assumed that for large particles capillary pressure was negligible. A finite-element scheme was used and fluid velocities were eliminated in favor of pressure and void fraction. A constant pressure condition was imposed at the porous layer-overlying liquid layer interface. Figure 14 shows the vapor and liquid flow patterns in an axisymmetric geometry analyzed by Tung and Dhir in which particles in the central region were volumetrically heated while surrounded

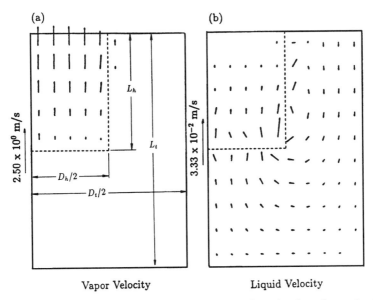

FIG. 14. (a, b) Geometrical configuration used in two-dimensional studies and velocity vectors at onset of dryout.

by glass particles. A liquid layer existed over both the heated and non-heated particles. The flow pattern shown in Fig. 14 is at the onset of dryout. Deficiency of liquid in the upper portion of the porous layer is evident. Dryout occurs because of liquid starvation in the heated region rather than by countercurrent flooding limitation.

Figure 15 shows a comparison of the prediction of the dryout heat flux with the data of Tsai [48]. In Tsai's experiments, D_h was varied parametrically, with D_t, L_t, and L_h (Fig. 14) being 15, 25, and 10 cm, respectively. The diameter of the particles in the heated and nonheated regions was 3.2 and 6.0 mm, respectively. While plotting the predictions and the data, the dryout heat fluxes have been nondimensionalized, with the dryout heat flux obtained under purely countercurrent flooding conditions. In Fig. 15 three distinct regions of dryout can be identified. At $D_h/D_t \cong 1$ a reduction in the diameter of the heated zone causes a flow channel to establish for the liquid and to flow downward in the nonheated region while vapor and liquid flow cocurrently upward in the heated region. Liquid friction is thus drastically reduced and the dryout heat flux increases rapidly as the diameter ratio is reduced from a value near unity. Once the flow channel for the liquid has been established in the nonheated region, liquid friction

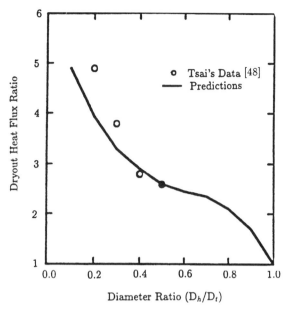

FIG. 15. Dependence of dryout heat flux on the ratio of diameter of heated to nonheated region.

no longer controls the amount of liquid which enters the heated region. The dryout heat flux then is essentially controlled by a balance of the pressure drops in the heated and nonheated channels, with a relatively mild influence of the side flow. Consequently, in the middle portion the dryout heat flux is relatively insensitive to the diameter ratio.

When the diameter ratio is near zero the side flow becomes dominant and the dryout heat flux is mostly controlled by the amount of liquid that enters through the side of the heated region. The principal flow path for liquid is thus very short and the dryout heat flux increases very rapidly with decreasing diameter of the heated region. Good agreement of the predictions with the data for this two dimensional situation highlights the merits of the mechanistically developed models for relative permeabilities and interfacial drag.

E. GENERAL REMARKS

One-dimensional dryout heat flux models based on hydrodynamic considerations alone appear to be adequate for engineering applications. However, for nuclear reactor applications large uncertainties in predictions result from the difficulty in a priori specification of the mean particle size and shape and the porosity of the porous layer. Deviations of the laboratory data from the predictions are a result of interactions of several ill-understood parameters. In porous layers of small particles these are the formation of vapor channels and the effect of porous media and fluid properties (e.g., wettability) on the relative permeabilities, and the capillary pressure. Spatial variations in the porous medium properties, and their effect on the constitutive relations, and coupling between the overlying liquid layer and the porous medium are considered to be important parameters for large particles. Extensions of one-dimensional models to multidimensional situations appear to yield good agreement with the data. Future studies should stress the hydrodynamic process at the interface between regions of widely differing particle sizes and porosities.

IV. Summary

1. A significant amount of data for natural convection in volumetrically heated layers contained in rectangular, cylindrical, and spherical geometries are available in the literature. Correlations based on these data cover Rayleigh numbers up to 10^{14} (spherical geometry). For applications to a pool of molten core material formed in the lower head of the reactor vessel, the interest is in heat transfer at Rayleigh numbers up to 10^{17}. The form of Nusselt number correlation for this geometry is such that the correlations can be extended to Ra up to 10^{17} without introducing much error. However, uncertainty in the thermophysical properties of the core material can be a source of significant error.

2. Limited data and correlations for convective heat transfer in superposed layers exist. The data suggest that two layers can be treated as independent layers, and correlations representative of each layer can be used to predict the rate of heat transfer. However, no data exist when the overlying layer has a very low Prandtl number, as is the case for a layer of steel overlying a layer of core material.

3. Dryout heat flux models based on hydrodynamic considerations are adequate for engineering applications. However, the effects of several variables, such as formation of vapor channels, coupling between the particulate bed and the overlying liquid layer, and the effect of system variables on the constitutive relationships for relative permeabilities are not understood.

Nomenclature

$C_1, C_2, C_3,$
$\quad C_4, C_5, C_6$ empirical constants
c empirical constant
D bed diameter
D_b bubble diameter
D_h heated zone diameter
D_p diameter of particles constituting the porous medium
D_p' dimensionless diameter of particles, $D_p \sqrt{g(\rho_l - \rho_v)/\sigma}$
F^* dimensionless force, per unit volume, $F/g(\rho_l - \rho_v)$
F_i vapor–liquid interfacial drag per unit volume
Fo   Fourier number, $\alpha t_{max}/L^2$
F_{pv}, F_{pl} particle drag per unit volume on vapor and liquid, respectively
f geometrical factor
Gr   Grashof number
g gravitational acceleration
$\bar{h}_{av}$ area and time-averaged heat transfer coefficient over the surface
h_{fg} latent heat of vaporization
J Leverett's function
j_l, j_v liquid and vapor superficial velocities
j_s slip velocity between phases
K viscous permeability
k thermal conductivity of liquid layer
k_l, k_v liquid and vapor thermal conductivities
L bed or liquid layer depth
L' dimensionless bed depth, $L\sqrt{g(\rho_l - \rho_v)/\sigma}$
M molecular weight
m exponent
Nu   Nusselt number
P pressure
P^* dimensionless pressure gradient, $(-dP/dz)/g(\rho_l - \rho_v)$
P_c critical pressure
$\bar{P}_c$ capillary pressure
P_l, P_v pressure in liquid and vapor, respectively

$\dot{Q}_v$ volumetric heat generation rate
q heat flux
q_d dryout heat flux
q_w wall heat flux
Pr   Prandtl number
$\mathscr{R}$ gas constant
R radius
R_e equivalent radius of a spherical cavity
Ra   Rayleigh number, $g\beta\dot{Q}_v(L$ or R or $R_e)^5/k\nu\alpha$
Ra_E external Rayleigh number, $g\beta L^3(T_1 - T_2)/\alpha\nu$
T_c critical temperature
T_{max} maximum temperature in the liquid layer
T_{sat} saturation temperature
T_w wall temperature
y coordinate normal to the surface
z vertical axis
z' dimensionless distance in the vertical direction, $z\sqrt{g(\rho_l - \rho_v)/\sigma}$

GREEK LETTERS

$\alpha, \alpha_e, \alpha_w$ area and time-averaged void fraction, equivalent void fraction, and wall void fraction, respectively
α_{eb}, α_{et} area and time-averaged void fractions at bottom and top of the porous layer, respectively
β coefficient of isobaric expansion
ΔT temperature difference
ΔP pressure difference
ϵ porosity
η fraction of energy transferred downward or inertial counterpart of viscous permeability
θ angular position measured from lower stagnation point
κ relative permeability
λ index of pore size distribution

μ	viscosity
ν	kinematic viscosity
ρ	density
σ	surface tension
ϕ	pool angle

SUBSCRIPTS

1	upper layer or upper bounding surface
2	lower layer or lower bounding surface
l	liquid
s	side wall
v	vapor

Acknowledgment

The author is grateful to Cindy Gilbert for her excellent typing of the manuscript.

References

1. Gluekler, E. L., and Baker, L., Jr. (1977). Post accident heat removal in LMFBR's. In *ASME Symposium on the Thermal Hydraulic Aspects of Nuclear Reactor Safety* (O. C. Jones and S. G. Bankoff, eds.), Vol. 2, American Society of Mechanical Engineers, New York, NY. pp. 285–324.
2. Kulacki, F. A., and Goldstein, R. J. (1972). Thermal convection in a horizontal fluid layer with uniform volumetric energy source. *J. Fluid Mech.* [2] 55, 271–287.
3. Mayinger, F., Jahn, M., Reineke, H. H., and Steinbrenner, U. (1975). *Examination of Thermohydraulic Processes and Heat Transfer in a Core Melt.* Federal Ministry for Research and Technology Final Report BMFT RS48/1.
4. Jahn, M., and Reineke, H. H. (1974). Free convection heat transfer with internal heat sources. *Proc. Int. Heat Transfer Conf., 5th*, Tokyo. Paper NC2.8.
5. Steinbrenner, U., and Reineke, H. H. (1978). Turbulent buoyancy convection heat transfer with internal heat sources. *Proc. Inter. Heat Transfer Conf., 6th*, Toronto. Vol. 2, pp. 305–310.
6. Emara, A. A., and Kulacki, F. A. (1980). A numerical investigation of thermal convection in a heat-generating fluid layer. *J. Heat Transfer* 102, 531–537.
7. Fiedler, H. E., and Willie, R. (1970). Turbulante freie Kovektion in einer horizontalen Flussigkeitsschicht mit volumen-warmequelle. *Proc. Int. Heat Transfer Conf., 4th*, Paris, Versailles. Paper No. NC4.5.
8. Kulacki, F. A., and Nagle, M. E. (1975). Natural convection in a horizontal fluid layer with volumetric energy sources. *J. Heat Transfer* 97, 206–211.
9. Kulacki, F. A., and Emara, A. A. (1977). Steady and transient thermal convection in a fluid layer with uniform volumetric energy sources. *J. Fluid Mech.* 83, 375–395.
10. Kulacki, F. A., and Goldstein, R. J. (1975). Hydrodynamic instability in fluid layers with uniform volumetric energy sources. *J. Appl. Sci. Res.* 31, 81.
11. Suo-Antilla, F. A., and Catton, I. (1976). An experimental study of a horizontal layer of fluid and volumetric heating and unequal surface temperatures. Paper presented at 10th National Heat Transfer Conference, St. Louis, MO.
12. Baker, L., Jr., Faw, R. E., and Kulacki, F. A. (1976). Post accident heat removal—Part I: Heat transfer within an internally heated non-boiling layer. *Nucl. Sci. Eng.* 61, 22–230.

13. Greene, G. A. (1982). *Experimental and Analytical Study of Natural Convection Heat Transfer of Internally Heated Liquids*. U.S. Nuclear Regulatory Commission Report NUREG/CR-2939.

14. Greene, G. A., Jones, O. C., Jr., Schwartz, C. E., and Abuaf, N. (1980). *Heat Removal Characteristics of Volume Heated Boiling Pools with Inclined Boundaries*. U.S. Nuclear Regulatory Commission Report NUREG/CR-1357.

15. Kymäläinen, D., Tuomisto, H., Hongisto, O., and Theofanous, T. G. (1993). Heat flux distribution from a volumetrically heated pool with high Rayleigh number. *Proc. Int. Topical Meeting on Nuclear Reactor Thermal Hydraulics, 6th*, Grenoble, France. Vol. 1, pp. 47–53.

16. Gabor, J. D., Ellison, P. G., and Cassulo, J. C. (1980). Heat transfer from internally heated hemispherical pools. Paper presented at 19th National Heat Transfer Conference, Orlando, FL.

17. Asfia, F. J., Frantz, B., and Dhir, V. K. (1996). Experimental investigation of natural convection in volumetrically heated spherical segments. *J. Heat Transfer* **118**, 31–37.

18. Asfia, F. J., and Dhir, V. K. (1996). Experimental study of natural convection in a volumetrically heat spherical pool bounded on the top with a rigid wall. *Nucl. Eng. Des.* **163**, 333–348.

19. Asfia, F. J. (1995). Experimental investigation of natural convection heat transfer in volumetrically heated spherical segments. Ph.D. Dissertation, UCLA, Los Angeles.

20. Kelkar, K. M., Schmidt, R. C., and Patankar, S. V. (1992). Numerical analysis of laminar natural convection of an internally heated fluid in a hemispherical cavity. *ANS Proc. Nat. Heat Transfer Conf.*, San Diego, pp. 355–366.

21. Schmidt, R. C., Kelkar, K. M., and Patankar, S. V. (1993). Numerical analysis of laminar natural convection of an internally heated fluid in a hemispherical cavity with an adiabatic flat surface. *ANS Proc. Nat. Heat Transfer Conf.*, Atlanta, pp. 331–338.

22. Min, J. H., and Kulacki, F. A. (1978). *Steady and Transient Natural Convection with Volumetric Energy Sources in a Fluid Layer Bounded from Below by a Segment of Sphere*. U.S. Nuclear Regulatory Commission Report NUREG/CR-0006.

23. Schramm, R., and Reineke, H. H. (1976). Natural convection in a horizontal layer of two different fluids with internal heat sources. *Proc. Int. Heat Transfer Conf., 6th*, Toronto, Canada, pp. 299–304.

24. Fieg, G. (1976). Experimental investigation of heat transfer characteristic in liquid layers with internal heat sources. *Proc. Int. Meet. on Fast Reactor Safety and Related Physics*. USERDA Conf. 761001, pp. 2047–2055.

25. Kulacki, F. A., and Nguyen, A. T. (1982). *Hydrodynamic Instability and Thermal Convection in a Horizontal Layer of Two Immiscible Fluids and Internal Heat Generation*. U.S. Nuclear Regulatory Commission Report NUREG/CR-2619.

26. Leverett, M. C. (1941). Capillarity behavior in porous solids. *Trans. Am. Inst. Min. Metall. Eng.* **142**, 152–169.

27. Hardee, H. C., and Nilson, R. H. (1977). Natural convection in porous media with heat generation. *Nucl, Sci. Eng.* **63**, 119–132.

28. Bau, H. H., and Torrance, K. E. (1982). Boiling in low-permeability porous materials. *Int. J. Heat Mass Transfer* **25**, 45–55.

29. Cornwell, K., Nair, B. G., and Patten, T. D. (1976). Observation of boiling in porous media. *Int. J. Heat Mass Transfer* **19**, 236–238.

30. Ogniewicz, Y., and Tien, C. L. (1980). Porous Heat Pipe. *AIAA Progress in Astronautics and Aeronautics* **70**, 329–345.

31. Udell, K. S. (1985). Heat transfer in porous media considering phase change and capillarity—the heat pipe effect. *Int. J. Heat Mass Transfer* **28**, pp. 485–495.

32. Chuah, Y. K., and Carey, V. P. (1985). Analysis of boiling heat transfer and two phase flow in porous media with non-uniform porosity. *Int. J. Heat Mass Transfer* **28**, 147–154.
33. Lu, S. M., and Chang, R. H. (1987). Pool boiling from a surface with a porous layer. *AIChE J.* **33**, 1813–1828.
34. Lipinski, R. J. (1984). A coolability model for post accident nuclear reactor debris. *Nucl. Technol.* **65**, 53–66.
35. Brooks, R. H., and Corey, A. T. (1966), Properties of porous media affecting fluid flow. *Journal of Irrigation Drainage Division of ASCE* **92**, 61–88.
36. Corey, A. T. (1977). *Mechanics of Heterogeneous Fluids in Porous Media*. Water Resources Publications, Fort Collins, CO.
37. Dhir, V. K. (1983). On the coolability of degraded LWR cores. *Nucl. Safety* **24**, 319–337.
38. Tung, V. X., and Dhir, V. K. (1988). A hydrodynamic model for two phase flow through porous media. *Int. J. Multiphase Flow* **14**, 47–65.
39. Tung, V. X. (1988). Hydrodynamic and thermal aspects of two phase flow through porous media. Ph.D. Dissertation, UCLA, Los Angeles.
40. Ishii, M., and Chawla, T. C. (1979). *Local Drag Laws in Dispersed Two-Phase Flow*. U.S. Nuclear Regulatory Commission Report NUREG/CR-1230.
41. Marshall, J. S., and Dhir, V. K. (1984). *Hydrodynamics of Counter-Current Two Phase Flow Through Porous Media*. U.S. Nuclear Regulatory Commission Report NUREG/CR-3995.
42. Schrock, V. E., Wang, C. H., Revankar, S., Wei, L. H., Lee, S. Y., and Squarer, D. (1984). Flooding in particle beds and its role in dryout heat fluxes. Paper presented at 6th Information Exchange Meeting on Debris Coolability, UCLA, Los Angeles.
43. Dhir, V. K., and Catton, I. (1977). Dryout heat fluxes for inductively heated particulate beds. *J. Heat Transfer* **99**, 250–256.
44. Schrock, V. E., Fernandez, R. T., and Kesavan, K. (1970). Heat transfer from cylinders embedded in a liquid filled porous medium. In *Heat Transfer 1970: Papers Presented at the Fourth Int. Heat Transfer Conference, Paris-Versailles*, Vol. VII, Paper CT 3.6. Elsevier, Amsterdam.
45. Catton, I., and Jakobsson, J. D. (1987). The effect of pressure on dryout heat flux on a saturated bed of heat generating particles. *J. Heat Transfer* **109**, 185–192.
46. Lienhard, J. H., and Schrock, V. E. (1963). The effect of pressure, geometry, and the equation of state upon the peak and minimum boiling heat flux. *J. Heat Transfer* **85**, 261–269.
47. Dhir, V. K. (1994). Boiling and two-phase flow in porous media. *Ann. Rev. Heat Transfer* (C. L. Tien, ed.) **5**, 303–350.
48. Tsai, F. P. (1987). Dryout heat flux in a volumetrically heated porous bed. Ph.D. Dissertation, University of California, Los Angeles.

Hydrogen Combustion and Its Application to Nuclear Reactor Safety

JOHN H. S. LEE

McGill University, Montreal, Canada

MARSHALL BERMAN

Sandia National Laboratories, Albuquerque, New Mexico

I. Introduction

In a severe loss-of-coolant accident (LOCA) in a nuclear reactor, large quantities of hydrogen can be generated from metal–water reaction. When the hydrogen is mixed with the air in the containment building, a large volume of combustible mixture may be formed. If ignited, the combustion of the hydrogen–air cloud could generate high pressure and temperature loads that could result in the breach of the containment and a release of radioactivity to the environment; this combustion could also damage safety equipment and thus jeopardize the management of the accident and the ultimate safe termination of the event. Such accident scenarios were considered as highly improbable events prior to the incident at the Three Mile Island Unit 2 (TMI-2) reactor on March 28, 1979. The TMI-2 accident confirmed that hydrogen combustion is a credible event that could threaten the integrity of the containment [1]. In the TMI-2 incident, a peak pressure transient of about 190 kPa (1.9 atm) was registered at about 10 hours into the accident. The magnitude of this pressure pulse corresponds to the constant–volume explosion pressure of an 8% hydrogen–air mixture, which is near the downward propagation limit. In any serious accident, it is doubtful that all the hydrogen generated could be uniformly mixed inside the containment prior to ignition. Thus, local concentrations of hydrogen could have been much higher than 8%. For

0065-2717/97 $25.00

mixtures above the flammability limit, the combustion rate can be much higher and, accordingly, higher peak pressures could be generated locally near the cloud. The time scale of the pressure records at TMI-2 can only reveal the averaged pressure—much higher pressure transients might have occurred during the hydrogen explosion near the hydrogen cloud. High local pressure transients could lead to severe damage of safety equipment, which might hinder the management of the accident itself. However, this seems not to have occurred at TMI.

In the wake of the TMI-2 incident, major research programs covering all aspects of the hydrogen issue (i.e., generation, mixing, transport, combustion, mitigation concepts) were initiated in various countries all over the world (e.g., the United States, Canada, Japan, Germany, the United Kingdom, France, Italy, the former Soviet Union). In this paper, we shall be restricting our discussion to the hydrogen combustion problem only, although the other issues are equally important. In fact some issues, such as the mixing problem, are an integral part of the hydrogen problem since the type of combustion (i.e., diffusion flame, slow deflagration, detonation) is a direct consequence of the mixing processes. The ultimate objective of all the hydrogen combustion research programs is to achieve realistic estimates of the pressure and temperature loads due to combustion in a given reactor for different postulated accident scenarios. These predictions are achieved via the use of computer codes, usually of the thermohydraulic type. As input to the code, one needs fundamental combustion data for hydrogen–air mixtures.

At the time of the TMI accident, fundamental combustion data (e.g., flammability limits, burning velocity, detonation cell size, and other dynamic detonation parameters such as initiation energy, detonability limits, and critical tube diameter) were quite limited and information on the various mixtures of hydrogen with different diluents (e.g., steam, CO_2, CO) and at various initial thermodynamic states (pressure, temperature, density) of interest to reactor safety was lacking. As a result, efforts were devoted to acquiring these fundamental data in most of the research programs. These fundamental studies were often carried out in laboratory–scale experiments. Combustion experiments have also been conducted at various institutions in different countries in fairly large vessels with more complex internal geometries to simulate prototypical conditions inside a containment building. Thus, over the past 15 years, a significant amount of combustion data has been accumulated. However, to date, little of this information has actually been directly applied to providing better input to the thermohydraulic codes in order to achieve a more realistic prediction of the combustion load for a given accident scenario.

The standard methodology for estimating combustion loads in the containment is to use a so-called "lumped-parameter" thermohydraulic computer code. These "lumped-parameter" codes are of "zero dimension" and do not resolve spatial variations of the dependent variables of interest within a defined control volume. In general, the containment is divided into a number of compartments (or nodes); typically, the number of compartments is of the order of 30 or 40. The conservation laws for the various species and the total energy are then written for each compartment. Because of the zero-dimensional nature of these codes, momentum conservation is accounted for only in connection with the mass transport between different compartments. The dependent variables of interest are assumed to be uniform within each compartment and thus the conservation equations are reduced to a set of ordinary differential equations, with time as the independent variable. The different compartments are linked via the transport coefficients between a compartment and its immediate neighbors.

The combustion information that these codes require is in the form of: (1) a limit mixture composition above which combustion can be "turned on"; (2) a burning rate law for the mixture; and (3) a degree of combustion completeness. The burning rate and the degree of combustion completeness are usually correlated with the mixture composition and the thermodynamic state in the compartment rather than more physically appropriate parameters like the turbulence intensities and the turbulence scales of the mixture itself. It is disappointing to note that, in spite of the vast amount of information that has been obtained in the past 15 years, the combustion models used in these codes are still somewhat of an arbitrary nature and quite often formulated without any physical basis. They are made to fit specific cases and hence lack generality.

The objective of this paper is to critically examine the problems of integrating the existing fundamental information on hydrogen combustion and large-scale test data into the thermohydraulic codes, and to suggest ways of implementing this integration. Our goal was not to provide a comprehensive review of the data obtained from the various hydrogen research programs to date. There are numerous reviews that have accomplished this task adequately (e.g., see [2–6]). However, to accomplish our objective, we also provide explanations of the various fundamental parameters of combustion and an extensive review of the literature on deflagrations, DDT (Deflagration to Detonation Transition), and detonations in order to discuss the implementation of existing and future experimental data into the current code models.

II. Combustion Parameters

In order to use the data accumulated from the various experimental programs, it is essential to understand what these combustion parameters mean and how they are measured. Generally speaking, gaseous combustion phenomena can be classified into three types: diffusion flames, premixed flames, and detonations. In a diffusion flame, the fuel (hydrogen) is not uniformly premixed with the oxidizer prior to combustion. Mixing occurs only along the boundary separating the fuel and the oxidizer. Upon ignition, the flame front locates itself in the mixing zone between the fuel and the oxidizer, where the composition is favorable for combustion. The fuel and oxidizer then diffuse to the reaction zone from both sides of the flame boundary. Diffusion flames occur when the fuel jet is ignited immediately before substantial mixing with the oxidizer has progressed. There are accident scenarios in which diffusion flames are important, for example, direct containment heating (DCH) in some large Pressurized Water Reactor containment buildings [7, 8] or in the region above the wetwell of certain Boiling Water Reactor containment designs [9]. These diffusion flames pose a turbulent mixing problem in fluid mechanics, which is relatively well understood. In a reactor accident, diffusion flames can overheat parts of containment, or degrade emergency equipment due to imposed thermal loads. This thermal loading represents a turbulent convective heat-transfer problem, which can be described quite adequately by current numerical simulations (at least from an engineering point of view). Diffusion flames do not involve high-pressure transients and are localized near the fuel jet itself. In term of high pressure transients that threaten containment integrity, we are more concerned with propagating premixed flames and detonations.

For premixed combustion, the fundamental combustion parameters can be classified into equilibrium and nonequilibrium (or dynamic) parameters. Equilibrium parameters are independent of the chemical kinetic and transport rates, and depend only on the energetics of the mixtures. Equilibrium parameters can readily be computed from a thermochemical calculation if the thermodynamic data (i.e., heat capacities, equilibrium constants, heat of formation) of the various chemical species are known. There are a number of standard computer codes (e.g., STANJAN [10], NASA Gordon-McBride Code [11]) that are available for that purpose. The equilibrium parameters that are of interest here are the adiabatic isochoric complete combustion (AICC) pressure (i.e., the pressure increase as a result of combustion in a closed volume without any heat losses), and the adiabatic flame temperature (i.e., the temperature rise across a constant pressure flame without heat losses). Detonation velocities, pressures,

TABLE I

EQUILIBRIUM COMBUSTION PARAMETERS

Fuel	Mole% fuel	Adiabatic flame temp. T_{ad} (K)	Sound speed at T_{ad} (m/s)	Constant-volume explosion pressure (atm)	C–J detonation velocity (m/s)	C–J detonation pressure (MPa)
CH_4	9.506	2225.72	894.5	8.810	1804.2	1.7428
C_2H_6	7.752	2260.11	892.0	9.204	1803.8	1.8233
C_3H_8	4.032	2267.01	889.8	9.343	1801.0	1.8515
C_4H_{10}	3.131	2270.42	888.6	9.520	1800.1	1.8873
C_2H_4	6.545	2370.37	899.5	9.399	1825.1	1.8621
C_2H_2	7.752	2539.39	914.8	9.777	1866.8	1.9383
C_3H_6	4.460	2334.26	893.3	9.458	1811.7	1.8744
H_2	29.586	2382.69	979.9	8.022	1971.3	1.5828

etc., can also be obtained via equilibrium thermodynamics calculations using the standard computer codes based on the Chapman–Jouguet (C–J) detonation model. Equilibrium combustion parameters for some common fuel–air mixtures at standard initial conditions (i.e., 1 atm and 298 K) and at stoichiometric composition are given in Table I: The *reflected* C–J detonation pressures will be more than twice as large as the C–J pressures shown in Table I [3, 4].

Note that the equilibrium combustion parameters are very similar for the various fuels; for example, the adiabatic flame temperatures are all of the order of 2200–2400 K, while the constant–volume explosion pressures are about 8–9 atm. The similarity is due to the fact that, for stoichiometric compositions, all the available oxygen in the air is assumed to be consumed. Thus, the amount of fuel burned is adjusted to suit the amount of oxygen available in air. The product species for all the hydrocarbon fuels are the same (i.e., H_2O and CO_2). Hence, the energetics of these mixtures are all similar, dictated by the total amount of oxygen available to make products. Similarly, the detonation velocities for all the mixtures are also more or less the same (i.e., detonation velocities are about 1800–1900 m/s and the detonation pressures are about 16–19 atm). Unlike the laminar burning velocity, the detonation velocity of a mixture can be calculated from the conservation equations using the C–J criteria of sonic conditions at the downstream equilibrium plane of the detonation front. Thus, the chemical kinetic and the transport rates are not required and the C–J detonation states are equilibrium parameters depending only on the energetics of the mixture. It is interesting to note that the detonation pressure is about twice the constant-volume explosion pressure.

Another parameter of interest is the sound speed in the product gases corresponding to the adiabatic flame temperature. This is also shown for the various cases in Table I. Note that their values are all of the order of 800–900 m/s. This sound speed in the product gases is an important parameter because experimentally it is found that the maximum deflagration speed for an accelerated flame in a confined obstacle field corresponds closely to the sound speed of the product gases [12]. It is also the maximum deflagration speed observed experimentally at the onset of detonation in a DDT process. Much insight can be gained from the equilibrium parameters of an explosive mixture.

It is of interest to compare the equilibrium parameters for hydrogen–air mixtures (at standard initial states) at different hydrogen concentrations. The chosen composition range corresponds to the rich and lean flammability limits (for upward flame propagation, this is 4–74% according to the classical values reported by Coward and Jones [13]. The equilibrium parameters for hydrogen–air mixtures are given in Table II below.

TABLE II

EQUILIBRIUM PARAMETERS FOR HYDROGEN–AIR MIXTURES

Mole% H_2	Adiabatic flame temp. T_{ad} (K)	Sound speed at T_{ad} (m/s)	Constant-volume explosion pressure (atm)	C–J detonation velocity (m/s)	C–J detonation pressure (MPa)
4	627.36	501.6	2.447	910.8	0.44528
10	1097.25	660.4	4.268	967.4	0.48760
15	1470.46	765.5	5.552	1516.7	1.0804
20	1828.19	856.0	6.643	1705.0	1.3030
25	2159.41	931.6	7.517	1861.2	1.4808
30	2391.11	983.7	8.044	1979.2	1.5873
35	2336.46	1024.4	8.026	2051.3	1.5862
40	2213.17	1048.1	7.772	2100.1	1.5348
45	2079.97	1068.8	7.445	2141.5	1.4673
50	1939.01	1088.3	7.070	2180.0	1.3910
55	1793.26	1107.6	6.656	2216.7	1.3060
60	1643.03	1126.5	6.202	2251.3	1.2131
65	1488.60	1146.3	5.709	2283.5	1.1125
70	1329.77	1167.0	5.177	2312.5	1.0041
74	1199.19	1183.8	4.721	2332.2	0.91136

Note that below 10% H_2, the adiabatic flame temperature is below 1000 K. Experiments indicate that there exists a minimum flame temperature for self-sustained propagation of a flame of about 1000 K. Thus, mixtures below 10% H_2 are *comparatively* benign in accord with experimental observations. Note that the equilibrium parameters provide some indication of the sensitivity of the various mixtures; for example, the adiabatic flame temperature, constant-volume explosion pressure, and C–J detonation pressure all peak near the stoichiometric composition of hydrogen in air (30%). However, sound speed and C–J detonation velocity continue to increase with increasing hydrogen mole fraction because of the decrease in the molecular weight of the mixture. These equilibrium data are quite informative regarding the explosive properties of the mixture.

Nonequilibrium or dynamic combustion parameters are dependent on the transport rates and the chemical kinetic rates of the reactions. For premixed mixtures, the dynamic parameters of interest for the deflagration mode are the flammability limits, minimum ignition energy, quenching distance, and the laminar burning velocity. For the detonation mode, the dynamic parameters of interest are the detonability limits, the critical initiation energy, the critical tube diameter, and the conditions for the transition from deflagration to detonation (i.e., DDT). Combustion science has not yet advanced to the stage where these dynamic parameters can be obtained theoretically from first principles. For the laminar burning velocity, the detailed chemistry and transport coefficients are sufficiently well established to permit the numerical calculation of the laminar burning velocity for most of the common fuel–air mixtures. However, in general, most of the dynamic parameters have to be determined experimentally. It should be pointed out that there is a certain degree of arbitrariness associated with the experimental measurement of these dynamic parameters. This stems from a lack of a unique operational definition for these parameters. There are also no universally accepted standard apparatus and procedures for the experimental determination of the dynamic parameters in general. We shall elaborate further when we discuss each of these parameters later.

A. FLAMMABILITY LIMITS

Flammability limits are defined as the critical compositions of the fuel–air–inert diluent mixture beyond which the mixture cannot sustain steady flame propagation. It is understood that the flammability limits also depend on the initial conditions of the mixture, which must be specified. For example, the flammability limits at high initial temperatures are, in general, wider than at room temperatures [14, 15]. Theoretically, the mechanisms responsible for the existence of the flammability limits are heat losses and flame stretching. Thus, the size and geometry of the apparatus, buoyancy, and convective flow play important roles in determining these limits. The more or less accepted standard apparatus and procedure for measuring the flammability limits are based on the pioneering work of Coward and Jones of the U.S. Bureau of Mines [13]. In essence, the Coward and Jones apparatus consists of a flame tube of 5-cm diameter and about 1.2 m long, which is filled with the premixed combustible mixture to be tested. Ignition is initiated at the bottom of the tube and flame propagation is observed visually in a darkened room. The flammability-limit criterion is based on an arbitrary decision that if the flame fails to propagate more than one third of the tube length, then the mixture is considered to be outside the limits. If ignition is at the bottom, then the limit is the upward flame propagation limit. If ignition is at the top of the tube, then the flame propagates down for the downward propagation limit. The horizontal flame propagation limit is determined with the flame tube in the horizontal position. Thus, there are three limits depending on the direction of the flame propagation and they differ because of buoyancy effects. Upward propagation is the widest because the flame is aided by buoyancy effects. The lean and rich limits for upward propagation correspond to the minimum and maximum fuel concentrations, compared to the downward and the horizontal limits. Because heat losses are the dominant mechanism responsible for these limits, there is a dependency on tube diameter, material, and geometry, as well as the initial thermodynamic state of the mixture (i.e., pressure and temperature). It is expected that increasing initial temperature widens the flammability limits; for example, at 500°C, the minimum concentration of hydrogen for upward propagation is only 2% as compared to 4% for room temperature mixtures [6].

Regarding the appropriate tube diameter that should be used for the flammability tests, experience indicates that when the tube diameter gets to be of the order 5 cm or larger, the dependency of the limits on tube diameter becomes insignificant. Similarly, the geometry of the cross-section of the flame tube is also not as important when the diameter of the tube is large. It should be noted that the thickness of laminar flames at atmospheric pressure is of the order of a millimeter and the quenching distance is about twice the laminar flame thickness. A tube dimension of the order of 5 cm corresponds to about 50 times the flame thickness. Thus, it is understandable that tube diameter is not as important above a certain minimum size as the 50 mm diameter of the classical Coward–Jones apparatus. It should be noted that the flammability-limits data reported in the literature are not always determined using the standard Coward–Jones apparatus and procedure. Closed combustion vessels of different sizes and geometries with the ignition source located at different positions are often used to determine the flammability limits. Sometimes fans are even installed inside the vessel to create convection currents and turbulence to see their influence on the limits. Using a closed vessel, a different criterion has to be used to define the limits, and quite often it is arbitrarily based on achieving a certain percentage of the total expected pressure rise from theoretical calculations. For hydrogen–air mixtures at standard initial conditions (i.e., atmospheric pressure and room temperature) and saturated with water vapor, the flammability limits as reported by Coward and Jones are as follows: upward propagation (4.1–74%); horizontal propagation (6.0–74%); and downward propagation (9.0–74%). In typical nuclear reactor accident scenarios, the hydrogen–air mixtures are at elevated temperatures and also with steam present in significant amounts. The flammability limits widen with elevated initial temperatures. The effect of high initial temperatures on the downward propagation limits of hydrogen–air mixtures is shown in Fig. 1.

Note that above a certain temperature (called the autoignition limit) the flammability limits no longer exist and any amount of hydrogen in the mixture will react spontaneously with the oxygen present. The autoignition temperature is not a well-defined parameter and depends on the method and the scale of the apparatus used for its determination. There is a lack of data at the lean limit and the results given in Fig. 1 should be considered as an approximate guideline for a self-ignition temperature of hydrogen–air mixtures.

When the initial temperature is of the order of 100°C and where the partial pressure of water vapor is about 1 atm, much higher steam concentrations can be obtained in the hydrogen–air mixtures. A compilation of flammability-limits data for hydrogen–air–steam at around 100°C was given by Berman and Cummings [3] and is shown in Fig. 2.

The scattering of the data, especially near the peak steam concentration, is quite large and can be attributed to the use of different vessels and igniters as well as the arbitrariness of the different criteria used for determining the limits themselves. From Fig. 2 it can be observed that the lean limit is rather insensitive to steam dilution and is primarily governed by the amount of hydrogen available for combustion. The rich limit drops more rapidly with steam dilution and is primarily governed by the amount of oxygen available.

Generally speaking, a mixture of hydrogen–oxygen–nitrogen will be combustible if the hydrogen concentration exceeds 4% and the oxygen concentration is above 5% approximately. Figure 2 indicates that it re-

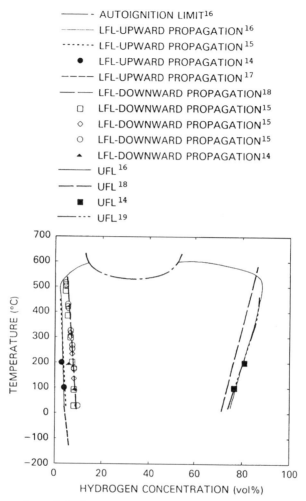

FIG. 1. Comparison of the Lower Flammability Limits (LFLs) and Upper Flammability Limits (UFLs) for hydrogen–air mixtures as a function of initial temperature. Note that for the symbols of [15], □ = no propagation, ◇ = partial propagation, and ○ = propagation (from [6]).

quires about 60% of steam to render the hydrogen–air mixture completely inert. At stoichiometric conditions, this corresponds to an oxygen concentration of about 5.6% which is slightly higher than the 5% required for self-sustained flame propagation in dry air. However, the higher heat capacity of steam as compared to nitrogen tends to lower the flame

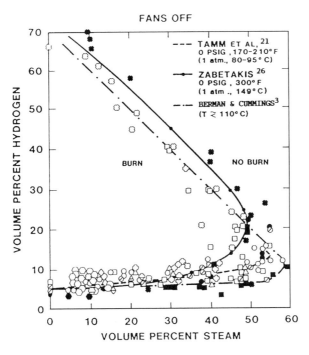

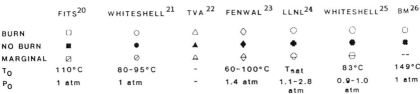

FIG. 2. Flammability limits for hydrogen–air–steam mixtures with fans off. FITS = Fully Instrumented Test System (Sandia); WHITESHELL = Whiteshell Nuclear Research Establishment (Canada); TVA = Tennessee Valley Authority; FENWAL = Fedwall, Inc.; LLNL = Lawrence Livermore National Laboratory; BM = Bureau of Mines (from [3]).

temperature and hence the burning velocity. Thus, steam is slightly more effective in suppressing flammability of the mixture than nitrogen.

Although not a rule of universal applicability, the concept of a minimum flame temperature appears to define the flammability limits. For mixtures of hydrogen with other fuels (e.g., CO), the flammability limits can be estimated using the LeChatelier rule [13], which is based on a consideration of the energetics of the mixture.

Mixtures of hydrogen with other gases (e.g., CO, CO_2, steam, air) at high initial temperatures are of interest in severe accidents involving molten core–concrete interactions. Flammability-limits data for those high-temperature mixtures are lacking. Approximate values can readily be estimated from empirical correlations based on the energetics of the mixtures. Experimental programs have been proposed to determine the flammability of these high-temperature mixtures. However, it is not clear to what degree of accuracy these flammability data are required for severe-accident analysis.

The experimentally determined flammability limits discussed above cannot be considered as fundamental properties of the mixture. The principal mechanism responsible for their existence is heat loss, which depends on the mode of heat transfer and that in turn depends on the boundary conditions of the system. Thus, it is in some ways apparatus- as well as scale-dependent. A further complication arises from the very nature of the propagation mechanism of the flame itself, in particular, the effect of selective or preferential diffusion of molecules of different molecular weights. Depending on the diffusivity of the fuel molecules as compared to oxygen, near-limit flames tend to be unstable and break up into cells; that is, lean mixtures are unstable when the fuel molecule is lighter than oxygen and rich mixtures are unstable if it is heavier). Thus, for hydrogen near the lean limit, the flame breaks up into cells and, due to preferential diffusion, hydrogen enrichment occurs at the convex flame cells. The local hydrogen concentration at the flame cell is no longer the global average concentration of the mixture itself. Thus, the lean limit when referred to the global concentration is not representative of the local combustion phenomenon where the local hydrogen concentration is actually higher due to the selective diffusional transport.

The flammability limits are a quite complex phenomenon involving convective and transport processes in addition to the chemistry of the reaction mechanism. It is not clear to what accuracy or degree of conservatism the flammability-limits data are needed for severe-accident analysis; combustion at leaner concentrations causes low pressures, while burning at higher concentrations would be a more conservative approach for the pressures induced in a particular combustion event. Employing low-flammability limits would be nonconservative in the sense that the burned hydrogen is no longer able to accumulate and build to higher concentrations at a later time if ignition were delayed. Thus, we do not really know the experimental methodology nor the scale of the experiment to use to determine the limits. The small-scale flame tube of Coward and Jones may be adequate and, if so, it simplifies considerably the task of obtaining the data.

Currently, reactor safety experiments are often designed with the philosophy of attempting to simulate as much as possible, the so-called realistic or prototypical conditions. As a result the data obtained in large vessels with complex geometries are neither of fundamental significance nor can they necessarily be generalized and scaled up to be of practical interest.

The flammability limits discussed earlier are based on an initially quiescent mixture. Quite often, it is claimed that under real accident conditions, the mixture is generally turbulent. Thus, the flammability limits should be obtained under the turbulent conditions expected to be encountered in a real accident. In a turbulent mixture, the flammability limits tend to be narrower due to the quenching effect of turbulence. Bradley and coworkers [27] have demonstrated that a mixture cannot sustain flame propagation when the ratio of the chemical to the turbulent eddy time scales exceeds a certain limiting valued. Thus, for turbulent mixtures, the flammability limits now involve specifying additional parameters to characterize the turbulence intensity and scale. It may be difficult to decide on an appropriate apparatus and procedure for its determination. Furthermore, it is not clear how the resulting information could be scaled to actual reactor conditions. For this reason, it is difficult to assess the validity and the usefulness of the existing data on the flammability limits of hydrogen–air–steam mixtures, which have been obtained under a variety of apparatus and conditions.

B. COMPLETENESS OF THE COMBUSTION

Associated with the study of the combustion of near-limit mixtures in closed vessels, it is observed that the total pressure rise is often only a fraction of the theoretical value obtained from thermodynamic calculations, especially when the mixture is below (i.e., leaner than) the downward propagation limit. For mixtures well above the downward propagation limit hydrogen concentration, the experimental value for the constant-volume explosion pressure generally agrees quite well with the theoretical value. Figure 3 shows a collection of data for the constant-volume explosion pressure versus the hydrogen concentration near the lean limit [24, 28–33]. The results are from vessels of different sizes and geometries, different igniter locations, with and without fans to generate turbulence inside the vessel, and even in the presence of water sprays used to promote heat losses as well as generating air recirculation and turbulence. In general, the data are highly scattered below the downward propagation limit of about 8% H_2, indicating a dependence of the combustion completeness on the vessel characteristics and turbulence. However, the agreement between the theoretical adiabatic constant-volume explosion pressures and

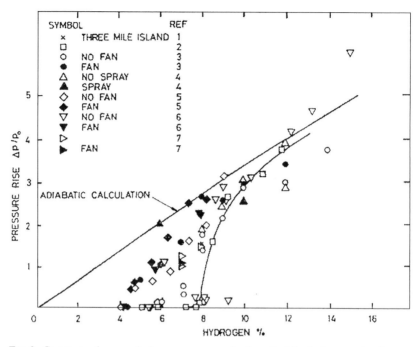

FIG. 3. Constant-volume explosion pressure near the lean limit for hydrogen–air mixtures.

experimental data is quite good for hydrogen concentrations above the downward limit.

Below the downward propagation limit, the experimental values deviate significantly from the calculated values. It is clear that the degree of combustion completeness depends on the vessel geometry, location of the igniter, and the turbulence, or, more appropriately, the forced convective flow inside the vessel. If the mixture is below (leaner than) the downward propagation limit, the flame cannot propagate below the igniter position in a quiescent mixture. Furthermore, buoyancy lifts the flame ball upward as it expands and hence only a fraction of the total volume of mixture (contained inside a cone whose half angle is given by the arc tangent of the flame speed to the buoyancy-induced velocity) is burned. If the mixture is turbulent, the flame speed will be higher and correspondingly the cone angle is larger and more of the mixture is burned. Similarly, the geometry of the vessel plays an important role. For example, if the vessel is cylindrical with a large L/D ratio, then once the flame reaches the tube wall, all the mixture will be burned as the flame continues to propagate upward, in contrast to a spherical or cubical vessel where only a conical volume is burned.

Near the upward propagation limit of about 4% H_2, turbulence does not influence the degree of combustion significantly. This is due to the quenching effect of turbulence which offsets its influence on increasing the flame speed via higher convective transport rate of the turbulent motion. For intermediate concentrations from 4% to the downward limit of about 8% H_2, the data suggest that turbulence does have an effect in increasing the total pressure rise by increasing the turbulent flame speed. Above the downward limit of about 8%, we see that the results approach the theoretical values and turbulence plays a decreasing role since the flame can now propagate downward to consume all of the mixture.

The effects of turbulence on the flame propagation near the lean limit are well illustrated by the results of the work of Al-Khishali et al. [34]. Figure 4 shows the flame contours for different mixtures below and above the downward propagation limit. At 6% H_2, the flame ball is lifted by buoyancy and the mixture below the igniter is not burned. Above the downward propagation limit of 8% H_2, the flame is almost spherical and symmetrical about the igniter. Figure 4 shows the flame contours for 6% H_2 at different turbulent intensities (as measure by the fan speeds).

For modest turbulence level, the flame speed for lean hydrogen–air mixtures increases. This in turn results in a more complete burning of the mixture inside the vessel. At too high a turbulence level, we see that the burning velocity decreases instead of increases due to quenching. Accordingly, the total pressure rise, which is associated with the degree of completeness, is reduced. The results of Al-Khishali et al. [34] are summarized in Fig. 5, where the maximum pressure rise is plotted against the concentration of hydrogen for different fan speeds (i.e., turbulence intensities).

Above the downward limit, turbulence has little effect on the total pressure rise. The effect of turbulence is most prominent in the range of hydrogen concentrations between 5 and 8% H_2.

To account for this pressure deficit near the lean limit, a combustion-completeness parameter is defined based on the experimental pressure deficit observed as shown in Fig. 3. This combustion-completeness parameter has been correlated with the fuel composition and used as an input in the lumped-parameter thermohydraulic code to estimate the pressure transients that might occur in a nuclear reactor containment for different hypothetical accident scenarios. It is clear that this so-called combustion completeness is predominately a buoyancy effect and depends on the position of the igniter, the flame speed, and the scale and geometry of the vessel. From the preceding discussion, it is clear that this combustion-

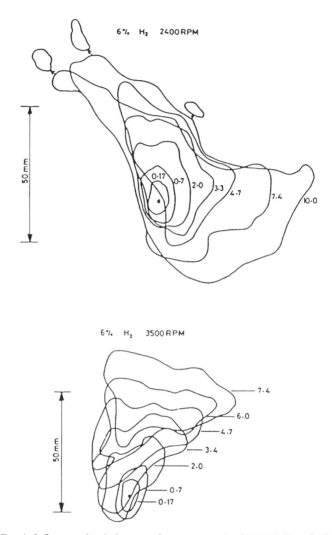

FIG. 4. Influence of turbulence on flame propagation (6% H_2) (from [34]).

completeness parameter is not a fundamental property of the mixture and depends on numerous parameters specific to the particular experiment apparatus. It is not meaningful to attempt to fit the data obtained from different vessels under different conditions and arrive at a generically applicable empirical relationship to use in the thermohydraulic code calculations. (Note that in conventional combustion science, the complete-

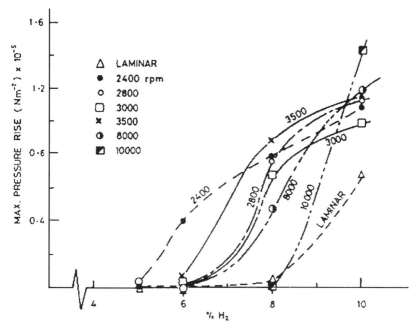

FIG. 5. Influence of turbulence on the constant-volume explosion pressure (from [34]).

ness of a combustion process generally refers to the degree of conversion of carbon to carbon dioxide rather than carbon monoxide.)

C. BURNING VELOCITY

Given a mixture at a specified temperature and pressure, there exists a unique value of the so-called laminar burning velocity, that is, the speed at which a laminar flame front advances *relative to the unburned gases ahead of it*. Relative to a fixed observer, the velocity of the flame front is referred to as the flame speed. The flame speed is in general greater than the burning velocity by a factor that depends on the boundary conditions. For a flame propagating from a closed-end tube toward the open end, the flame speed is greater than the burning velocity by a factor equal to the ratio of the specific volume of the hot burned products to the cold unburned mixture ahead of the flame front. This is due to the displacement of the unburned gases by the expansion of the combustion products (since the pressure across the flame is more or less constant, the increase

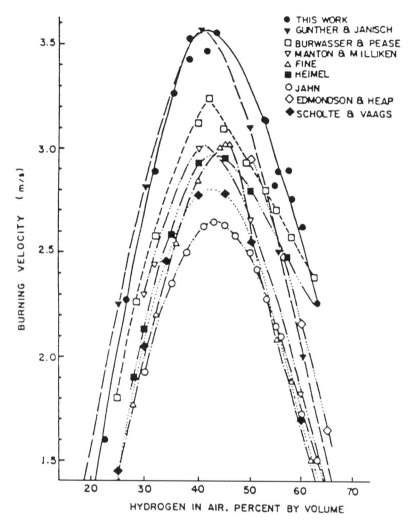

FIG. 6. Laminar burning velocity of hydrogen–air mixtures at 20°C (from [35]).

in the specific volume of the products corresponds to approximately the temperature increase across the flame). The actual rate of burning (energy release) is based on the burning velocity and not on the flame speed. For the case of hydrogen–air mixtures at an initial temperature of 20°C, the laminar burning velocity as measured by different investigators is shown in Fig. 6. One can observe a considerable scatter of the data. This is perhaps due to the different experimental methods used in the determination of the burning velocity.

The laminar burning velocity is a fundamental property of the mixture and depends on the chemical kinetic rates of the reaction as well as the transport coefficients. Thus, it should have a unique value for a given mixture composition and initial condition. The scatter shown in Fig. 6 is indicative of the difficulty involved in its precise determination. The current status in combustion science permits one to theoretically calculate the laminar burning velocity for most of the common fuels; for the case of hydrogen–air mixtures, laminar burning velocity has been computed by Mass and Warnatz, Dixon-Lewis, Egolpopoulos and Law, among others [36–38]. A meaningful comparison between the calculated and experimental values requires great care in the choice of the method used in the determination of the burning velocity itself. It is important to note that flame stretch and preferential diffusion can cause significant deviation of the experimental values from the theoretically computed ones. This is particularly important for hydrogen flames where the diffusivities of the hydrogen atoms and molecules are high. This leads to thermal and diffusional instabilities of lean hydrogen flames, accounting for the fact that most of the experimental data seldom go below 20% H_2, corresponding to an equivalence ratio of about 0.6. For leaner mixtures, cellular instabilities become important and the experimental results are no longer accurate for comparison with theory.

The development of the counterflow twin flame technique by Law and his students [39, 40] has provided an accurate means of determining the laminar burning velocity free from stretch and diffusional instabilities. Thus, it is now possible to make a meaningful comparison with theoretical values and investigate the detailed kinetic mechanisms of lean hydrogen flames using this technique. A comparison between the experimental and numerically determined laminar burning velocity for hydrogen flames using the counterflow twin flame technique is shown in Fig. 7 [38].

Now that a method has been developed for the precise measurement of the laminar burning velocity free from extraneous effects, it may be said that the laminar burning velocity is truly a fundamental parameter of a given combustible mixture. As a result, knowledge of the detailed kinetic mechanisms of the reactions in the flame zone can now be obtained via numerical modeling guided by the experimental results. For most practical purposes, the data reported are adequate.

For reactor safety analyses, mixtures of hydrogen, steam, and air are of interest. Figure 8 shows the results for hydrogen–air mixtures over their flammability range with different amounts of steam dilution. The initial temperature of the mixture is 100°C. The method used for determining the burning velocity is the nozzle–burner/schlieren cone angle using a 5-mm-diameter nozzle. The unburned gas flow velocity is measured by particle tracking with a dual-beam laser Doppler anemometer.

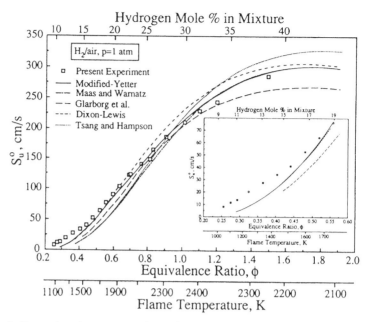

FIG. 7. Comparison between experimental and numerically determined laminar burning velocity for different kinetic schemes (from [38]).

Note that at elevated initial temperatures the burning velocity is increased through the increase in transport and reaction rates. Steam serves as an inert diluent and decreases the burning rate. The laminar burning velocity, although not realized in any real accident scenarios, serves as an important fundamental combustion parameter for the particular mixture and is useful in modeling the turbulent burning rates.

D. TURBULENT BURNING RATES

In any large-scale combustion event, the flame is invariably turbulent. In general, the turbulent burning velocity is higher than the laminar value. The turbulent burning velocity is not a fundamental property of the combustible mixture, and depends strongly on the turbulence parameters (i.e., turbulent intensity and scales). Turbulent burning velocity is usually expressed as a function of the laminar burning velocity and the turbulence parameters. The correlation is obtained from experiments. A typical experiment consists of the use of a number of fans to generate turbulence inside a closed vessel; the turbulent burning velocity is measured by the double

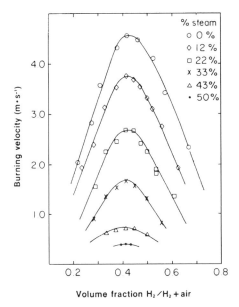

FIG. 8. Laminar burning velocity of hydrogen–air–steam mixtures at 373 K (from [41]).

flame kernel method. The turbulence intensity can be controlled via the rotational speed of the fan, and the turbulence scale by the pitch of the fan blades. Typical results for hydrogen–air mixtures over a range of hydrogen concentrations from 12 to 60% are shown in Fig. 9 [42].

It can be observed that, in general, the turbulent burning velocity increases with the turbulent intensity. The influence of the scale of the turbulence is not shown in Fig. 9. It should be noted that for very high turbulent intensities, the turbulent burning velocity decreases and eventually the flame is quenched. For lean hydrogen–air mixtures, Figs. 10 and 11 show the influence of turbulent intensities on the burning velocity of the flame kernel in the sideways and downward directions.

The results in Figs. 10 and 11 show considerable scatter due to the higher irregular shape of the flame near the limit as shown previously in Fig. 4. In both Figs. 10 and 11 we note that the turbulent burning velocity increases to a peak value and then starts to decrease with further increase in the turbulent intensity until, eventually, the flame is quenched. Thus, for near-limit mixtures, where the flame speed is low, turbulence can easily quench the flame instead of promoting its propagation.

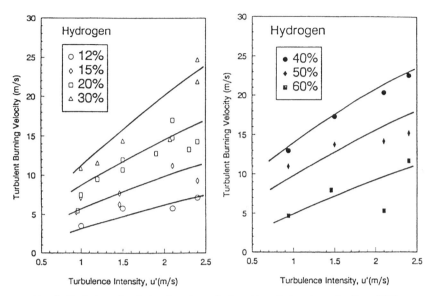

FIG. 9. Variation of turbulent burning velocity with turbulent intensity (from [42]).

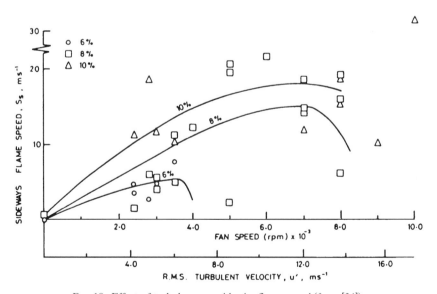

FIG. 10. Effect of turbulence on sidewise flame speed (from [34]).

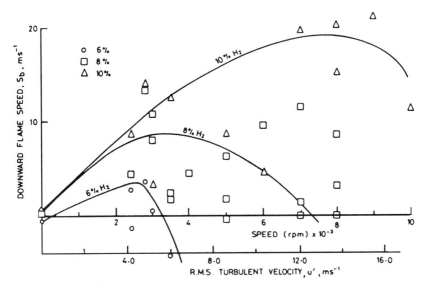

FIG. 11. Effect of turbulence on downward flame speed (from [34]).

Recognizing the importance of turbulence on the flame propagation rate, it is understandable that numerous attempts have been made to correlate the turbulent burning velocity with the parameters that characterize the turbulence of the system. Very large combustion vessels with complex internal geometries have been used to simulate realistic conditions. If is of interest to discuss the value of such complex large-scale experiments. It should first be noted that for the near-limit lean mixtures of interest, the burning velocity is, in general, quite low. Thus, the flame is convected in the mean flow without significant perturbation of the mean flow field itself. The global shape of the flame will be dependent on the mean flow field structure, which can be very complex and specific to the geometry and boundary conditions of the system. Locally the flame may be wrinkled according to the smaller-scale turbulence, which is still large compared to the laminar flame thickness, which is of the order of a fraction of a millimeter.

The rate of the pressure rise is dependent on the total energy release rate inside the vessel, hence on the surface area of the flame as it is being convected in the complex mean flow. The flame also advances into the unburned mixture ahead of it and thus the local burning velocity is also important in governing the rate of the pressure rise in the vessel.

The key problem is to determine the influence of turbulence on this local burning velocity. This local burning velocity may in fact be just the laminar burning velocity of the mixture, influenced perhaps by preferential diffusion and stretch associated with the local flame curvature and straining rate. The wrinkling of the flame is of the order of the integral scale of the turbulence. It is only when the Kolmogorov scale of the turbulence is smaller than the laminar flame thickness that the flame structure itself becomes modified and turbulent diffusivity plays a role in the burning velocity. The thickened flame is still wrinkled according to the integral scale. Detailed discussions on the classification of turbulent premixed flames are given by Borghi [43]. For the lean mixtures relevant to most severe accident analyses, the turbulent combustion regime is in the wrinkled laminar flame regime. In considering the combustion in a closed vessel, it is important to distinguish between mean motion from the turbulence by careful analysis of the length scales involved in the turbulent flow. The pressure–time history depends only on the total energy release inside the vessel, and that depends on both the total flame surface area as well as the local burning rate. Numerous tests have been carried out in vessels of intermediate to very large volumes for lean hydrogen–air mixtures in connection with reactor safety programs. However, the data have not been interpreted in a meaningful way for severe-accident analyses.

E. IGNITION SENSITIVITY

From a practical point of view, it is important to know the ignition sensitivity or an explosive or combustible mixture. In general, this is determined by minimum value of the ignition spark energy of the mixture. The standard method for determining the minimum ignition energy of a mixture is to use a low-inductance spark from a capacitor discharge. Different electrode geometries (e.g., spherical or flanged electrodes) are used and the spark energy differs for different electrode configurations. The RLC (resistance, inductance, capacitance) parameters of the discharge circuit also play a role, but since the time scale of the ignition phenomenon is long compared to that of the electrical discharge (i.e., milliseconds versus microseconds), the spark energy is not influenced much by slight variations in circuit parameters. The minimum spark energy for ignition is usually taken as the total energy stored in the capacitor (i.e., $\frac{1}{2}CV^2$); the actual energy deposited in the gas is only a small fraction of this total energy. However, as a relative measure, attempts are not usually made to obtain the actual energy released to the gas. Figure 12 shows the minimum spark energy for ignition of hydrogen–air mixtures at different hydrogen concentrations for a range of initial pressures.

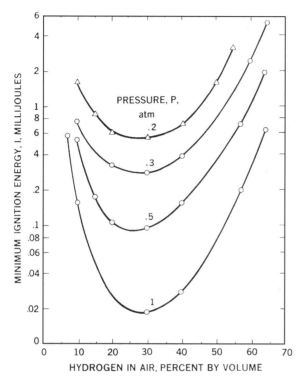

FIG. 12. Minimum spark ignition energy for hydrogen–air mixtures (from [44, 45]).

Note that the ignition energy is, in general, very small, of the order of a fraction of a millijoule even near the flammability limits. By comparing the minimum ignition energy of hydrogen to those of other fuels, one can obtain a relative measure of the ignition sensitivity of hydrogen–air mixtures. The minimum ignition energy is a useful fundamental parameter that can be easily determined. It can also be correlated with other fundamental combustion parameters of the mixture (e.g., quenching distance). If one wants to assess the influence of steam dilution or the sensitivity of hydrogen–air mixtures at high initial temperatures, the minimum ignition energy can be a useful parameter. Due to the fact that spark ignition is generally not considered relevant in severe accident scenarios, the minimum ignition energy for hydrogen–air mixtures has largely been ignored as a combustion parameter of interest. Instead, ignition by hot surfaces and by glow plug igniters has been carried out. Ignition by hot surfaces is essentially a convective-heat-transfer problem specific to the geometry and boundary conditions. The results lack generality and are of little fundamental interest.

F. AUTOIGNITION TEMPERATURE

In direct containment heating (DCH) analyses, a fundamental combustion parameter is involved, namely, the autoignition temperature. The autoignition temperature of an explosive gas mixture is the bulk temperature in which spontaneous reactions occur throughout the entire volume of the mixture without an ignition source. The autoignition temperature arises from the critical balance between heat (and free radicals) losses from the surface and the heat generation in the volume of an explosive mixture. Since the rate of heat and radicals lost is proportional to the surface-area-to-volume ratio of the mixture (i.e., inversely proportional to the length scale), the autoignition (or runaway) temperature depends on the scale. The experimental determination of the autoignition temperature is not unique, and depends on the method and scale used. Stamps and Berman [6] have reviewed the experimental data available and the results are illustrated in Fig. 13.

The data in Fig. 13 were based on a technique where the mixture is uniformly heated throughout its volume and where the autoignition threshold is established from a slight pressure rise when combustion occurs. Note that the autoignition temperature is a minimum at the stoichiometric composition and increases sharply toward the limits. For stoichiometric hydrogen–air mixtures, autoignition is observed at about 800 K. Conti and Hertzberg [46] reported an autoignition temperature of 873 K when the hydrogen concentration was 6%, near the lean limit. Sheldon [47] reported an autoignition temperature of about 1200 K using a hot surface, while Tamm et al. [48] reported that hydrogen–air–steam mixtures (5 to 55% H_2, 10 to 55% steam) can be ignited with a glow plug heated to about 975–1150 K. In the recent studies at Brookhaven National Laboratory (BNL) of high-temperature hydrogen–air detonations, it was found that self-ignition of the mixture occurs at about 700 K [49]. In the BNL facility, the "cold" hydrogen–air mixture is injected into the heated detonation tube. Autoignition corresponds to the heating via a hot surface (i.e., the heated wall of the tube). Thus, there exists a discrepancy between the different experiments on autoignition by a hot surface. As mentioned previously, the ignition by a hot surface is a heat-transfer problem and the convection flow around the hot surface plays an important role in the ignition process because of the difference in the residence time of the mixture near the hot surface.

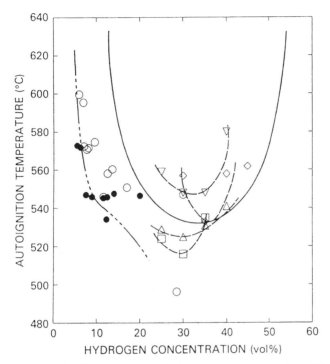

Fig. 13. Autoignition temperatures for hydrogen–air mixtures (from [6]).

Shock tubes have also been used to determine the autoignition temperature and the induction time for hydrogen–air mixtures via adiabatic shock
compression. The shock tube heats the mixture up very rapidly due to the
passage of the shock wave and its reflection from the end wall of the shock
tube. The time scale of the shock heating process is too fast for convective
effects to play a role. However, heat and radical losses to the tube wall
become important particularly for a small diameter tube, and also when
the induction time is long near the ignition limit. The interaction of the
reflected shock with the boundary layer also produces a highly nonuniform
flow field. The autoignition temperature as determined in a shock tube
differs from that determined from slow heating with a hot surface.

Hot jets of hydrogen at temperatures of 900 to 1000 K [26, 50] would
also ignite spontaneously when injected in air. In this case, the hydrogen is
not premixed with the oxidizer (i.e., air). Mixing occurs at the jet boundary,
and gradients of concentration as well as temperature exist in the mixing
zone of the two scalar quantities. It would not be meaningful to attempt to

determine a critical autoignition temperature for this case. The scale of the jet and its turbulence parameters play a strong role in determining the critical ignition condition.

Steam dilution tends to increase the autoignition temperature. A critical temperature beyond which hydrogen–air–steam mixtures would spontaneously ignite is an important parameter in severe accident analysis. However, it is difficult to define a precise autoignition temperature based on the properties of the mixture alone without reference to the particular experimental method and scale used in its measurement.

III. Fast Flames and Detonations

Experiments have shown that a combustible mixture within the flammability limits can, in general, also support a fast turbulent deflagration or undergo a transition to a detonation under "favorable" conditions. In safety analyses, it is important to know the concentration limits (for a given geometry, scale, and boundary conditions) under which a fast flame or transition to a detonation can occur. A fast flame propagates at a few hundred meters per second (subsonic) and can generate high-pressure transients. For the slow flames near the limits that have been discussed so far, significant pressures can only be generated under highly confined conditions without or with only negligible venting. However, when the flame speed is high, high local pressure transients around the flame region can be obtained even without any confinement. Since high-speed turbulent flames depend strongly on geometry, scale, and boundary conditions (i.e., degree of confinement) it is not possible to define a maximum turbulent flame speed that is a characteristic only of the mixture itself.

However, research in the late seventies and early eighties at McGill University [51] identified some optimum conditions for flame acceleration. Following the work of Schelkhin [52], the McGill researchers demonstrated that the use of regularly spaced orifice rings inside a confined tube (closed only at the ignition end or at both ends) can lead to rapid flame acceleration and transition to detonation (DDT). The optimum spacing of the orifice plates was about one tube diameter apart and the blockage ratio (where blockage ratio = $1 - (d/D)^2$) was about 0.43, corresponding to an orifice diameter "d" of about 75% of the tube diameter "D." The flame acceleration was not too sensitive around the preceding stated conditions. However, when the blockage ratio is too large (i.e., the orifice diameter is too small), the flame propagation is essentially one of a constant-volume explosion in the chamber between orifice plates, with the combustion products vented through the small orifice opening to ignite the mixture downstream of the orifice plate. Thus, the flame propagation is one of sequential jet ignition.

When the orifice was less than a certain critical size (characteristic of each mixture composition), quenching occurred after the flame had advanced a few chambers and the explosion pressure had built up sufficiently to result in a high-speed jet (i.e., a choked jet). This is referred to as the quenching regime and is not of practical interest in general. On the other hand, when the blockage ratio was too small, insufficient turbulence (i.e., randomization of the mean flow kinetic energy) was generated to result in the optimum flame speed. Thus, one notes that there exists an optimum blockage ratio to yield the maximum turbulent flame speed. Similarly, if the orifice plate spacing is too close, the individual effect of each orifice in blocking the flow is not fully manifested. If the plates are too far apart, the turbulence generated as insufficient to affect the combustion. Thus, the optimum condition should occur at a minimum spacing of the order of at least about one jet or orifice diameter. Under this optimum flame-acceleration condition, the flame, in general, accelerates rapidly to reach some steady-state value after a flame travel of about 5 or 6 tube diameters. Figure 14 shows the terminal steady-state flame velocity for hydrogen–air mixtures in a rough (i.e., orifice-plate-filled) tube.

With a blockage ratio at the optimum value of about 0.43, the orifice diameter is about $3/4$ of the tube diameter and quenching is not observed. From Fig. 14 we note that the first transition occurs at about 10% H_2. Less than 10% H_2, the maximum flame speed is about 50 m/s. Pressure transients associated with flame speeds of this order of magnitude are negligible. For concentrations greater than 10% H_2, the maximum flame speed jumps from 50 to about 800 m/s, which corresponds closely to the sound speed of the burned products. The sound speed of the combustion products in an adiabatic constant-pressure flame (denoted as the isobaric sound speed in Fig. 14) is tabulated in Table II and plotted in Fig. 14. The experimental flame speeds agree quite well with this sound speed. Streak photographs [53] of the flame in this regime confirm that the flame speed is indeed close to the sound speed in the products. This regime is referred to as the "choking regime." Theoretical analysis by Chue et al. [54] also demonstrated that the flame speed in this choking regime corresponds to the C–J deflagration speed (i.e., the lower tangency point of the Rayleigh line with the equilibrium Hugoniot). The C–J deflagration solution represents the maximum deflagration speed possible for a mixture. For higher propagation speeds, the combustion wave has to be a detonation wave and the solution is now switched to the upper branch of the Hugoniot curve. It should also be noted that the C–J deflagration is not a constant-pressure flame. However, the different between the sound speeds in the products

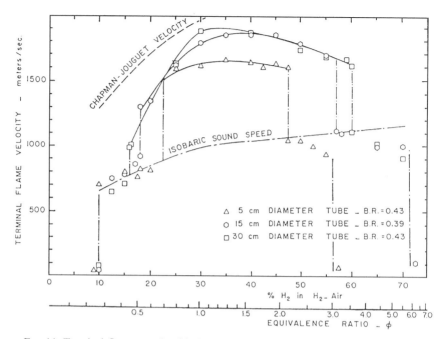

FIG. 14. Terminal flame speeds of hydrogen–air mixtures in a rough (orifice-plate-filled) tube (from [51]).

for different modes of combustion (i.e., constant pressure, constant volume, C–J detonation is insignificant because the sound speed is proportional to the square root of the temperature. The difference in the temperature of the products for the different modes of combustion arises from the difference in the concentration of the different species.

Pressure measurements of the high-speed deflagration in this choking regime indicate high-pressure spikes associated with the propagating flame. The averaged pressure, however, corresponds closely to the constant-volume explosion pressure of the mixture itself. The results of Fig. 14 indicate that the transition limit of 10% is quite independent of the tube diameter (three tubes of 5-, 15-, and 30-cm diameters were used) suggesting that the kinetic rates and the energetics of the mixture control the transition to the choking regime. As the mixture sensitivity is increased (i.e., increasing hydrogen concentration), a second transition is observed. The deflagration velocity undergoes a sudden increase of a few hundred meters per second from about 800 m/s to about 1300 m/s. The critical concentration where the transition occurs is found to depend on tube diameter (or more precisely, on the orifice diameter). For the 30-cm-diameter tube, transition occurs at about 17% H_2, while for the 5-cm-diameter tube, transition occurs at about 23% H_2 instead. This transition from the choking regime to the so-called quasi-detonation regime is found to correspond to a criterion based on the detonation cell size "λ" of the mixture. For transition to occur, it is found that the cell size should be less than or equal to the orifice diameter (i.e., $d/\lambda \geq 1$) [51]. In other words, transition requires that the orifice opening must be sufficient to accommodate at least one detonation cell. This transition criterion is in accord with the detonability limits criterion suggested by Schelkhin [55]. It should be noted that the regime is called quasi-detonation because the detonation velocity is substantially lower than the normal C–J velocity of the mixture (1500–1800 m/s). For quasi-detonation, the obstacle plates provide strong frictional losses resulting in a significantly lowered velocity. On the other hand, shock reflections and diffractions around the orifice plates produce "hot spots" for autoignition. Without the orifice plate for hot spots generation, the maximum detonation velocity deficit is typically 15% when the detonation fails. Thus, the orifice plates on the one hand promote propagation via the generation of hot spots, and on the other hand decrease significantly the detonation velocity via large momentum losses. Note that as the hydrogen concentration increases toward stoichiometry, the quasi-detonation velocity increases toward the normal C–J value as the cell size decreases when the mixture becomes more sensitive.

When the number of cells across the orifice diameter increases, the boundary condition at the wall (i.e., friction) has a lesser influence on the propagation of the detonation. In the limit when the orifice diameter is of the order of the critical tube diameter (i.e., $d_c = 13\lambda$), the wall has no influence on the detonation propagation at all. Similar behavior occurs on the rich side of stoichiometry (i.e., transition from slow deflagration to the choking regime; namely, C–J deflagration), and the second transition to the quasi-detonation regime. It is important to interpret this result as some fundamental property of an explosive mixture and not try to correlate directly the various transition limits with realistic accident analyses. However, the results in Fig. 14 do indicate that for lean mixtures less than 10% H_2 and for rich mixtures beyond 70% H_2, flame acceleration to high-speed deflagrations and DDT is highly unlikely.

Since high-temperature mixtures of hydrogen–air–steam are of interest in reactor safety analysis, flame acceleration and DDT tests have also been carried out in the high-temperature detonation tube (HDT) facility of Sandia by Slezak [56]. The blockage ratio of the orifice plates and their spacing are similar to those of the McGill experiments. Without steam dilution, flame speeds of about 825 m/s were observed at a hydrogen concentration of about 11% in agreement with the McGill results shown in Fig. 14. With 28% steam dilution, fast flames of the order of 600 m/s were observed in mixtures of an equivalence ratio of 0.5 (i.e., 17.3% H_2 on a dry basis). With higher steam dilution of 47%, fast flames were observed at an equivalence ratio of one. Flame acceleration and DDT experiments were carried out at still higher temperatures of 650 K at BNL recently and these results will be available in the near future.

A. DYNAMIC PARAMETERS OF DETONATION

If detonation occurs, very high peak pressures of the order of 15 bars or more are generated at the detonation front. Upon reflection, the pressure increases further by approximately a factor of two or more. At the onset of detonation in DDT, the initially overdriven detonation that is formed can produce much higher pressure transients depending on the degree of precompression of the mixture prior to the onset of detonation. Thus, detonation is a highly destructive mode of combustion and it is difficult to design containments to withstand detonation loading. For a given explosive mixture, it is therefore essential to know the sensitivity of the mixture to detonation (how lean would the concentration have to be before it cannot sustain a detonation, i.e., the detonability limits), the relative ease in which detonations can be initiated (i.e., the critical initiation energy), the minimum dimension of an opening for detonation to be transmitted (i.e., critical tube diameter), etc.

The sensitivity of a mixture to detonation is governed by a set of so-called "dynamic parameters," which depend on the kinetic rates of the chemical reactions. However, unlike flames, the detonation velocity and the detonation states (i.e., pressure, density, temperature, concentration of the various species in the products) can be calculated from equilibrium thermodynamics without knowledge of the reaction rates. However, the detonation state does not provide information on the sensitivity, for it depends only on the energetics of the mixture. The current status in detonation theory is that there exist a number of semi-empirical correlations and theories whereby the dynamic parameters can be predicted from a given characteristic length scale that describes the detonation process [57]. Classical one-dimensional theory gives a length scale from the kinetic rates, for example, the Zeldovich–Von Neumann–Doering (ZND) reaction zone length [58]. However, for most of the mixtures of practical interest, the detonation is unstable and a cellular structure forms from the interaction of an ensemble of transverse waves with the leading shock front; this cellular structure is observed instead of the classical one-dimensional ZND

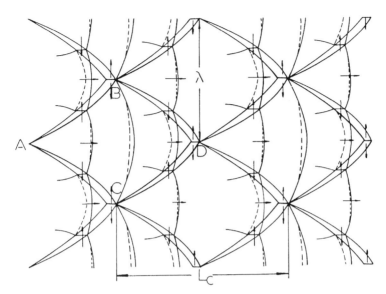

FIG. 15. Schematic of the propagation of a cellular detonation front showing the trajectories of the triple points (from [57]).

structure. Figure 15 shows a schematic of a cellular detonation and the trajectories of the triple points formed from the Mach interactions between the leading shock front and the transverse waves. The trajectories of the triple points are what is recorded on the smoked foil when the detonation propagates past it.

For an unstable detonation, the averaged cell size is the appropriate length scale that characterizes the detonation front. The cell size is a consequence of the nonlinear instability of the coupling between the chemical and the fluid mechanical processes. Although numerical simulation can reproduce the cellular structure of a two-dimensional detonation wave [59], the computation of the cell size of a three-dimensional multicellular detonation with real chemistry is still in the distant future. Fortunately, the cell size can readily be measured experimentally using the smoked foil technique pioneered by the Russians in the late 1950s [60]. The smoked foil records the trajectories of the triple points of the Mach interactions between the transverse waves and the leading shock front, and the cell size is determined from the average spacing between the triple point trajectories. A collection of the detonation cell size data for most of the common duel–air mixtures (at atmospheric pressures and room temperature initially) is shown in Fig. 16.

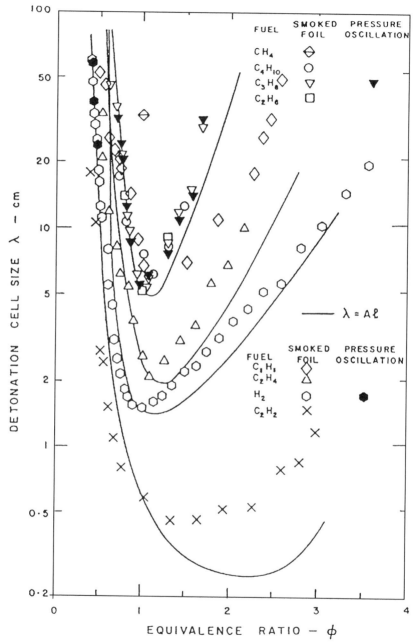

FIG. 16. Detonation cell size for common fuel–air mixtures initially at atmospheric pressure and room temperature (from [57]).

Note that the smaller the cell size, the more sensitive is the mixture. For a given mixture, the minimum cell size corresponds, in general, to the stoichiometric composition. From Fig. 16, we note that acetylene (C_2H_2) is the most sensitive mixture to detonation with a minimum cell size of the order of half a millimeter. Hydrogen is next with a minimum cell size at stoichiometric composition of about 15 mm. The alkenes (e.g., ethylene, C_2H_4) and the alkanes (e.g., propane, C_3H_8) follow. Note that methane (CH_4) of the alkane family is particularly insensitive with a cell size almost an order of magnitude larger than other members of the same alkane family. Propylene oxide (C_3H_6O) and ethylene oxide (C_4H_4O) have sensitivities in between acetylene and hydrogen. Although the cell size cannot be predicted theoretically like the ZND reaction length of an ideal one-dimensional detonation wave, we note that dimensional considerations show that the cell size "λ" should be proportional to the ZND reaction length "L." Assuming a linear dependence, that is, $\lambda = AL$, the constant of proportionality can be obtained by fitting with one experimental data point. The solid lines in Fig. 16 are based on fitting the linear relationship to the stoichiometric composition of the fuel. The linear relationship provides an adequate approximation for cell sizes on the lean side of stoichiometry, if the experimental point is chosen as the stoichiometric composition. For fuel-rich mixtures, however, the linearly extrapolated predictions underpredict the cell size in many cases. The error could be in the linear extrapolation itself or in the experimental determinations of cell size, which might be more prone to error for rich mixtures. For hydrogen oxidation, the kinetics are sufficiently well known to permit cell size to be estimated from the ZND reaction length when one or more experimental data points are available.

The detonation sensitivity of high-temperature mixtures of hydrogen–air–steam has been studied at Sandia National Laboratories up to 373 K and recently extended to 650 K at Brookhaven National Laboratory [49]. The detonation cell size as determined by the standard smoked-foil method was used in these studies. Without steam dilution, the BNL results at 650 K are summarized in Fig. 17.

For mixtures near the stoichiometric composition, the influence of high initial temperatures on the detonation cell size is small. However, for lean mixtures in the range of 10–15% H2 (and for rich mixtures), Fig. 17 shows that high initial temperatures can have a much stronger effect, significantly reducing the cell size and hence increasing the sensitivity of the mixture to detonation. The concentration range between 10–15% is of particular importance since many loss-of-coolant accident scenarios yield hydrogen concentrations in this range. The continuous curves shown in Fig. 17 are based on the assumption of a linear dependence of the cell size on the

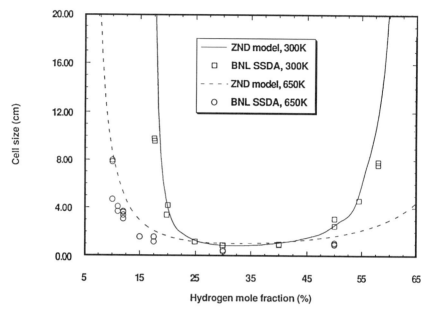

FIG. 17. Comparison between the detonation cell size of hydrogen–air mixtures at 650 and 273 K (from [49]).

ZND reaction zone length with the constant of proportionality obtained by fitting with one experimental point (in this case, the stoichiometric composition). Note that this kind of ZND modeling provides a fairly good estimate of the cell size, even far from the data point where the fitting takes place.

Figure 18 shows the dependence of the cell size on the degree of steam dilution for stoichiometric hydrogen concentrations.

Again we note that for sensitive mixtures with small amounts of steam dilution, the sensitivity of the mixture is not influenced much by higher initial temperatures. However, for the same hydrogen–air relative concentrations diluted with large amounts of steam, elevated initial temperatures tend to increase the overall sensitivity of the mixture.

B. CRITICAL ENERGY FOR INITIATION

Perhaps the most important dynamic parameter that characterizes the detonation sensitivity of an explosive mixture is the critical energy for direct initiation of an unconfined spherical detonation. Experimentally it has been observed that the direct (instantaneous) initiation of a spherical

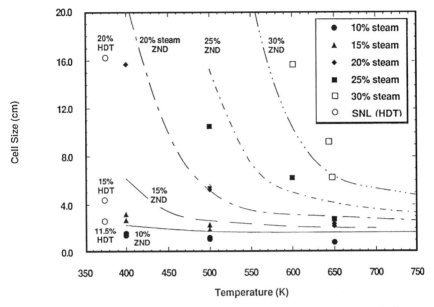

FIG. 18. Cell size for hydrogen–air–steam mixtures at 650 K and 1 atm (from [49]).

detonation requires the explosive release of a well-defined amount of energy. Above this critical value of the initiation energy, the strong spherical blast wave generated by the igniter will decay asymptotically to a C–J detonation. If the igniter energy is below this critical value, the reaction zone will decouple from the decaying blast and the reaction will propagate subsequently as a deflagration wave or flame, whose propagation mechanism depends on transport of heat and mass. In the absence of strong turbulence-generation mechanisms in a pure unconfined geometry, it is very difficult for the spherical deflagration to accelerate and undergo a transition to detonation. Thus, generally speaking, spherical detonations can only be formed via direct initiation with a powerful ignition source.

For fuel–air mixtures in general, the critical energy for direct initiation is typically of the order of megajoules and hence solid explosive charges are the most convenient energy sources with the concentrated energy of the order of magnitude required (1 g of condensed explosive releases about 4.2 MJ of energy). A typical procedure to measure the critical energy for direct initiation is to use a bare charge of condensed explosive (e.g., C4 plastic explosive, which can easily be shaped into a sphere), and place it inside a large plastic bag containing the premixed explosive mixture to be tested. The event can be photographed and pressure transducers can also be used to monitor the pressure-time history to indicate if a detonation has been initiated. In general, one can readily tell if direct initiation has been successful or not just from the manner in which the plastic bag breaks. If detonation has been initiated, the plastic bag fragments and tears into small pieces whose dimension is typically of the cell size of the mixture. If only deflagration results, the plastic bag is torn into large irregular pieces. In general, it is easier to change the mixture concentration and keep the initiation charge weight constant rather than changing the charge weight by small increments to arrive at the critical condition. Figure 19 shows the critical energy (i.e., charge weight) for many common fuel–air mixtures.

From Fig. 19, we note that the critical charge weight for direct initiation of a stoichiometric hydrogen–air mixture is about 1 g of tetryl (i.e., 4.27 MJ). Comparing Fig. 19 with the cell size data shown in Fig. 16, we note that a direct correlation between the critical energy E_c and the cell size λ exists. Zeldovich *et al.* were the first to point out the dependence of E_c on the cube of the cell size λ [61]. Since then, this dependence has been well demonstrated experimentally and numerous semi-empirical theories exist linking the critical energy to the cell size (for example, see [62]). The solid lines in Fig. 19 are based on one such analytical theory developed by Lee *et al.* [63]. Given the cell size data, the critical energy can be estimated quite adequately from these empirical theories.

Note that the critical energy is a much more accurate measure of the detonation sensitivity of an explosive mixture than the cell size. This was first proposed by Matsui and Lee [64]. It is also a much more sensitive parameter than the cell size due to its cubic dependence. In general the cell size is difficult to estimate and requires experience in selecting the appropriate scale as being the characteristic cell size from the spectrum of sizes inscribed on the smoke foil. For relatively insensitive fuel–air mixtures, very large initiation energies are required, necessitating the use of condensed explosive charges. The source characteristics (i.e., charge dimension and energy release time) are not important, and we can consider it as an ideal instantaneous point energy source. The disadvantage is that the experiment is of fairly large scale and, in general, has to be carried out in the field. The cell size can be measured in relatively smaller-scale detonation tubes in the laboratory. It should be noted that for certain mixtures, the dynamic parameters are not correlated with the cell size via the established empirical relationships (e.g., mixtures with high argon dilution). For these cases, the dynamic parameters have to be determined individually instead of being derived from the cell size.

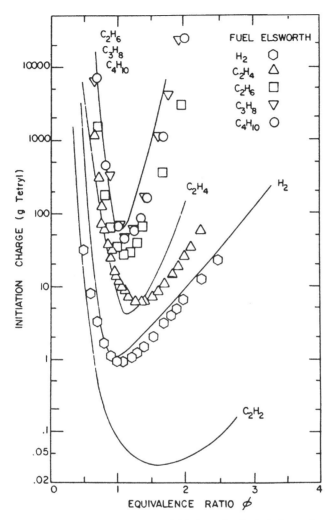

FIG. 19. Critical charge weight for direct initiation of spherical detonation in fuel–air mixtures (from [57]).

C. CRITICAL TUBE DIAMETER

A planar detonation propagating in a confined tube will be quenched by the lateral expansion waves if it is subjected to an abrupt area increase. However, if the diameter of the tube is sufficiently large, the detonation will reinitiate itself near the charge axis and continue to propagate as a spherical detonation. For a given explosive mixture, there exists a critical value of the tube diameter below which the detonation will fail. The critical tube diameter is one of the more important dynamic parameters of the mixture. Dimensional considerations indicate that the critical tube diameter must be correlated to the characteristic length scale of the detonation wave, that is, the cell size or the ZND reaction zone length. Mitrofanos and Soloukhin [65] first demonstrated that critical tube diameter is about 13 times the cell size for acetylene–oxygen mixtures, Edward et al. [66] later suggested that the 13λ correlation should be valid for other mixtures as well. A systematic study was later carried out by Knystautas et al. [67] in which cell size and the corresponding critical tube diameter were measured simultaneously for all the common fuels with oxygen as well as with air. The 13λ correlation was confirmed and thus critical tube diameter served as an alternative length scale that could characterize the detonation front. Figure 20 shows the experimentally measured critical tube diameters for some fuel–air mixtures and their comparison with the 13λ correlations (solid lines). The values of the cell sizes came from several independent measurements.

As can be observed from Fig. 20, the empirical correlation appears to be quite adequate for fuel–air mixtures in general. It should be noted that the critical tube diameter can be determined much more precisely than the cell size and is more preferable as a length scale for characterizing the detonation. This was suggested by Moen et al. [68]. However, the scale of the experiment to determine the critical tube diameter is an order of magnitude larger than that required for the cell size. Thus, the cell size remains as an important fundamental parameter from which other dynamic parameters can be derived.

The 13λ correlation was later found to be invalid for mixtures highly diluted with argon [69]. It was found that the critical diameter is of the order of 25λ, about twice that for the normal fuel–air mixtures. For mixtures highly diluted with argon, the cell patterns are, in general, much more regular. Thus, it was thought that cell regularity has something to do with the breakdown of the 13λ rule.

In a recent paper by Lee [70], the critical tube diameter problem was clarified by pointing out that there exists a dual failure mechanism for detonation. For stable detonation, the propagation of the detonation is governed by the classical shock ignition mechanism of the ZND model. Transverse waves (hence cell patterns on smoked foils) are always present but they are weak and do not contribute to the ignition mechanism. This is in accord with experimental observations that for mixtures that yield regular cell patterns, the transverse wave trajectories are much weaker.

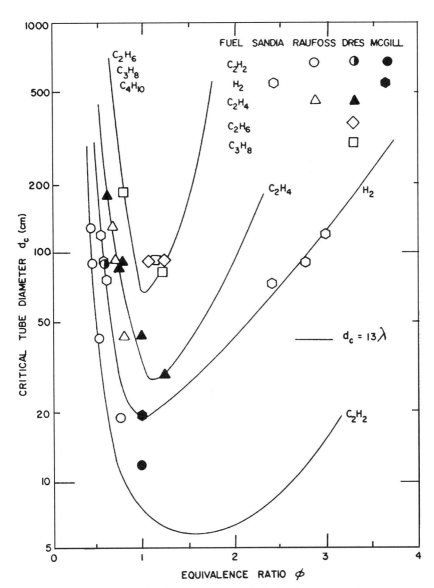

FIG. 20. Critical tube diameters for some fuel–air mixtures (from [57]).

From theoretical stability analysis, detonations are stable when the parameter E/RT (E refers to the activation energy and T the postshock temperature) is below a certain critical value. For mixtures highly diluted by argon, the shock temperature is high, thus resulting in a smaller value of the stability parameter E/RT, and rendering the wave more stable. For stable detonation waves the failure can be credited to an excessive curvature of the detonation front. For large values of the stability parameter E/RT, the detonation front is unstable and the propagation mechanism now depends on the generation of local hot spots for ignition. The hot spots are formed by the collision of the transverse waves associated with the Mach stems. In a stable detonation, the transverse waves are weak acoustic waves sustained by the chemical energy release. They do not perturb significantly the essentially one-dimensional ZND structure. Thus, the coupling of the transverse waves to the ZND structure is weak, resulting in a more regular cellular pattern. In an unstable detonation, the transverse waves are strong and form an integral part of the three-dimensional cellular structure of the detonation front. The cellular patterns are more random structures due to the strong temperature dependence of the reaction rate for mixtures with high activation energies. Small perturbations grow rapidly, resulting in the continuous formation and destruction of transverse waves leading to a higher irregular cell pattern. For unstable detonations, failure is due to the inability to develop local instabilities for hot spot ignition to occur. Thus the failure mode for an unstable detonation would be due to failure to form local hot spots from instabilities. A more complete discussion of the dual failure mechanisms is given by Knystautus et al. [67]. For the limits of detonation in tubes, the study by Dupre and Lee [71] indicate that for unstable detonation, the failure appears to correlate with the criterion that the ratio of cell size to tube diameter is of the order of unity. For stable detonations in mixtures with high argon dilution, failure occurs when the tube diameter is many times the cell size, very similar in nature to the critical tube diameter problem. For hydrogen–air mixtures at high initial temperatures, recent studies at BNL [49] indicate that transition to detonation requires a much higher value for the tube-diameter-to-cell-size for high initial temperature mixtures. This is in accord with the explanations given above for stable and unstable detonations.

D. DETONATION LIMITS

The detonability limits refer to the critical compositions beyond which the mixture cannot sustain steady propagation of a detonation wave. It is clear that these limits depend on the initial and the boundary conditions of the system. For example, for the case of spherical unconfined detonations, there are no boundary conditions in the problem, and hence only initial conditions (i.e., the initiation energy) control the limits. From the U-shaped curve of the critical initiation energy versus composition, one can arbitrarily define an upper bound for the initiation energy beyond which the mixture is considered too insensitive to be detonable. This provides an operational definition for the limits of spherical detonations. However, whether there exists a true composition limit for unconfined detonations has not yet been established. In principle, there should be such a limit when the reaction rate is just too slow to maintain the gas dynamic instability processes of the cellular detonation. No experimental attempts have been made to determine the detonability limits for unconfined detonations due to the large scale involved. The direct initiation of methane–air mixtures at stoichiometric composition requires in excess of 20 kg of condensed explosives [72]. This was perhaps the largest scale initiation experiment for unconfined fuel–air detonations that has been conducted to date.

For confined detonations in tubes, both the initial and boundary conditions become important. Assuming that the specific conditions of the initiation process of a detonation can be "forgotten" when a very long tube is used, the propagation limits should then be governed by the tube dimension and geometry. For simplicity, consider a circular tube. The tube diameter should then be correlated to the characteristic length scale (i.e., cell size) of the detonation wave itself. If we ignore the highly fluctuating "galloping mode" of propagation, then the lowest mode of propagation corresponds to that of a spinning detonation [73]. Since spinning detonations can be observed for a range of compositions, the onset of the spinning mode can be chosen to define the detonation limit in a given tube. The onset of spinning detonation is chosen to correspond to $\lambda = \pi D$, that is, when the tube circumference corresponds to a cell dimension. For a two-dimensional planar channel, the limits are defined by $W = \lambda$, where W is the smaller dimension (width or height) of the channel [74]. Thus with a knowledge of the cell size as a function of the mixture composition, the detonation limits in a tube or channel can be estimated.

For stable detonations where the propagation mechanism is not governed by instability, the cell size may not be the appropriate length scale to use to correlate with the physical dimension of the system. For stable detonations where shock ignition as described by the classical ZND model applies, the failure in a tube may be governed by excessive curvature of the detonation front. The curvature results from the divergence of the streamlines arising from the negative displacement thickness of the boundary layer [75].

For hydrogen–air mixtures, the classical value for the lean limit at room temperature is 18.5% H_2. This is based on the experiment of Breton [76] using a tube of diameter of about 25 mm. In the 300-mm-diameter HDT facility of Sandia National Laboratories, detonations have been initiated in hydrogen–air mixtures at room temperatures initially at concentrations as low as 11.5% H_2 [77]. At a higher initial temperature of 370 K, detonations have been observed at a concentration of only 9.5% H_2. It is not clear if the Sandia results for the limits are based on the onset of single-headed spin or not. The Sandia results may be based on the fact that a single-headed spin detonation is observed at the end of their tube. As mentioned previously, after the onset of spin, the spinning mode persists for a range of even lower fuel concentrations. However, the spinning detonation is no longer stable and when subjected to a finite perturbation, the spinning detonation fails. This has been demonstrated by Donato [73].

In the more recent study at BNL [78], the limits are 16 and 14% in the 100- and 270-mm-diameter tube, respectively, for room temperature mixtures. A short length of a more sensitive acetylene–oxygen mixture is used to initiate the near-limit hydrogen–air mixtures. For higher initial temperatures, the limits are 12 and 9% in the 100-mm tube for 500 and 650 K, respectively. For the larger tube of 270-mm diameter, the limits are found to be 10 and 7% at 500 and 650 K, respectively. The BNL limits are based on the onset of single-headed spin. Thus, we note that the detonability limits are dependent on the criterion used to define them as well as on the initial thermodynamic states and the boundary conditions of the system itself. It is not meaningful to assign a single value to a given fuel without reference to the criterion used and the initial and boundary conditions.

IV. Application to Severe Accident Analysis

Assuming that all the relevant combustion parameters discussed previously have been determined for hydrogen–air–steam mixtures (and perhaps with other components such as CO and CO_2) at different initial thermodynamic states, it is not at all clear how this information can and should be used in a meaningful way to assess the risk for postulated severe-accident scenarios. It should be noted that the hydrogen release is usually highly localized with respect to the typical 40,000-m^3 containment volume. The hydrogen mixing and distribution inside the complex internal configuration of the reactor containment is a formidable problem.

The nature of the combustion events and their consequences depend on the hydrogen distribution in space and time. Thus, it is essential to obtain an accurate description of the hydrogen distribution prior to analyzing the combustion phenomena. The hydrogen distribution is a turbulent mixing problem in fluid mechanics. Even without chemical reactions and shock waves, a detailed description is beyond current computing capabilities. For local regions of simple geometry and well-defined initial conditions, the turbulent mixing of scalar quantities may be adequately described in a variety of computer codes of different degrees of accuracy. However, for the hydrogen distribution within the entire reactor containment, again thermohydraulic codes have to be used. These codes do not resolve the hydrogen distribution within a cell and only an averaged value for the entire cell is computed as a function of time. Note that the time step must also be taken to be sufficiently large for local variations of the hydrogen concentration to homogenize over the cell. If we divide the containment volume of say, 40,000 m^3 into 40 cells, then the typical dimension of a single cell would be of the order of 10 m. The length scales associated with the combustion phenomena are at least three to four orders of magnitudes less. Thus, to be compatible with thermohydraulic code description of the temporal and spatial states of a cell, the effects of combustion inside a cell have to be averaged out according to the appropriate length and time scales involved. This problem has not been addressed properly in the currently available lumped-parameter codes.

Leaving the hydrogen distribution problem aside and assuming that the hydrogen concentration is known, let us examine the manner in which the combustion parameters can be of use in risk analysis. Perhaps the most straightforward parameter to consider is the flammability limits. According to the definition, flammability limits define the critical composition outside of which the mixture is nonflammable. Therefore, in a severe-accident scenario, if the estimated hydrogen concentration does not fall within the flammability limits, we simply do not have a combustion problem. The flammability limit curves of Figs. 1 and 2 define a map of flammable and nonflammable regions. Given a hydrogen–air–steam composition, a point can be located on the map and, depending on which region it is in, one may or may not have a combustion issue to worry about. Similarly, if the detonability limits are dependent only on the mixture composition, then a smaller region can be defined within the flammable region of Fig. 2. If the mixture composition falls inside the detonable region, then the detonation mode should be considered. In between the flammability and the detonability limits, one may further divide this region into one where burning is benign and one where flame acceleration and high-speed deflagrations are possible. For example, the results of Fig. 14 indicate that for hydrogen–air mixtures below 10% H_2, high-speed deflagrations are not obtained at room temperature. Thus, if similar experiments are carried out for a

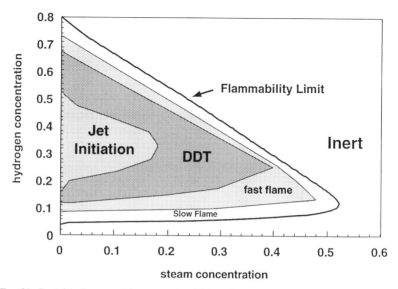

FIG. 21. Postulated composition range in which various combustion modes can be realized (from [78]).

different amount of steam dilution, the flame acceleration boundary can also be defined between the flammability and the detonability limits.

Other important combustion events may similarly be defined. For example, if the composition limits for hot jet initiation of detonations can be obtained experimentally, then another region can be defined within which the possibility of detonation initiation by a turbulent jet of combustion products can be addressed. An attempt to create such a map has been made by Shepherd [78] and is shown in Fig. 21.

Shepherd's map is based on an analysis of existing experimental data and empirical correlations for flame acceleration, DDT, jet initiation, etc. The situation for the containment is steam-saturated conditions at 100°C. Since detonation events are scale dependent, Shepherd arbitrarily (and very conservatively) assumed a characteristic length of 10 m for the channel width for the estimation of the detonability limits. The detonation-limit criterion is based on the ratio of the channel width to the detonation cell size to be of the order of unity. For the spontaneous initiation of a detonation by a jet of hot product gases, a length scale of 1 m is assumed for the jet diameter. Experiments have shown the critical condition for the direct initiation of detonation by a hot jet of combustion products is when the jet diameter is at least of the order of $10–13\lambda$ with a minimum jet velocity of at least about 600 m/s. Since available experimental data for the detonation cell size do not cover the entire range of steam and hydrogen concentrations, nor all the appropriate initial temperatures and pressures, Shepherd estimated the required cell size theoretically from ZND modeling. The data necessary to complete a map such as Fig. 21 are far from complete at the present time. It should be noted that Shepherd's map is only qualitative. However, it provides an idea as to how the experimental data on hydrogen combustion could be applied to severe-accident analysis.

Assuming that such a map becomes available, we must still caution against its use on a quantitative basis. We have discussed in previous sections the problems associated with the definitions and measurements of these combustion parameters. We shall now comment on their significance when used to construct a map like that shown in Fig. 21.

Consider first the flammability limits. There are three limits according to the direction of flame propagation, namely, upward, downward, and horizontal. For hydrogen, the difference between upward and downward propagation lean flammability limits is quite substantial, that is, from about 4 to 9% H_2 at STP in dry air. In constructing the map, one may simplify matters by taking the wider upward propagation limits. This may be a conservative or a highly nonconservative choice, depending on the accident scenario. For containment buildings too weak to survive such low-pressure transients (and there may be no such containments anywhere), any combustion event, even at lean concentrations, may increase risk. Thermal threats to the survival of safety equipment may also increase if they are subjected to a large number of small burns that would be predicted based on wide limits, when compared to fewer burns for the narrower limits. For overpressure threats, it is much more likely that a conservative assumption would be to choose the narrower flammability limits. This allows the hydrogen to accumulate to higher levels, and produce pressures that might threaten stronger containment buildings. Hence, the *conservative* assumption depends on the accident scenario and the analyst's safety considerations.

The next question is whether the method used in determining the limits is compatible with realistic conditions, since the map is being used for actual accident analysis. If the flammability limits are obtained by the standard procedure and the flame tube apparatus of Coward and Jones [13], one must recognize that the limits data may be incompatible with the actual conditions found inside a containment, where the volume is large with a dense population of equipment and the atmosphere is highly turbulent. If one deviates from the standard Coward and Jones apparatus and procedure and attempts to measure the limits in a large vessel, then one faces the problem of deciding on an appropriate criterion for defining the limits as well as resolving the problem of scaling the results afterward to even larger vessels of volumes comparable to the actual reactor containment.

Similarly, the influence of turbulence on the flammability limits may need to be taken into consideration to achieve even more realistic conditions; recall that very high levels of turbulence can quench the flame and allow higher concentrations of hydrogen to accumulate. Taking all these factors into consideration, we see that the flammability limits would no longer be a simple fundamental parameter characteristic of the mixture and its thermodynamic state that could be put on the simple map of Fig. 21. From the preceding discussions, we see that the flammability-limits data should be used only as guidelines to provide a qualitative idea as to how combustible the mixture would be for a given hydrogen–air–steam composition for given initial temperatures and pressures. These limits should not be used as precise values for regulatory or design purposes. There is also no point in attempting to obtain flammability limits under prototypical conditions. The small-scale standard apparatus is sufficient for the purpose of providing a relative measure of the flammability of the mixtures under different initial and boundary conditions.

The same preceding comments on the flammability limits also apply to the detonability limits. Different methods are used for determining the limits for unconfined detonations and for confined detonation. Due to the large scales involved in reactor accidents, the confined detonation limits are usually chosen. For confined detonations, a long detonation tube of a certain diameter is used. Again, a criterion has to be chosen to define the limits; for unstable cellular detonations where instabilities and hot-spot ignition plays the dominant role in the propagation mechanism, the onset of spinning detonation (i.e., the lowest steadily propagating unstable mode) is usually chosen to define the limits in a circular tube (i.e., $\lambda = \pi d$). Since the detonation limits are scaled according to the cell size, knowledge of the cell size permits one to determine the limits for a given physical dimension. As previously mentioned, Shepherd arbitrarily chose an extremely conservative value of 10 m as an appropriate length scale and the detonation limit boundary is then defined by evaluating the corresponding composition of the hydrogen–air–steam mixture that has a cell size of 10 m!

It is clear that detonation limits do not simply define a mixture composition beyond which detonations cannot occur. The limits depend on scale, geometry (planar or spherical), and other boundary conditions. Hence a map that defines a detonable region as shown in Fig. 21 is very qualitative in nature, and useful only as a guideline. Besides, the detonation problem is further complicated by the existence of so-called "quasi-detonations" in a complex obstacle environment typical of a congested equipment-filled volume of a containment building. Unlike normal C–J detonations that have a unique velocity for a given mixture composition, quasi-detonations have no unique velocity and depend on the boundary conditions (obstacle and confinement) as well as the mixture composition.

Similarly, the other regions in the map of Fig. 21 that define the last deflagration, DDT, and jet initiation regions are, at best, only qualitative in nature. The "Shepherd map" permits the existing information on hydrogen-combustible research to be visually collected in a manner useful for qualitative severe-accident analysis. However, the map should not be used in a quantitative manner, since the combustion parameters are not uniquely defined and measured, and the parameters themselves are not simply functions of the mixture composition, but depend also on numerous other factors outside of the mixture composition space.

A. Pressure Transients from Combustion

It is important to estimate the pressure loading on the containment in the case of a combustion event. For the detonation mode, it is possible to estimate the pressure loading on a containment structure due to a global detonation if one assumes fairly idealized conditions, that is, a uniform mixture of a given composition in a vessel of fairly simple internal geometry and direct initiation at a given location with the subsequent detonation propagation corresponding to its ideal C–J conditions throughout [79]. Such a calculation does a good job of estimating the pressure loads due to geometry and the pressure and impulse amplifications resulting from reflection and superposition of detonation waves. However, this type of calculation represents neither the worst-case scenario of precompression of the unburned mixture prior to detonation, nor a necessarily physically realistic situation. In reality, the detonation fails and reinitiates as it undergoes diffraction around bends and emerges from discontinuous openings. To describe the diffraction process even in the simplest geometry (e.g., the critical tube diameter problem) in a realistic way is beyond the computational capability currently available. One can make estimates of detonation pressure and impulse corresponding to various conditions, and use them as a guideline for safety analysis [80]. If a detonation were to occur, it would do so locally and the pressure and impulse estimates can be of use for evaluating the possible local damages in the event of a local detonation. Flame acceleration, DDT, and jet initiation of detonations are beyond current computational capability.

B. Slow Burn (Deflagration)

Perhaps the only combustion event that can be described analytically in a meaningful manner is the case of a very slow burn of near-limit mixtures between about 4 to 8% H_2 (i.e., between the upward and downward limits). Combustion of these near-lean-limit mixtures are of interest since this is the regime in which deliberate igniters are designed to operate for mitigation purposes. The prediction of the pressure development as a result of combustion is affected by a lumped-parameter thermohydraulic code. There are a number of them in existence currently (e.g., CONTAIN, MELCOR, MAPHY BURN, HECTR, MAAP); however, they all have essentially the same approach. The reactor containment is first divided into a number of compartments (referred to also as cells or nodes), typically of the order of 10 to 40. The compartments are arbitrarily chosen to roughly correspond to major containment building units, but they may have different geometries and volumes. The conservation laws of mass (total mass in the cell as well as for each chemical specie and phase), momentum, and energy are then written for each compartment as a set of ordinary differential equations in time. No spatial variation within a compartment is considered and the quantities of interest (i.e., the dependent variables) are assumed to be uniform within a cell. Transport between compartments is assumed to be due to pressure differences between compartments, convection, buoyancy, as well as molecular diffusion.

We shall elaborate more on the structure of these thermohydraulic codes by considering the specific example of HECTR (Hydrogen Event Containment Transient Response), which was developed by Sandia National Laboratories [81]. (All the HECTR models have been incorporated into the CONTAIN [82] code, and HECTR is no longer available. However, the following equations are still illustrative of the key models employed.) The independent variables of HECTR are N_{ij}, E_i, and F_i. N_{ij} denotes the number of moles of the jth specie in the ith compartment. The gaseous species considered usually are steam, oxygen, nitrogen, and hydrogen; and other gases such as carbon monoxide and carbon dioxide can readily be added. E_i is the total internal energy in the ith compartment and can be expressed as

$$E_i = \sum_{j=1}^{N_{\text{gases}}} N_{ij} u_{ij}, \qquad N_{\text{gases}} = 4,$$

where u_{ij} is the internal energy of the jth specie in the ith compartment. F_i is the volumetric flow rate in the ith flow junction given by the product of the flow area of the junction and the gas-flow velocity at the junction.

The conservation laws for mass, momentum, and energy are given below:

$$E_i = \sum_{j=1}^{N_{\text{gases}}} N_{ij} u_{ij}, \qquad N_{\text{gases}} = 4.$$

Conservation of Molar Content:

$$\frac{dN_{ij}}{dt} = - \sum_{l, F_l > 0} \frac{F_l}{V_i} N_{ij} - \sum_{l, F_l > 0} \frac{F_l}{V_k} N_{kj} + \left(\frac{dN_{ij}}{dt} \right)_{\substack{\text{external} \\ \text{source}}}$$

$$+ \left(\frac{dN_{ij}}{dt} \right)_{\text{chemical}} + \left(\frac{dN_{i, \text{H}_2\text{O}}}{dt} \right)_{\substack{\text{evaporation} \\ \text{condensation}}} + \left(\frac{dN_{ij}}{dt} \right)_{\text{fans}}.$$

Conservation of Momentum:

$$\frac{dF_l}{dt} = \frac{l}{\dfrac{\rho_i L_i}{a_i} + \dfrac{\rho_k L_k}{a_k}}$$

$$\times \left[(P_i - P_k) + \left(\frac{\rho_i + \rho_k}{2} \right) g(z_i - z_k) - \frac{K_l(\rho_i + \rho_k) F_l |F_l|}{4 A_l^2} \right].$$

Conservation of Energy:

$$\frac{dE_i}{dt} = \sum_{j=1}^{N_{\text{gases}}} \left[- \sum_{l, F_l > 0} \frac{F_l}{V_i} N_{ij} h_{ij} - \sum_{l, F_l > 0} \frac{F_l}{V_k} N_{kj} h_{kj} \right.$$

$$\left. + \left(\frac{dN_{ij}}{dt} \right)_{\substack{\text{external} \\ \text{source}}} h_{j, \text{source}} \right] + \left(\frac{dN_{i, \text{H}_2\text{O}}}{dt} \right)_{\substack{\text{evaporation}, \\ \text{condensation}}} h_{i, \text{H}_2\text{O}}$$

$$- \left(\frac{dQ}{dt} \right)_{\substack{\text{radiation}, \\ \text{convection}}} - \left(\frac{dQ}{dt} \right)_{\text{sprays}} + \left(\frac{dE_i}{dt} \right)_{\text{fans}},$$

where V_i is the volume of compartment i (m^3); $(dN_{ij}/dt)_{\text{external source}}$ is the rate of addition into compartment i from an external source (steam or hydrogen), (moles/s); $(dN_{ij}/dt)_{\text{chemical}}$ is the chemical molar rate of change due to combustion (moles/s); $(dN_{i, \text{H}_2\text{O}}/dt)_{\text{evaporation, condensation}}$ is the rate of change of water vapor due to evaporation from or condensation on the wall surfaces and the spray droplets (moles/s); $(dN_{ij}/dt)_{\text{fans}}$ is the mass in compartment i due to intercompartment fans (moles/s); h_{ij} is the enthalpy of the jth gas in compartment i (J/mole); $h_{j, \text{source}}$ is the enthalpy of the jth gas evaluated at the source temperature (J/mole); $(dQ/dt)_{\text{radiation, convection}}$ is the sum of the rates of radiative and convective heat transfer from the gas (J/s); $(dQ/dt)_{\text{sprays}}$ is the rate of heat transfer from the gas to the spray droplets (J/s); $(dE_i/dt)_{\text{fans}}$ is the energy addition to compartment i due to intercompartment fans (J/s); ρ_i is the gas density in compartment i (kg/m^3); L_i is the gas path length in compartment i (m); a_i is the effective gas flow area through compartment i (m^2); P_i is the pressure in compartment i (Pascals); g is the acceleration due to gravity (m/s^2); z_i is the elevation of compartment i (m); K_l is the loss coefficient due to flow through the lth junction; A_l is the interconnection (flow) area of the lth junction. Separate expressions for the various heat and mass transfer source terms in the preceding equations are independently specified. We shall not give further details of these source terms here but concentrate subsequent discussions on the modeling of the combustion processes.

Combustion is initiated in a compartment when the mixture composition reaches some prescribed value. This may or not necessarily correspond to the flammability limits as given in Fig. 2. For example, the default values of flammability limits used in CONTAIN 1.12, which employs the combustion model of HECTR 1.8, are given as

$$X_{\text{H2}} + BX_{\text{CO}} \geq X_c^{\text{crit}}$$

Condition	B	X_c	X_s	X_0
Spontaneous Ignition	0.541	0.07	0.55	0.05
Propagation				
Downward	0.600	0.09	0.55	0.05
Horizontal	0.435	0.06	0.55	0.05
Upward	0.328	0.041	0.55	0.05

where X_c and X_0 are the concentration thresholds for ignition, and X_s is the steam concentration threshold for inerting.

Thus, we note that combustion is initiated when the fuel concentration (i.e., hydrogen + 0.541 carbon monoxide) exceeds 0.07, and the oxygen concentration exceeds 0.05, if the total concentration of inert species (steam, carbon dioxide, and nitrogen) is less than 0.55. The default values for the diluent and oxygen are consistent with the data of Tamm *et al.* [21]. Similarly, flame propagation (upward, horizontal, and downward) is allowed when the gas concentrations of the mixture satisfy the prescribed limiting values. A comparison between the above correlation and the flammability-limits curve (without CO) as given by Fig. 2 is illustrated in Fig. 22.

When ignition and flame propagation occur, HECTR does not model flame propagation as a moving flame front. Rather, it calculates the rate in which hydrogen and oxygen in the compartment are consumed (burned) to form steam. The procedure is to first calculate a burn time and the final mole fraction of hydrogen in the compartment. The final mole fraction of

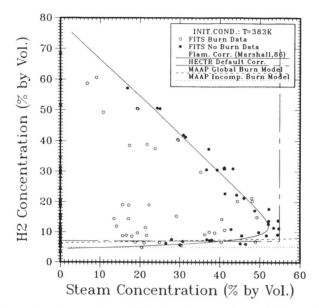

FIG. 22. Comparison between flammability-limits data and the correlation used in HECTR (from [83]).

hydrogen is computed on the basis of a combustion-completeness correlation, while the burn time is based on a specified length scale for the compartment and an averaged flame speed. The combustion-completeness correlation is derived from experimental data like that shown in Fig. 3, where we note that for mixtures below the downward propagation limits (i.e., leaner than about 9% H_2), the total pressure rise falls short of the theoretical adiabatic, constant-volume explosion pressure. Since turbulence generated by fans and sprays influences the degree of combustion completeness, different correlations are used when sprays are on or off. An example of the burn-completeness correlation with sprays off and on is shown in Fig. 23. The HECTR correlation for burn completeness is defined by:

Burn completeness = $f(X_c, X_d, \text{spray})$
  with sprays off:
  Completeness = $\min[(30.499X_c - 1.2827)\exp(A), 1.0]$
  with sprays on:
  Completeness = $\min[(28.638X_c - 1.0463)\exp(A), 1.0]$
  where $X_c = H_{H2} + 0.6X_{CO}$
    $A = X_d(-4.1966 + 3.3985X_d)$,
    $X_d = $ concentration of diluents.

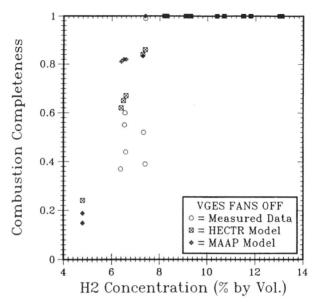

FIG. 23. Comparison of combustion completeness between experimental data from VGES and the correlation used by HECTR (from [83]).

Comparison of the above HECTR correlation for the burn completeness with the experimental data of VGES [83] is shown in Fig. 23.

The flame speed correlations are also derived from fitting to experimental data. Experimental flame speeds are obtained from direct video records of the flame front motion, from thermocouple arrays in some experiments, or from the pressure-time records. Apart from composition, turbulence has a dominant influence on the flame speed. Accordingly, different correlations are obtained for different ranges of the fuel concentration and for whether fans and sprays are on or off. For example, for fuel concentrations of less than 10%, the flame speed correlations for sprays on and off are given below.

Flame speed $= V = f(\text{spray}, L, X_d, X_c)$; $X_c \leq 10\%$
with sprays on:
$$V = L^{1/3}(59.65X_c - 1.248)\exp(A)$$
with sprays off:
$$V = L^{1/3}(23.70X_c - 0.862)\exp(A)$$
where $X_c = X_{H2} + 0.541X_{CO}$
$A = X_d(-4.1877 - 3.008X_d)$
$X_d = $ concentration of diluents.

Similar correlations are obtained for different ranges of fuel concentrations up to the rich limit [84]. However, it should be noted that for mixtures below 10% hydrogen, flame acceleration and DDT are highly improbable. Besides, the short time scale of fast flames is incompatible with the basic assumptions of the thermohydraulic code, which would render meaningless the code predictions of fast combustion events in a compartment.

In the codes, combustion can propagate from one compartment to neighboring compartments after a specified delay time, provided that the mixture concentrations in the neighboring compartments satisfy the propagation criteria. The time delay is specified as a fraction of the total burn time which varies between 0 and 1. The default value is usually taken as 0.5 corresponding to a flow junction located halfway along the compartments. It should be noted that this time delay cannot really be specified arbitrarily. In a lumped-parameter code, the mixture composition at each time step is averaged between burned and unburned mixtures in a compartment where combustion is taking place. At the same time, gas transport is allowed through the flow junctions. Thus, if the time delay is taken to be too long, the mixture in the adjacent compartments may be rendered nonflammable due to dilution with the burned products transported in from the burning compartment. On the other hand, if too short a delay time is assumed, many compartments are burning simultaneously and, thus, the containment pressure and temperature may be overestimated. The time sequence for ignition of different compartments requires that the flame front be tracked and its transmission through flow junctions be calculated. However, this is incompatible with the assumption of the lumped-parameter codes, which do not identify a propagating flame front.

The thermohydraulic codes can also handle other modes of combustion. For example, in a DCH event, the temperature of the mixture may exceed the autoignition temperature and, hence, would undergo a spontaneous reaction. This is referred to as bulk spontaneous recombination (i.e., BSR); this mode of combustion can readily be included into the code via a criterion that allows the fuel and oxygen to recombine when the bulk gas temperature exceeds a prescribed value for a given mixture composition (e.g., as given by Fig. 13). BSR can also occur when the temperature and mass concentrations of debris exceed specified limits. The HECTR default value for autoignition limit is 773 K; for debris ignition, a minimum debris temperature of 773 K and a mass concentration that exceeds 1 kg/m^3 are adopted without experimental justification.

Similarly, diffusion flame burning can also be included into the code when specified conditions are met. For example, the fuel coming into a cell will be burned as a diffusion flame when the following three conditions are met: (1) an ignition source is present or the bulk temperature exceeds 773 K; (2) sufficient oxygen is available in the receiving cell; (3) the receiving cell inert gas concentration is below a certain threshold value (e.g., 55%); and (4) the mass inflow of combustible gas is above a certain threshold. Other similar conditions can also be specified and when these conditions are met, the diffusion flame mode of combustion can be assumed in the particular cell. Since the criteria for deflagration, bulk spontaneous recombination, and diffusion flame overlap, an order of preference for these combustion events must be specified. The assumed order is diffusion flame, deflagration, and bulk spontaneous recombination; however, the ordering can readily be changed according to user preference.

C. LIMITATIONS OF THERMOHYDRAULIC CODES

Perhaps the most important point to consider in assessing the limitation of the thermohydraulic codes is the time scale associated with the variation of the dependent variables. The code is essentially of zero dimension and the spatial variations of the dependent variables within the reactor containment are defined by the actual locations of the different compartments themselves. Within a compartment, the variables take on a single averaged value for each time step. Thus, strictly speaking, the time steps used in the code must be long enough for averaged values of the variables to be meaningful. In other words, the equilibration time of the variables within the compartment must be short compared to the time step. The equilibration time for different variables can be orders of magnitude different and, strictly speaking, the time step of the code should correspond to the slowest equilibration time. For example, the pressure in a compartment is equalized in a time scale of the order the characteristic length scale (i.e., $V^{1/3}$) of the compartment divided by the local sound speed. If we consider a length scale of 10 m and a sound speed of 340 m/s, then the characteristic time for pressure equilibration is of the order of 30 ms. Thus the time step used to calculate the pressure variation by the code should be long compared to 30 ms if a single averaged value for the pressure in the compartment were to be meaningful.

For the concentrations of the different species to equilibrate, the time scale required is much longer and depends on the transport mechanism. If we assume molecular diffusion as the transport mechanism, then the time scale would be of the order of the square of the characteristic length divided by the diffusivity (i.e., L^2/a). Taking a typical value for the diffusivity of hydrogen to be of the order of 0.26×10^{-5} m^2/s and a characteristic length scale of the compartment of 10 m, we see that it will take a time of the order of 4×10^7 s, or over a year! This is clearly too long, and under prototypical accident conditions, one would expect a lot of turbulence and recirculation flow inside the containment. So if we consider sufficiently small-scale turbulence for rapid mixing and a large-scale recirculating, convective flow of the order of 2 m/s, the mixing time would be reduced to the order of 2.5 s, considering the characteristic length scale of the compartment to be reduced by a factor two for the recirculation. Thus, for code predictions of the averaged gas concentrations of the cell to be meaningful, the flow inside the cell must be highly turbulent to reduce the characteristic mixing time required. Furthermore, we note that equilibration of the mixture-composition still involves time scales of the order of a few seconds even in a highly turbulent recirculating flow.

Similarly for the combustion process, the models used in current thermohydraulic codes are not compatible with the basic approximations of the code itself. Strictly speaking, these lumped-parameter codes do not describe combustion within a compartment, since flame propagation involves spatial variation within the cell itself. To be compatible with the assumptions of these codes, we may only consider the "sudden" conversion of reactants to products in one time step. Hence, the time step must at least be greater than the "burn-out" time of the compartments. If we consider a typical value of the flame speed to be of the order of 3 m/s for the near limit mixtures as observed experimentally, then the burn-out time is of the order of 3.3 s (i.e., of the same order as the equilibration time for gas mixing). Within the burn-out time period, it is clear that the gas concentrations are not uniform within the cell and that the flame front divides the mixture into unburned reactants and burned products. Similarly the temperature equilibration would also require a time scale at least of the order of the burn-out time. Since pressure equilibration takes only about 30 ms, we may assume that the pressure within the cell is uniform within the burn-out time. The validity of this assumption is well demonstrated by experiments in closed vessels. Pressure equilibrium is not satisfied only for very fast flames and detonations.

The method of handling combustion within a compartment in the current models in thermohydraulic codes it not rigorously based on scientific considerations. The consequences of averaging burned and unburned gases at each time step during the combustion process itself has not been properly assessed. It is clear that the gas transport to the adjacent compartments during combustion is not correctly handled by the codes. Without discerning the flame front, it is not possible to describe when unburned gases and products are being transported out of the flow junction. Although this problem has been previously recognized [85], no serious effort has been made thus far to analyze the assumptions used in these codes and their consequences.

In view of the continuing efforts in conducting expensive large-scale hydrogen-combustion tests, we should comment on the application of experimental test data as inputs to the thermohydraulic codes. Let us just concentrate on the deflagration mode where three experimental correlations are required: the flammability limits, the flame speed, and the combustion completeness. The current default values used in HECTR and CONTAIN are all based on experimental data from single-chamber premixed combustion tests carried out at Sandia, EPRI, NTS, and AECL at White Shell in the 1980s. Analysis of these experimental data to obtain the correlations was carried out by C. C. Wong and others [83, 84]. Intermediate- and large-scale combustion tests are currently still being carried out by NUPEC in Japan (e.g., Hitachi, Takasago). The validity of these empirical fits (derived from limited experimental data from single closed vessels of different scales and geometries) as universal inputs to the lumped-parameter codes has not been carefully examined.

Consider first the flammability-limits data. It should be noted that ignition and flame propagation are local phenomena within the compartment. They depend on local conditions. Hence, we need the spatial variation of the mixture concentration within a compartment. To see if ignition can occur in a compartment, we need to know if the fuel concentration at the specific location of the igniter is within the limits. Similarly, we need to know the concentration at the flame front to know the flame speed. If only global averages are known, it is difficult to assess the significance of an ignition limit or a flame speed based on an averaged mixture composition. In any case, there is really no need to obtain precise flammability-limits data. The code calculates an averaged value for the entire compartment (of a length scale typically of the order of 10 m). Thus, this averaged mixture composition may not represent at all the local concentrations within the compartment where ignition and flame propagation take place. The averaged value of the composition taken over such a large length scale may involve local concentrations that can vary over the entire flammability range from the lean to the rich limit. Thus, in accordance with the approximation of the code itself (where the mixture composition is averaged over the entire cell), it is perhaps sufficient to use the flammability-limits data from small-scale apparatus (e.g., a Coward and Jones flammability tube) to give a qualitative idea of the flammability of the mixture. Any attempt to improve the accuracy to simulate actual conditions by using larger vessels and more complex internal geometries and including turbulence will only confuse the database by generating an even larger scatter of the experimental data. It is clear that the problem comes from the limitations of the codes and not from the data. The accuracy required is dictated by the coarseness of the grid (nodalization) used in the thermohydraulic code. Thus far, this has not been pointed out explicitly and carefully examined.

Regarding the flame speed correlations, it should be noted first of all that only very slow burns of near-limit mixtures can in principle be modeled by the lumped-parameter codes. Thus, it is not valid to consider mixtures that can sustain flame acceleration and DDT. For hydrogen–air mixtures, one may reasonably consider the flame acceleration lower limit to be about 10% H_2. As already pointed out previously, the flame speed is the velocity of the front relative to the fixed laboratory coordinate system. Depending on the flow velocity of the unburned mixture ahead of the flame, the flame speed can take on any arbitrary value. In contrast, the burning velocity is measured relative to the unburned mixture and hence is a fundamental quantity that depends on the mixture composition, the thermodynamic state, and the turbulence parameters in the unburned mixture. For the same mixture and the same thermodynamic and fluid mechanical states, the flame speed can be different for different boundary conditions that control the flow of the unburned mixture (e.g., a vented vessel). Even for a closed vessel, the flame speed changes as the flame propagates toward the wall even though the burning velocity itself may be constant throughout.

The flame speed correlation used in HECTR and CONTAIN are derived from fitting to experimental data from closed vessels of different sizes and geometries. These correlations cannot be applied universally to a compartment of different scale and geometry, particularly with a distribution of flow obstacles (e.g., equipment, doors, ladders) inside the compartment which would make the unburned gas flow very complex. Furthermore, the compartments are usually vented to neighboring compartments, permitting mass transport through the flow junctions. For vented vessels, the flow velocity of the unburned mixture may be quite high (as compared to a closed vessel). Accordingly, the flame speed is also high since the flame front is convected in the unburned gas flow. The correlations in the combustion models of HECTR and CONTAIN obtained from closed vessels of different scale and geometries certainly cannot be expected to be universally valid when applied to open compartments of different geometries, scale, and internal complexities. Large-scale combustion tests are currently still being carried out. The codes are used to predict the outcomes and the correlations are modified to fit the experimental data. This may not be a fruitful direction to pursue.

Similar arguments can be made concerning the combustion-completeness correlations. Their very existence is a consequence of nonuniform burning within the compartment due to buoyancy. The combustion models used in the thermohydraulic codes do not describe nonuniform burning inside a compartment. Combustion completeness is computed at the start of a burn based on the initial preburn mixture composition. In reality, the burn completeness depends on the mixture composition and flame speed throughout the burn phase and cannot be predetermined initially. It should be noted that there are also source terms and mass transport through the flow junctions, which can further modify the mixture composition during the burn phase. Thus, at best, the burn-completeness computation as incorporated in the codes at present may provide only a very qualitative correction to the total amount of burning that occurs inside a compartment for near-limit mixtures. It is not clear at all whether it is even meaningful to do so within the assumptions of the code itself.

V. Recommendations

The ultimate objective of hydrogen combustion studies is to achieve realistic predictions of the consequences of combustion events in hypothetical severe accidents in nuclear reactors, and these predictions are effected via thermohydraulic codes. Therefore, it is important to discuss the modifications that have to be implemented in these codes. One must recognize that these codes predict the time variation of dependent variables (e.g., pressure, temperature, species concentrations) within a chosen compartment (i.e., control volume or open system). For these variables to be physically meaningful, one must ensure that the time steps are chosen to permit a meaningful average to be defined. This implies that the characteristic equilibration times for all the variables should be properly evaluated and that the time steps chosen to compute the time variation of the variables are sufficiently large to permit the variable to equilibrate within the compartment. Note that one could always use arbitrary time steps and ignore the equilibration times for the different variables, but the results obtained may not correspond to reality.

A. SUBSCALE MODELS

Since the equilibration times for different variables may differ by many orders of magnitude, one has to use subscale models to compute the spatial variations of the variables that have longer relaxation times within the compartment itself. This is best illustrated by considering a simple problem of the combustion of a homogeneous gas mixture inside a spherical closed vessel. We may consider central ignition and that the flame propagates outward subsequently at the laminar burning velocity of the mixture. For simplicity, we may ignore the pressure and temperature dependency of the laminar burning velocity and assume it to be constant throughout. Considering the vessel as a compartment, the thermohydraulic code will compute the time variation of the pressure, temperature, and species concentrations (assume reactants and products) in the vessel. As discussed previously, the pressure equilibrates rapidly in a time of the order of milliseconds (i.e., the radius of the vessel divided by an appropriate sound speed averaged between cold reactants and hot products). Thus, the time step for the pressure variation can be chosen to be, say, an order of magnitude longer (e.g., 10 ms). However, the time scale to equilibrate the species concentrations is at least two orders of magnitude longer (since the sound speed is typically a hundred times the flame speed). By choosing a time scale long enough to permit the species concentrations to equilibrate (order of the burn-out time), the combustion event in the vessel cannot be described. In one time step, the reactants will all turn to products. If one insists on modeling the combustion process by choosing time steps less than the burn-out time, then one has to average the amount of reactants remaining and products produced at each instant during the combustion process over the entire volume of the vessel, to obtain the averaged values at that instant. Hence, these averaged concentrations of reactants and products are no longer real quantities.

This would be unacceptable if one were to calculate mass transport to adjacent compartments during the combustion process or to evaluate the change in the flame speed as the flame propagates into mixtures of varying composition. To describe the combustion process within the scale of the burn-out time, it is clear that a subscale model is required in which the compartment is now subdivided into two with a moving boundary (flame front) separating unburned mixture and products. Thermal equilibrium involves an even longer time scale. After the flame has reached the vessel wall and the combustion is completed, there exists a radial temperature gradient due to the adiabatic compression during the combustion process. The characteristic time for the temperature to equilibrate is the square of the vessel radius divided by the thermal diffusivity of the products if molecular diffusion is the mechanism. Temperature equilibration involves a time four to five orders of magnitude longer than the species concentration equilibration time. One would have to solve the heat conduction equation for the temperature equilibration. One could always compute an averaged temperature in the vessel at any instant of time during the combustion process as is done in the current codes. However, this averaged temperature is not a physical quantity. Serious errors are involved if this averaged temperature were to be used to calculate local heat transport to equipment inside the vessel.

From the preceding discussions we note that, as a consequence of the existence of a spectrum of equilibration time scales that can differ by many orders of magnitude for the various dependent variables, a proper description in a lumped-parameter code must use the proper subscale models to compute the variables corresponding to different time scale. Thus a *hierarchy of subscale models* is required for a proper description.

This procedure is well established in statistical physics. For example, a macroscopic system is described by a set of thermodynamic state variables (i.e., pressure temperature and the specific volume or density). The time scale for describing the variation of this set of thermodynamic state variables of the system must necessarily be large enough to permit spatial nonuniformities of these variables to equilibrate within the system. The next level of description determines the equilibration of the spatial variations of the state variables within the system itself. This is usually referred to as the hydrodynamic regime because the equations that govern the equilibration process are the hydrodynamic equations. In the hydrodynamic regime, local thermodynamic state variables can still be defined as moments of the local Maxwellian distribution. The time step of the hydrodynamic equations must be long enough to permit local Maxwellian equilibrium to be established and thus permit the local thermodynamic state variables to be defined.

Further up in the hierarchy of subscale model is the so-called kinetic regime. The kinetic regime is governed by the Boltzmann Equation, which describes the evolution of the distribution function toward local equilibrium (i.e., Maxwellian). The time scale for the kinetic regime is of the order of the collision time. Even higher levels of description involve the time evolution of the n-particle distribution function (where n is greater than 2) and the equation that governs the time evolution of the n-particle distribution function is called the BBGKY hierarchy [86]. At the top of the hierarchy is the Liouville equation that describes individual molecular interactions. In general, the computation time escalates exponentially for each higher level of description and one seldom goes beyond the kinetic regime. The number of variables is reduced as one goes down the hierarchy from the kinetic regime to thermodynamics. This is referred to as "contracted description" by Uhlenbeck [86] and the reduction of the number of variables is a consequence of the averaging procedure.

The use of lumped-parameter codes to describe the transients inside a nuclear containment is identical to the fundamental problem of statistical physics. For further progress, it is essential to develop a hierarchy of subscale models and establish the proper averaging procedure for the dependent variables involved. It appears that pressure equilibration within a compartment requires the shortest time scale (order of milliseconds).

Thus, the time step for computing the pressure variation must be at least of the order of this pressure equilibration time. Subscale models are required for all the other variables of longer relaxation times. This is the major deficiency of current thermohydraulic codes and must be remedied if existing experimental data accumulated over the past 15 years are to be properly analyzed.

It should be noted that experimental data also correspond to different levels of description. If the data do not belong to the same level of description, then they cannot be used directly as input to the code. For example, consider the case of a combustion test in a single closed vessel. Since the pressure equilibrates rapidly, the single value measured represents the averaged value over the entire volume at each instant. However, if one measures the flame shape as a function of time (e.g., via high-speed video), then this information represents a higher level of description since it provides a spatial variation of states inside the vessel. Therefore, the flame shape data cannot be used directly in a code that only describes the entire vessel as a compartment. To use to flame speed data, some averaging procedure is needed to derive an averaged value of the flame speed, and from this averaged value, a burn-out time can be calculated. This burn-out time can then be used in the lumped-parameter code. If a subscale model exists for describing the flame propagation process inside the vessel, then the experimental data on flame shape could be used directly. Without proper subscale models, it is not possible to use the various experimental data in an intelligent way.

Subscale flame models can also eliminate a lot of the large-scale tests with complex geometries. The complexities of the flame propagation inside a real compartment are mostly associated with the mean recirculating flow. The flame speed of near-limit mixtures is slow and combustion process does not significantly perturb the mean flow. The flame front is essentially being convected with this large-scale mean flow. A subscale model can readily compute this mean flow and the flame front can be superimposed on it as a slowly moving boundary. If one wishes to obtain a local turbulent burning velocity to describe the rate at which the front advances into the unburned mixture ahead of it, one could carry out relatively small-scale studied such as those of Bradley [27] and obtain correlations of the ratio of the turbulent to laminar burning velocity as a function of the turbulence parameters. The combustion tests in large-scale vessels with complex internal geometries such as those carried out currently by NUPEC at Takasago are extremely difficult to interpret. These tests provide neither basic information of general applicability nor can they simulate actual reactor conditions.

B. QUALITATIVE ASSESSMENT OF THE POSSIBILITY OF
LOCAL DETONATIONS

Although the thermohydraulic codes attempt to quantify some aspects of combustion, they cannot deal with some events such as the timing and location of random ignition, the concentration of gases and diluents at the time of ignition, or many aspects of accelerated flames and transition to detonation. Sherman and Berman [87] reviewed the available experimental data and models on transition to detonation. They developed a set of qualitative criteria that, taken together, could be used to estimate the likelihood of a detonation. The criteria included mixture class and geometry class. The two classes were then conflated into a results class. The mixture classes ranged in sensitivity from *Class 1* = *extremely detonable* to *Class 5* = *unlikely to undergo DDT*. *Class 1 geometries* were conducive to DDT, while *Class 5 geometries* were highly unfavorable to flame acceleration. The resulting classes included:

Results class 1:   DDT is highly unlikely.

Results class 2:   DDT is likely.

Results class 3:   DDT may occur.

Results class 4:   DDT is possible but unlikely.

Results class 5:   DDT is highly unlikely or impossible.

These qualitative criteria were applied to the conditions estimated for hypothetical accidents in the Bellefonte plant using code calculations of the transport and mixing of gases together with the geometric description of the plant. Under the assumed conditions of a degraded-core accident, the qualitative methodology indicated that DDT was unlikely or impossible in all but a few cases of accident conditions and certain compartments. This methodology has been used extensively, especially in Europe, to determine potential problems.

VI. Concluding Remarks

Since hydrogen explosions pose a real threat to reactor containment integrity in a severe accident, it is essential to fully understand the combustion properties of hydrogen mixtures (with CO, CO_2, and steam) at different initial conditions. This is achieved by a systematic measurement of the various equilibrium and dynamic combustion parameters. It should be noted that many of these combustion parameters can be determined in small laboratory-scale experiments with the exception of perhaps certain detonation parameters at near-limit conditions (e.g., critical tube diameter, cell size, initiation energy). With these fundamental parameters, one could identify the conditions in which the different combustion events can take place (e.g., the Shepherd map as shown in Fig. 21). It should be emphasized that knowledge of the combustion parameters can only provide a qualitative estimate of the possibility of different combustion events. One should not use them for quantitative assessments. One should also refrain from attempting to extend these combustion parameters to more realistic prototypical conditions. By doing so, one tends to increase the scatter of the data and generate more confusion regarding their meaning.

Regarding the prediction of the transient development in the containment for different accident scenarios, it should be noted that it is only meaningful to consider slow combustion events at near-limit conditions. Under that restriction, lumped-parameter codes can provide a useful qualitative description of the combustion events. Perhaps the most important improvement that has to be made in these thermohydraulic codes is the development of a hierarchy of subscale models that describe properly the events of different time scales. With these subscale models, experimental data can be used properly as inputs to render the code prediction more meaningful.

Acknowledgments

The authors would like to thank George Greene for inviting them to prepare this article. Martin P. Sherman and Martin Pilch reviewed the draft of this paper. This work was supported in part by the U.S. Department of Energy under Contract DE-AC04-94AL85000.

References

1. Rogovin, M. (1980). *Three Mile Island: A Report to the Commissioners and to the Public.* NUREG/CR-1250, U.S. Nuclear Regulatory Commission, NTIS.
2. Sherman, M. P. (1984). Hydrogen combustion loads in nuclear power plants and associated containment loads. *Nucl. Eng. Des.* **82**, 13–24.
3. Berman, M., and Cummings, J. C. (1984). Hydrogen behavior in light-water reactors, *Nucl. Safety* **25**(1), 53–74.
4. Berman, M. (1986). Hydrogen behavior and nuclear reactor safety. In *Recent Developments in Hydrogen Technology* (K. D. Williamson, Jr. and Frederick J. Edeskuty, eds.), Vol. II, Chapter 2, CRC Press, Boca Raton, FL.
5. Berman, M., (1986). A critical review of recent large-scale experiments on hydrogen–air detonations. *Nucl. Sci. Eng.* **93**, 321–347.
6. Stamps, D. W., and Berman, M. (1991). High-temperature hydrogen combustion in reactor safety applications. *Nucl. Sci. Eng.* **109**, 39–48.

7. Pilch, M. M., Yan, H., and Theofanous, T. G. (1994). *The Probability of Containment Failure by Direct Containment Heating in Zion.* NUREG/CR-6075, SAND93-1535, Sandia National Laboratories, Albuquerque, NM.
8. Williams, David C. (1995). CONTAIN code analyses of direct containment heating (DCH) experiments: Model assessment and phenomenological interpretation. In *Trans. Am. Nucl. Soc.*, Vol. 73, pp. 495–497. ANS Thermal Hydraulics Division, San Francisco.
9. Cummings, J. C., Camp, A. L., and Sherman, M. P. (1983). *Review of the Grand Gulf Hydrogen Igniter System.* NUREG/CP-2530, SAND82-0218, Sandia National Laboratories, NTIS, Albuquerque, NM.
10. Reynolds, W. C. (1986). *Implementation in the Interaction Program*, Vol. 3. Stanford University Report, Stanford Univ. Mech. Eng. Dept., Palo Alto, CA.
11. Gordon, S., and McBride, B. J. (1976). *Computer Program for Calculation of Complex Chemical Equilibrium Compositions, Rocket Performance, Incident and Reflected Shocks, and Chapman–Jouguet Detonations.* NASA SP-273, Interim Revision N78-17724.
12. Lee, J. H. S., Knystautas, R., and Chan, C. F. (1984). Turbulent flame propagation in obstacle filled tubes. In *20th Combustion Symposium International*, Vol. 99, pp. 1663–1672. The Combustion Institute, Pittsburgh, PA.
13. Coward, H. F., and Jones, G. W. (1952). *Limits of Flammability of Gases and Vapors.* U.S. Bureau of Mines Bulletin 503.
14. Kumar, R. K. (1985). Flammability limits of hydrogen–oxygen–diluent mixtures, AECL-8890. *J. Fire Sci.* **3**, 245–262.
15. Hustad, J. E., and Sonju, O. K. (1988). *Combust. Flame* **71**, 283.
16. DeSoete, G. G. (1975). *Rev. Combust.* **29**, 166.
17. Karim, G. A., Wierzba, I., and Boon, S. (1985). *Int. J. Hydrogen Energy* **10**(1), 117.
18. Shapiro, Z. M., and Moffette, R. T. (1957). *Hydrogen Flammability Data and Application to PWR Loss-of-Coolant Accident.* WAPD-SC-545, Westinghouse Electric Corp.
19. Bunev, V. A. (1972). *Fiz. Goreniya Vzryva* **8**(1), 82.
20. Roller, S. F., and Falacy, S. F. (1982). Medium-scale tests of H_2 : air : steam systems. In *Proc. 2nd Int. Conf. on the Impact of Hydrogen on Water Reactor Safety* (M. Berman, J. Carey, J. Larkins, and L. Thompson, eds.), NUREG/CP-0038, EPRI RP 1932-35, SAND82-2456. pp. 683–708. Sandia National Laboratories, Albuquerque, NM.
21. Tamm, H. (1982). A review of recent experiments at WNRE on hydrogen combustion. In *Proc. 2nd Int. Conf. on the Impact of Hydrogen on Water Reactor Safety* (M. Berman, J. Carey, J. Larkins, and L. Thompson eds.), NUREG/CP-0038, EPRI RP 1932-35, SAND82-2456, pp. 633–650. Sandia National Laboratories, Albuquerque, NM.
22. Renfro, D. (1982). Development and testing of hydrogen ignition devices. In *Proc. 2nd Int. Conf. on the Impact of Hydrogen on Water Reactor Safety* (M. Berman, J. Carey, J. Larkins, and L. Thompson eds.), NUREG/CP-0038, EPRI RP 1932-35, SAND82-2456. pp. 1029–1044. Sandia National Laboratories, Albuqerque, NM.
23. Tennessee Valley Authority, Sequoyah Nuclear Plant (1982). *Research Program on Hydrogen Combustion and Control.* Quarterly Report No. 3, Appendix A.2, 16 June.
24. Lowry, W. E. (1982). *Final Results of the Hydrogen Igniter Experimental Program.* NUREG/CR-2486, UCRL-53036.
25. Liu, D. D. S. (1980). *Canadian Hydrogen Combustion Studies Related to Nuclear Reactor Safety Assessment.* Atomic Energy of Canada Ltd. Report AECL-6994.
26. Zabetakis, M. G. (1956). *Research on the Combustion and Explosion Hazards of Hydrogen–Water Vapor–Air Mixtures.* U.S. Atomic Energy Commission Report AECU-3327.

27. Ramzy, G. Abdel-Gayed, Bradley, Derek, and McMalion, Michael (1979). Turbulent flame propagation in premixed gases: Theory and experiment. In *17th Combustion Symposium International*, pp. 245–254. The Combustion Institute, Pittsburgh, PA.

28. Henrie, J. O., and Postma, A. K. (1983). In *Proc. 2nd Int. Topical Mtg. on Nucl. Reactor Thermal-Hydraulics*, 11–14 January, Santa Barbara, CA, Vol. 2, pp. 1157–1170. ANS, Washington, DC.

29. Furno, A. L. Cook, E. B., Kuchta, J. M., and Burgess, D. S. (1971). In *13th Symposium Combustion International*, p. 593. The Combustion Institute, Pittsburgh, PA.

30. Hertzberg, M. (1981). In *Proc. Workshop on the Impact of Hydrogen on Water Reactor Safety* (M. Berman, J. Carey, J. Larkins, and L. Thompson eds.), NUREG/CR-2017, Vol. 3, pp. 13–65. Sandia National Laboratories, Albuquerque, NM.

31. Liparulo, N. J., Olhoeft, J. E., and Paddleford, D. F. (1981). Westinghouse Nuclear Corporation Report WCAP-5909.

32. Berman, M. (1981). U.S. Nuclear Regulatory Commission Report NUREG/CR-2163/1.

33. Tamm, H., Harrison, W. C., and Kumar, R. K. (1981). Atomic Energy of Canada Ltd. Report AECL-7468.

34. Ali-Khishali, K. J., Bradley, D., and Hall, S. F. (1983). Turbulent combustion of near limit hydrogen–air mixtures. *Combust. Flame* **54**, 61–70.

35. Liu, D. D. S., and MacFarlane, R. (1983). Laminar burning velocity of hydrogen–air and hydrogen–air–steam flames. *Combust. Flame* **49**, 59–71.

36. Mass, U., and Warnatz, J. (1988). *Combust. Flame* **74**, 53.

37. Dixon-Lewis, G. (1979). *Phil. Trans. Royal Soc. London A* **292**, 45.

38. Egolpopoulos, F. N., and Law, C. K. (1990). In *23rd Combustion Symposium International*, pp. 333–340. The Combustion Institute, Pittsburgh, PA.

39. Wu. C. K., and Law, C. K. (1984). In *20th Combustion Symposium International*, pp. 1941–1949. The Combustion Institute, Pittsburgh, PA.

40. Zho, D. L., Egolpopoulos, F. N., and Law, C. K. (1988). In *22nd Combustion Symposium International*, pp. 1537–1545. The Combustion Institute, Pittsburgh, PA.

41. Koroll, G. W., and Mulpuru, S. R. (1986). In *21st Combustion Symposium International*, pp. 1811–1819. The Combustion Institute, Pittsburgh, PA.

42. Koroll, G. W., Kumar, R. K., and Bowles, E. M. (1993). *Combust. Flame* **94**, 330–340.

43. Borghi, R. (1985). On the structure of morphology of turbulent premixed flames. In *Recent Advanced in Aerospace Sciences* (C. Casci, ed.), pp. 117–138. Plenum, New York.

44. Drell, I. L., and Belles, F. E. (1958). *Survey of Hydrogen Combustion Properties*. NACA R1383, National Advisory Committee for Aeronautics.

45. Lewis, B., and von Elbe, G. (1961). *Combustion Flames and Explosion of Gases*, 2nd ed. Academic Press, New York.

46. Conti, R. S., and Hertzberg, M. (1988). *J. Fire Sci.* **6**, 348.

47. Sheldon, M. (1984). *Fire Eng. J.* **43**, 27.

48. Tamm. H., Ungurian, M., and Kumar, R. K. (1987). *Effectiveness of Thermal Ignition Devices in Rich Hydrogen–Air–Steam Mixtures*. EPRI NP-5254, Electric Power Research Institute, Palo Alto, CA.

49. Cicarelli, G., Ginsberg, T., Boccio, J., Econonmos, C., Finfrock, C., Gerlach, L., and Sato, K. (1994). *High Temperature Hydrogen–Air–Steam Detonation Experiments in the BNL Small-Scale Development Apparatus*. NUREG/CR-6213, Brookhaven National Laboratory, Upton, NY.

50. Shepherd, J. E. (1985). *Hydrogen Steam Jet Flame Facility and Experiments*. NUREG/CR-3638, SAND84-0060, Sandia National Laboratories, Albuquerque, NM.

51. Peraldi, O., Knystautus, R. K., and Lee, J. H. S. (1986). In *21st Combustion Symposium International*, pp. 1629–1637. The Combustion Institute, Pittsburgh, PA.

52. Schelkhin, K. J. (1956). *Sov. Phys—JETP* **2**, 296–300.
53. Wagner, H. G. (1982). Some experiments about flame acceleration. In *Fuel–Air Explosions* (J. H. S. Lee and C. M. Guirao, eds.), pp. 77–100. University of Waterloo Press, Waterloo, Ontario, Canada.
54. Chue, R., Lee, J. H. S., and Clarke, J. (1993). Chapman–Jouguet deflagrations. *Proc. Royal Soc. London, Ser. A* **441**(1913), 607–623.
55. Schelkhin, K. I. (1959). Two cases of unstable combustion. *Zh. Eksp. Teor. Fiz*, **36**, 600–606.
56. Slezak, S. E. (1990). Flame acceleration in hydrogen–air–steam mixtures. Presented at the *Severe Accident Partners Meeting*, Upton, NY.
57. Lee, J. H. S. (1984). Dynamic parameters of gaseous detonations. *Ann. Rev. Fluid Mech.* **16**, 331–336.
58. Shepherd, J. E. (1986). Chemical kinetics of hydrogen–air–diluent detonations. *Prog. Astronaut. Aeronaut.* **106**, 263–293.
59. Oran, E. S., Boris, J. P., Yonge, T., Flanagan, M., Burks, T., and Picone, M. (1981). In *18th Combustion Symposium International*, pp. 1641–1649. The Combustion Institute, Pittsburgh, PA.
60. Denisov, Y. H., and Troshin, Y. K. (1959). *Dokl. Akad. Nauk SSSR* **125**, 110. [Transl. *Phys. Chem. Sect.* **125**, 217 (1980)].
61. Zeldovich, Y. B., Kogarko, S. M., and Simonov, M. N. (1956). An experimental investigation of spherical detonation of gases. *Sov. Phys. Tech.* **1**(8), 1689–1713.
62. Benedick, W. B., Guirao, C. M., Knystautas, R. K., and Lee, J. H. S. (1986). Critical charge for the direct initiation of detonation in gaseous fuel-air mixtures. *Prog. Astronaut. Aeronaut.* **106**, 181–202.
63. Lee, J. H. S., Knystautas, R. K., and Guirao, C. (1982). In *Fuel–Air Explosions* (J. H. S. Lee and C. M. Guirao, eds.), pp. 157–187. University of Waterloo Press, Waterloo, Ontario, Canada.
64. Matsui, H., and Lee, J. H. S. (1979). In *17th Combustion Symposium International*, pp. 1269–1280. The Combustion Institute, Pittsburgh, PA.
65. Mitrofanos, V. V., and Soloukhin, R. I. (1964). The diffraction of multi-front detonation. *Sov. Phys. Dokl.* **9**, 1055–1965.
66. Edward, D. H., Thomas, G. O., and Nettleton, M. A. (1979). The diffraction of a planar detonation wave at an abrupt area change. *J. Fluid Mech.* **95**, 79–96.
67. Knystautas, R. K., Lee, J. H. S., and Guirao, C. M. (1982). The critical tube diameter for detonation failure in hydrogen–air mixtures. *Combust. Flame* **48**, 63–83.
68. Moen, I. O., Thibault, P., Funk, J., Ward, S., and Rude, G. M. (1995). Detonation length scales for fuel–air explosives. *Prog. Astronaut. Aeronaut.* **94**.
69. Moen, I. O., Sulainistras, A., Thomas, G. P., Bjerketvedt, D., and Thibault, P. A. (1995). Influence of cellular regularity on the behavior of gaseous detonations. *Prog. Astronaut. Aeronaut.* **6**, 220–243.
70. Lee, J. H. S. (1995). On the critical tube diameter problem. In *Dynamics of Exothermicity* (J. R. Bowen, ed.), pp. 321–335. Gordon and Breach, Singapore.
71. Dupre, G., and Lee, J. H. S. (1986). Near-limit propagation of detonations in tubes. *Prog. Astronaut. Aeronaut.* **106**, 244–259.
72. Bull, D. C., Ellsworth, J., Hooper, G., and Quinn, C. P. (1976). A study of spherical detonation in mixtures of methane and oxygen diluted by nitrogen. *J. Phys. D: Appl. Phys.* **9**, 1991–2000.
73. Moen, I. O., Donato, M., Knystautas, R., and Lee, J. H. S. (1981). In *18th Combustion Symposium International*, pp. 461–469. The Combustion Institute, Pittsburgh, PA.
74. Vasilieu, A. A. (1991). The limits of stationary propagation of gaseous detonations. *Dynamic Structure of Detonation in Gaseous and Dispersed Media* (A. A. Borisov, ed.), pp. 27–49. Kluwer Academic Press, Dordrecht, The Netherlands.

75. Ray, J. A. (1952). Two gaseous detonations: Velocity deficit. *J. Phys. Fluids* **2**(3), 283–289.
76. Breton, J. (1936). Recherches sur la detonation des malanges gazeau. Doctoral Thesis, Université de Nancy, France.
77. Tieszen, S., Sherman, M. P., Benedick, W. B., and Berman, M. (1987). *Detonability of H$_2$–Air–Diluent Mixtures*. SAND85-1263, NUREG/CR-4905, Sandia National Laboratories, Albuquerque, NM.
78. Shepherd, J. E. Hydrogen combustion and explosion during severe accidents in nuclear power plants. In preparation.
79. Byers, R. K. (1982). *CSQ Calculations of H$_2$ Detonations in Zion and Sequoyah*. SAND81-2216, NUREG/CR 2385, Sandia National Laboratories, Albuquerque, NM.
80. Tieszen, S. R. (1993). Effect of initial conditions on combustion-generated loads. *Nucl. Eng. Des.* **140**, 81–94.
81. Camp, Allen L., Wester, Michael J., Dingman, Susan, E., and Sherman, M. P. (1982). HECTR: A computer program for modeling the response to hydrogen burns in containments. In *Proc. Workshop on the Impact of Hydrogen on Water Reactor Safety* (M. Berman, J. Carey, J. Larkins, and L. Thompson, eds.), pp. 827–842, Sandia National Laboratories, Albuquerque, NM.
82. Washington, K. E. (1991). *Reference Manual for the CONTAIN 1.1 Code for Containment Severe Accident Analysis*. SAND91-0835, NUREG/CR-5715, Sandia National Laboratories, Albuquerque, NM.
83. Wong, C. C. (1988). *A Standard Problem for HECTR-MAAP Comparison: Incomplete Burning*. NUREG/CR-4993, SAND87-1858, Sandia National Laboratories, Albuquerque, NM.
84. Wong, C. C. (1988). *HECTR Analysis of the Nevada Test Site (NTS) Premixed Combustion Experiments*. NUREG/CR-4916, SAND87-0956, Sandia National Laboratories, Albuquerque, NM.
85. Geller, Antony, S., and Wong, C. C. (1991). *A One-Dimensional Material Transfer Model for HECTR Version 1.5*. SAND88-0974, UC-505, Sandia National Laboratories, Albuquerque, NM.
86. Uhlenbeck, G. E., and Ford, G. W. (1963). In *Lectures in Statistical Mechanics*, Chapters I, IV, V, VI, VII. American Mathematical Society.
87. Sherman, Martin P., and Berman, Marshall (1988). The possibility of local detonations during degraded-core accidents. *Nucl. Tech.* **81**, 63–77.

ADVANCES IN HEAT TRANSFER, VOLUME 29

Heat Transfer and Fluid Dynamic Aspects of Explosive Melt–Water Interactions

D. F. FLETCHER

Department of Chemical Engineering, University of Sydney, Australia

T. G. THEOFANOUS

Departments of Chemical and Mechanical Engineering, Center for Risk Studies and Safety, University of California, Santa Barbara, California

I. Introduction

If a hot liquid (melt) contacts a cooler volatile liquid, in some circumstances the energy transfer rate can be so rapid and coherent that an explosion results. Such explosions can present a hazard in any industry where there is the potential for contact between a hot liquid at a temperature well above the saturation temperature of a cold volatile liquid. Before we continue with the description of these explosions it is important to note that they are known by a variety of names. They are sometimes referred to as *physical explosions*, as they are physical rather than chemical in nature. They are often referred to as *vapor explosions* because of the rapid generation of vapor, but this term is very misleading as they are then confused with chemical gaseous explosions. In the hydrocarbon transport industry they are known as *Rapid Phase Transitions* (RPTs). In the nuclear industry they are known as *Fuel Coolant Interactions* (FCIs) or *Molten Fuel Coolant Interactions* (MFCIs) since they have been studied extensively in the context of severe nuclear accidents. In the water reactor context they are known as *Steam Explosions* (SEs)—this term is probably the oldest and perhaps the least objectionable, therefore it is adopted here in line with trends in the recent past.

Such explosions are a well-known hazard in the metal casting industry [1]. They are particularly common in the aluminum industry, where ways of preventing them have been investigated for the last 30 years [2]. The move to the use of aluminum lithium alloys in the aircraft industry has increased the severity of the problem as these can explode, chemically augmented, with greater violence [3]. With the increase of recycling of scrap metal, the major hazard in this industry has shifted from casting operations to scrap loading in the furnace, where water contained within the scrap poses a significant hazard [4]. They are also common in the steel industry and result when water accidentally contacts molten steel either in the blast furnace (if wet scrap metal is loaded) or during the transfer of melt from the blast furnace to a processing plant [5]. The tragic accident at Appleby-Frodingham in the UK was a result of the accidental entry of water into the "torpedo" used to transport molten steel [6].

Steam explosions are of concern in the transportation of liquefied natural gas over water [1, 7], where the potential hazard arises if (cold) LPG or LNG contacts (hot) water either because of a spill as a ship is being loaded or as the result of an accident. Although these explosions are not as violent as those observed in the metal industry, they still represent a significant hazard, especially as they can act as a very efficient means of dispersing flammable gases. Observations show that the explosivity of this system is very sensitive to the composition of the LNG, which changes as the more volatile species evaporate.

In the paper industry a molten salt composed mainly of Na_2CO_3, called "smelt," may contact water within a furnace [1], as for example from the accidental failure of a water pipe used for head recovery. The explosions can also occur in dissolver tanks when smelt is poured into water to convert "black liquor" into "green liquor" [1]. Again the explosivity of this system depends very strongly on the composition of the smelt. Explosions can also occur when molten glass contacts water [8] and are of concern in the vitrification process for highly radioactive waste. These examples show that the high thermal conductivity of metallic melts is not a prerequisite for an explosion.

It is believed that steam explosions also occur in submarine volcanism [9]. Here the scale is very large, with thousands of tons of molten lava mixing with water. The explosion which destroyed the island of Krakatau is the largest explosion of this type to occur in recent history [10]. The mechanisms into which water and lava came are still poorly understood but there is little doubt that melt–water contact played a role in this enormous explosion.

They are studied in the nuclear industry to assess the consequences in the unlikely event that a severe accident occurs, molten material contacts residual coolant, and such an explosion results [11]. In the Light-Water Reactor (LWR) context the main concern has been to show that if an explosion occurred in the reactor pressure vessel it would not lead to early containment failure via a sequence of events known as α-mode failure. With the resolution of the α-mode failure issue close at hand, attention has recently focussed on other possible melt-water contact modes, such as ex-vessel explosions and accident management issues [11]. It is important to realize that many other events besides the possibility of an energetic steam explosion must be considered in any severe accident analysis. The steam explosion is set in context with these other events in the review of severe accident phenomenology produced by Theofanous and Saito at the time of the Zion-Indian Point study [12].

Finally, and although not a liquid, finely divided ash from power station boilers is also known to result in similar (at least in some respects) explosions [13]. Ash collects around furnace walls and superheater banks. When large lumps break loose they fall into the ash collector that contains water and can explode if the agglomerate fragments. Such explosions represent a severe problem in the Australian power generation industry and result in significant economic loss.

We should also note that there are certain industrial operations where melt and water are deliberately mixed in order for the melt to be rapidly cooled [14]. This process is used to produce very fine particles, with the very high cooling rates ($> 10^6$ K/s) resulting in amorphous solids. By controlling the cooling rate the crystallographic structure, and hence the properties of the solid, can be varied. Here the aim is to work in a certain regime, by which we mean melt temperature, water temperature, contact mode, mass scale, etc., such that explosions are avoided. Clearly here it is important to have a good understanding of the regime under which explosions occur and to be able to calculate the explosion loading in the event that something goes wrong.

Having recognized that this type of explosion is indeed widespread, we will show that the different scenarios share many common features. As is apparent from the above description, the need arises to be able to calculate the likely energy release and pressure fields generated in an explosion. These quantities determine the extent of damage, to people and plant, and are required if active steps are taken to design against such explosions. In the next section we describe the explosion process in more detail, with the aim of trying to distill common elements in the explosion process for the above events.

II. Description of the Explosion Process

It is widely accepted that steam explosions progress through a number of distinct phases. If we take the example of melt poured into water, the stages are as follows:

1. Initially, the melt and water mix on a relatively slow timescale (~ 1 second). During this stage the melt and water zones have a characteristic dimension larger than the order of 10 mm. This size is determined in part by the initial length scale of the melt and in part by the degree of subsequent hydrodynamic breakup which results from Rayleigh–Taylor and Kelvin–Helmholtz instability of the system. Because of the high temperature of the melt, a vapor blanket insulates the melt from the water and there is relatively little heat transfer. This stage is usually referred to as *coarse mixing* or *premixing*,

2. If the vapor blanket is collapsed in some small region of the mixture, high heat transfer rates result and there is a rapid rise in the pressure locally. The vapor film collapse can occur because either the melt cools or some event forces the water and melt into physical contact locally [15, 16]. This is known as the *triggering* stage.

3. If the circumstances are favorable this pressure pulse can cause further vapor film collapses, so that it propagates and escalates through the mixture, causing coherent energy release. The propagating pressure pulse (which steepens to form a shock wave) has two main effects. First, it collapses the vapor blankets around the melt droplets initiating rapid heat transfer. Second, it initiates a variety of processes which can cause the melt droplets to fragment, resulting in a large increase in the heat transfer rate from the melt to the water. For a large-scale explosion to occur this *propagation* stage is essential to ensure that the energy transfer from the hot to the cold liquid is coherent.

4. As the energy of the melt is transferred rapidly to the water, the explosion mixture expands with the potential to cause damage to any surrounding structures. The damage may be caused either by the shock-wave itself that results from the propagation process or by a slug of water or melt driven by the explosion in the latter stages as the hot pressurized coolant flashes into steam. The latter process acts to integrate the kinetic energy generated, which can thus be focused upon impact with a structure. This is usually referred to as the *expansion* stage, but we note here that the effect of the shock wave must also be taken into consideration.

Figure 1 shows a schematic representation of the above sequence of events. It is important to realize that the above description applies to one particular contact mode between the melt and water. It is one that has

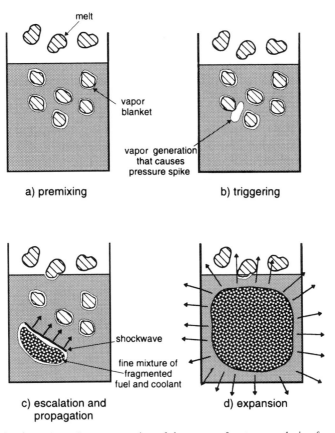

FIG. 1. (a–d) A schematic representation of the stages of a steam explosion for a pouring mode of contact.

been studied extensively because of its importance in the accident se-quences studied in the nuclear industry. Other modes of contact could also be important in a given application. An alternative mode of contact is that which results when the two fluids are initially separated in stratified geometry, as for example in the case when an explosion was avoided in the above scenario. Also, this situation occurs naturally, because of the density differences between the two fluids, if water is poured slowly onto melt. In this case there is no initial premixing, and when the vapor film collapses the propagation event must mix the fluids as it goes. Such explosions are discussed in detail in Sect. 6.4.

In industrial accidents and vulcanology applications an alternative contact mode is important, in which water becomes entrapped and is subsequently heated until it rapidly flashes to vapor and generates significant pressures [17]. Coherent explosions of this type would require superheating to spontaneous nucleation levels, and a feedback mechanism involving water that has not yet reached this level. Small-scale tests [18] reveal such possibilities, but very little is known about large scales. Alternatively, explosion-like behavior can occur in a highly constrained system, by means of a cascade of contacts that build up the pressure rapidly, but not on sonic time scales, until the constraint fails. Such may be the mechanism of the ash explosion mentioned above, or perhaps some hydrovulcanologic situations, and should be kept clearly distinct from our main subject here, which is *propagating* explosions.

In any accident situation it is likely that a combination of the above contact modes occurs, with some melt being premixed, some being strati-

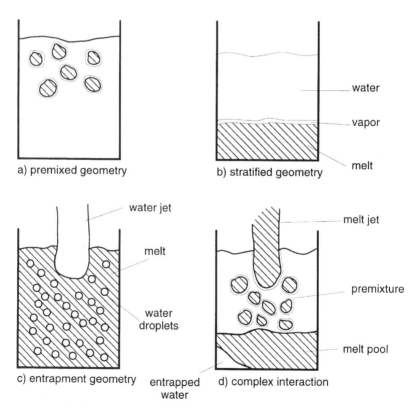

FIG. 2. (a–d) A schematic representation of various melt–water contact modes.

fied (either on the base of the vessel or because the melt jet is unfragmented) and some being mixed as the explosion progresses. An example of this complex mode of interaction is shown in Fig. 2. Thus the analysis of industrial incidents and steam explosion experiments must always begin with an investigation of the initial contact mode.

We should also point out here that although the explosions we are describing are physical in nature, they can also be combined with chemical explosions. The best-known example of this type is the highly energetic interaction of molten aluminum with water. This explosion can be coupled with an extremely exothermic chemical reaction if the aluminum fragments (which have a very large surface area) oxidize rapidly producing hydrogen. An added concern in these cases is the explosivity of the resulting hydrogen-steam-air mixture, following the expansion phase. Explosions of this type have been observed in a number of aluminum/water interactions. They are discussed in detail in Sect. 7.

III. A Very Brief History

A. Early Studies

The occurrence of steam explosions in nature goes back to the earliest days of this earth. No doubt as man began to work with metals he found that the contact of molten metal with water posed a significant hazard. In the fourteenth century the hazard of metal casting is recounted in the Canon's Yeoman's Tale in the *Canterbury Tales* of Chaucer. To quote:

> And how, d'you think? It happens, like as not,
> There's an explosion and goodbye the pot!
> These metals are so violent when they split
> Our very walls can scarce stand up to it.

In the 1950s it was the aluminum industry that first sought to limit the occurrence of this type of explosion. In a now classic study, Long [2] performed as extensive series of tests in which he poured a stream of molten aluminum into water. Long noted that explosions occurred only when molten metal spread across the bottom surface of the tank. He postulated that the melt trapped a layer of water against the tank, and that it was the sudden vaporization of this water that triggered an explosion. He proposed, based on this picture, that if the container walls were coated with grease or an asphalt-based paint that explosions would be prevented. He was correct, and casting pits in the aluminum industry are to this day painted with bituminous paint, called Tarset, to prevent such explosions.

At about the same time the nuclear industry started to get interested in steam explosions. Planned or accidental power excursions in a number of small test reactors resulted in violent explosions. These are listed in Table 2 of the review by Cronenberg [19]. He distinguishes between true steam explosions, in which shock pressurization occurred (e.g., SL-1, BORAX-1, and SPERT-1D) and milder events due to rapid boiling (e.g., NRX and EBR-1). These incidents initiated the massive investment in research which the nuclear industry has subsequently made.

B. THERMODYNAMIC AND PARAMETRIC MODELING

While explosions continued to happen in the aluminum industry and in the paper industry, the licensing procedure for sodium-cooled fast reactors required quantification of the maximum possible energy release from such an explosion if the core were to melt and contact sodium coolant. Hicks and Menzies [20] proposed that a limit on the energy release could be placed by representing the explosion as a simple two-stage process. In their model, the melt and coolant initially mix adiabatically at constant volume and come into thermal equilibrium. This mixture is then assumed to expand isentropically to ambient pressure. This model, which has been modified to use more sophisticated equations of state and to allow for vapor blanketing (by assuming that the coolant expands adiabatically and is thermally insulated from the melt) is still used to provide bounding calculations, especially for situations where the contact mode and mixing conditions are poorly understood.

As expected, the above model predicted large conversion from thermal to mechanical energy, which when expressed as a ratio gives ∼40% under optimal premixture compositions (assumption of thermodynamic equilibrium and reversible expansion). Experimentally determined values are much lower. Corradini et al. [21, Table B.1] have summarized data on conversion ratios determined in large-scale experimental programs and show that these lie in the range from 0.3 to 3%. More recently, values as great as ∼8% have been estimated, while the literature contains some indication that near-thermodynamic conversions may have been reached in some accidents [22]. Incidentally, efficiency is also used to express the energetics of steam explosions, it being the ratio of the obtained work to the maximum thermodynamically possible.

In reality, heat transfer rates are finite and transient, and the system expands as the pressure increases. In order to determine typical pressure–time curves for coolant channels, a number of parametric models were developed. A typical model was that of Cho et al. [23] which used the geometry shown in Fig. 3 and allowed for finite-rate heat transfer effects

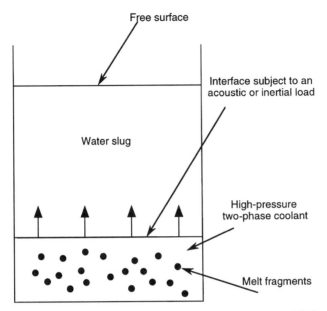

FIG. 3. The geometry used in the parametric model of Cho *et al.* [23].

and the compressibility of the surrounding coolant. A similar model developed by Berthoud and Newman [24] proved very useful in the interpretation of results from the CORRECT II facility. Details of the many variants of this type of model are tabulated in the review by Corradini *et al.* [21]. This modeling approach set a trend, in that it highlighted the need for an understanding of the detailed physics mechanisms, so that the rate-controlling processes could be identified and modelled.

At about this time studies of the basic mechanisms were started. These included investigations of rapid boiling, fragmentation, and explosion efficiency. A summary of these experiments can be found in the reviews by Cronenberg [19] and Corradini *et al.* [21].

On the theoretical side two important concepts appeared. One aimed at identifying explosivity thresholds is due to Fauske and is known as the Spontaneous Nucleation model [25]. The other sought to quantify the feedback that sustains an explosion already assumed to propagate is due to Board *et al.* and is known as the Thermal Detonation model [26]. These two models are discussed in the remainder of this section.

C. THE SPONTANEOUS NUCLEATION MODEL

First we need to recall that if two liquids (or solids) are brought into intimate contact, then the interface temperature (T_i) is given by the following well-known expression [27]:

$$T_i = \frac{T_h(k\rho c_p)_h^{1/2} + T_c(k\rho c_p)_c^{1/2}}{(k\rho c_p)_h^{1/2} + (k\rho c_p)_c^{1/2}}, \tag{1}$$

where k is the thermal conductivity, ρ is the density, c_p is the heat capacity at constant pressure, and the subscripts h and c refer to the hot and cold fluid, respectively.

Second we must discuss nucleation and metastable liquids. A liquid superheated above its saturation temperature for the given pressure is in a metastable state. It is stable to small thermodynamic fluctuations, but if a sufficiently large perturbation occurs it partially vaporizes to produce vapor and residual liquid (the system is adiabatic on the time scale of vaporization, so the heat of vaporization is extracted from some of the liquid). In order to form a vapor cavity in a bulk liquid, work must be done to overcome surface tension forces, which seek to collapse the bubble, and to "push back" the liquid. Once a bubble exceeds a critical size it grows and the effect of surface tension becomes unimportant. In everyday boiling from a solid object, such as a kettle element, surface imperfections provide small sites which trap permanent gas and vapor, allowing bubbles to grow. Such sites are very apparent in the early stages of boiling a kettle when the water is subcooled, and bubbles stream from such nucleation sites. However, in the case of a liquid in contact with a second liquid no such sites exist. Bubbles can only grow when the random molecular fluctuations are sufficiently large to provide critical-sized embryos. The rate of formation of such nuclei is given by Volmer's rate equation [28]:

$$J \propto \exp(-W/kT), \tag{2}$$

where J is the bubble nucleation rate per unit volume, k is the Boltzmann constant, and W is the reversible work of formation of an embryo, given by

$$W = \frac{16\pi\sigma}{3(p_v - p_l)^2}, \tag{3}$$

where σ is the surface tension of liquid, and p_v and p_l are the vapor and liquid pressures, respectively. If the bubble nucleation rate is plotted against the temperature the graph shows an exponential increase. The following example taken from Reid [1] illustrates the sensitivity of the bubble nucleation rate on the temperature. If one had 1 m^3 of ethyl ether at 410 K one would have to wait (on average) 10^{21} years to observe a bubble nucleate, but at a temperature of 420 K the average waiting time would be 10^{-14} s. The temperature at which very rapid nucleation occurs within the body of the liquid is known as the *homogeneous nucleation temperature* or the *superheat limit*, denoted T_{hn}. The increase in nucleation rate with temperature is so rapid that this value can be calculated to within a degree. Note that in many articles (e.g. [19, 21]) the nucleation rate is shown approaching an asymptote at a critical temperature. This is a wrong interpretation of the above equation.

The above superheat limit temperature applies to nucleation in the bulk of the fluid. As a good approximation, the value of T_{hn} is given by [1]

$$T_{hn} = [0.11(p/p_{crit}) + 0.89]T_{crit},\tag{4}$$

where the subscript "crit" denotes the value at the critical point.

At the interface between two different fluids the nucleation temperature may be reduced because of imperfect wetting. This reduced nucleation temperature is known as the *spontaneous nucleation temperature*, denoted T_{sn}. If the contact angle between the two materials is θ then the work of formation of a bubble given in Eq. (3) must be multiplied by the following factor [19]:

$$f(\theta) = \tfrac{1}{4}(2 + 3\cos\theta - \cos^3\theta).\tag{5}$$

From the above equation we see that if $\theta = 0°$, corresponding to perfect wetting, then $f(\theta) = 1$ and $T_{sn} = T_{hn}$. In the limit of imperfect wetting, for which $\theta = 180°$, $f(\theta) = 0$ and $T_{sn} = T_{sat}$.

Based on the above concepts Fauske [29] proposed that for a large-scale steam explosion to occur the following criteria must be satisfied:

1. The two fluids must be separated from each other by a vapor layer, that is, they must be in film boiling, in order to prevent excessive heat transfer.
2. There must be direct liquid–liquid contact due to vapor film collapse to allow rapid energy transfer.
3. The interface temperature upon film collapse must exceed the spontaneous nucleation temperature so that near instantaneous vapor formation occurs, which causes fragmentation of the liquids and mixing.
4. There must be adequate inertial constraint so that a shock wave can form and sufficient energy transfer can occur before the liquids are 'thrown apart'.

The above conditions provided the first conceptual picture of a steam explosion. Henry and Miyazaki [30] and Henry and Fauske [18] developed the concept into a mechanistic model and provided upper limits on the pressure at which explosions could occur. The model, known as the *Droplet Capture* model, is restricted to a pouring mode of contact and a well-wetted system. It considers that spontaneous nucleation cannot occur until the thermal boundary layer in the cold fluid has become sufficiently thick to contain a critical-sized vapor cavity and that significant bubble growth requires an established pressure gradient in the cold liquid. A consequence of their model, and the feature that gave it its name, is that small cold liquid drops would be "captured" by the hot surface and rapidly vaporized, whereas large droplets would remain in film boiling. The model also predicts that explosions are eliminated by elevated pressure (0.9 MPa for water) or a supercritical contact temperature [18].

The enduring legacy of this work is in the importance of film boiling and the spontaneous nucleation idea as a requirement for essentially instantaneous film boiling in a liquid–liquid system, thus allowing interpenetration and premixing. Also, the pressure cutoff has been supported experimentally, although the model itself has been criticized. Perhaps the most significant criticism can be found in recent data that show unambiguously highly supercritical pressures at the front of propagating explosions [31], whereas according to the droplet capture model peak pressures should be limited to $p_{sat}(T_{hn})$. On the other hand, it is not yet known (it may well be) whether the rapidity of nucleation in the spontaneous nucleation regime plays a role in the rapid feedback needed in the early stages of escalation. A discussion of these and other issues is given in the reviews by Reid [1], Bankoff [32], Cronenberg [19], and Corradini *et al.* [21].

D. The Thermal Detonation Model

At about the same time a competing theory, called the *Thermal Detonation* model, was developed by Board and Hall. This model laid the framework for most of the ongoing work in this subject and so will be reviewed here in detail. Based on their observations of tin–water explosions [33], they proposed an analogy between chemical detonations and vapor explosions [34, 26]. In the classical picture of a chemical detonation [35, 36], a shock wave passes through a homogeneous mixture of chemical reactants. The adiabatic compression across the shock wave increases the temperature, leading to extremely rapid chemical reaction. The chemical energy released generates high pressures, thus maintaining steady propagation of the shock wave through the reactants. Experimental measurements show that the chemical reaction is essentially complete in a narrow zone immediately behind the shock wave.

In a steam explosion, the propagating shock wave causes fragmentation and mixing of the two liquids, leading to transfer of thermal energy from the hot liquid to the cold volatile liquid. This produces high pressures, which drive the shock forward. Board and Hall proposed an analogy between chemical detonations, where the temperature jump across the shock wave leads to chemical energy release, and vapor explosions, where the jump in interfacial area and the collapse of the vapor film surrounding the hot fluid leads to rapid thermal energy transfer. They extended the analogy further by assuming that the two liquids reached thermal and mechanical equilibrium in a narrow zone immediately behind the shock wave. This assumption meant that they could ignore the detailed complexities of the fragmentation, mixing, and heat transfer processes and allowed them to base their model on only two states: the initially separated liquids and the final completely mixed equilibrium condition. This results in simple equations that were used to predict the propagation behavior.

Figure 4 shows the conceptual picture of a propagating detonation together with the variation of pressure and velocity through the front. Assuming that the detonation front is propagating at a steady velocity, it is possible to work in a frame of reference moving with the shock front, as shown in Fig. 5. In this coordinate system, conservation of mass, momentum, and energy across the reaction zone give

$$v_1 \rho_1 = v_2 \rho_2 \equiv j, \tag{6}$$

$$p_1 + \rho_1 v_1^2 = p_2 + \rho_2 v_2^2, \tag{7}$$

and

$$e_1 + \frac{p_1}{\rho_1} + \tfrac{1}{2} v_1^2 = e_2 + \frac{p_2}{\rho_2} + \tfrac{1}{2} v_2^2, \tag{8}$$

respectively. These equations can be combined to produce the following equation:

$$\tfrac{1}{2}(p_1 + p_2)(V_1 - V_2) = e_2 - e_1, \tag{9}$$

which provides a thermodynamic relationship between the initial conditions and all possible end-states. Given an equation of state for the reacted material (e.g., $p_2 = p(V_2, e_2)$), Eq. (9) gives the locus of all possible end-states on a P–V diagram and is known as the detonation adiabatic. Figure 6 shows a typical P–V diagram. Also shown on the diagram (as a dotted line) is the shock adiabatic, which represents all possible end-states in the absence of "combustion."

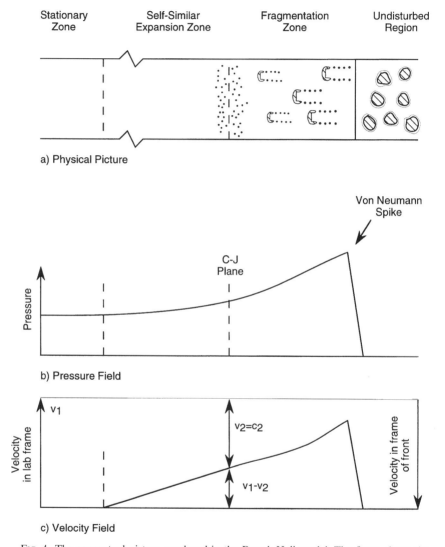

FIG. 4. The conceptual picture employed in the Board–Hall model. The figure shows the (a) physical picture on which the model is based and the (b) pressure and (c) velocity fields within the interaction zone.

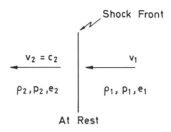

a) Shock Frame of Reference.

b) Laboratory Frame of Reference.

FIG. 5. (a–b) The detonation front in moving and stationary frames of reference.

Considering the jump conditions for mass and momentum alone, we obtain

$$j = \sqrt{\frac{(p_2 - p_1)}{(V_1 - V_2)}} \, , \tag{10}$$

where j is the mass flux through the detonation front. The above equation shows that region A–B of the detonation adiabatic is unphysical because it implies an imaginary mass flux. The point A corresponds to "combustion" at constant volume.[1] Equation (10) also shows that all end-states must also lie on a straight line through point O, with a slope determined by the mass flux through the front. This line is known as the Rayleigh line. Point C,

---

[1] A steady-state propagation causes the material entering the front to be compressed before it is "burned," so that pressures always exceed the constant volume mixing case. Thus the well-known thermodynamic model of Hicks and Menzies [20] discussed earlier is not a true upper bound, because a steady-state propagation leads to even higher pressures prior to material expansion.

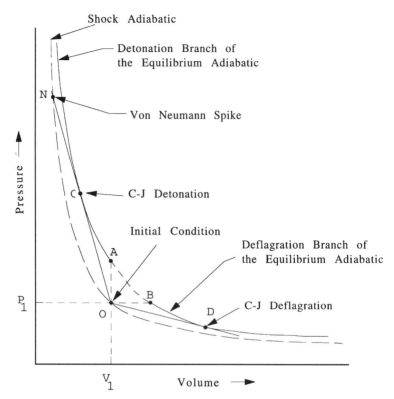

FIG. 6. The detonation trajectory in the P-V plane. The figure shows the shock and detonation adiabatics and the location of the CJ point.

where this line just touches the detonation adiabatic, is known as the Chapman–Jouguet (CJ) point and corresponds to the minimum mass flux. It is possible to show that in this case the material leaving the front is just sonic in the frame moving with the shock front. Physically the CJ point gives the minimum propagation velocity, which does not allow disturbances originating behind the detonation front to catch up with the front. It is also possible to show that this is the only steady-state solution [35]. Using Eq. (10) it is a simple matter to show that the fluid velocity behind the detonation front is given by

$$v_1 = V_1 \sqrt{\frac{(p_2 - p_1)}{(V_1 - V_2)}} \, . \tag{11}$$

This velocity is always supersonic relative to the conditions ahead of the detonation wave.

Returning to Fig. 4 we can now explain the shape of the pressure profile. Initially the pressure rises to point N on the shock adiabatic (see Fig. 6) as material entering the front is compressed without any "reaction" taking place. The point of peak pressure is known as the Von Neumann spike. The pressure than falls, as the reaction takes place, until the pressure reaches point C on the P–V diagram. At this point the steady-state reaction zone is connected to a time-dependent expansion zone. The pressure falls in the expansion zone in a manner determined by the far-field boundary condition.

Board and Hall used equations (6)–(11) to determine the CJ state for the tin–water system. For a mixture consisting of equal volumes of tin at 1000°C, water, and steam, they showed that the CJ pressure was of the order of 100 MPa and the propagation velocity was ~300 m/s. They also noted that if there was little or no vapor present initially the pressures would be much higher. Using the fragmentation correlation of Simpkins and Bales [37] they showed that tin droplets with an initial diameter of 10 mm would be fragmented into micron-size particles in a timescale of ~200 μs, giving a "combustion zone" thickness of ~100 mm. In addition, they predicted that pressures of the order of 1500 MPa could occur in the UO_2–Na system. However, it was recognized that the model only applied to situations where there was complete energy transfer from the hot to the cold fluid that the theory was only applicable to one-dimensional propagation fronts. Also the existence of a homogeneous mixture ahead of the detonation front is an essential part of the model.

In addition to their underlying conceptional picture that vapor explosions are analogous to chemical detonations, Board and Hall made four simplifying physical assumptions in deriving their mathematical mode [38]:

1. A pseudo-homogeneous geometry exists in the system before the shock waves arrives. This is a very idealized geometry which is not typical of that seen in "real" accidents or experiments. The disparity between reality and the assumed model geometry may therefore prevent meaningful comparison between model results and experimental data.

2. A shock wave propagates through the system at a steady speed. This assumption bypasses the need to model the complex process of escalation from a small trigger pressure to a steady-state detonation.

3. All of the hot liquid interacts with all of the cold liquid. As we will see later, there is experimental evidence that this assumption is incorrect. It may be incorrect for most "real" world explosions.

4. The two fluids reach thermal and mechanical equilibrium inside the reaction zone that moves along with the shock wave. The idealized picture allows the model to calculate the behavior of the system on

the basis of a completely reacted end-state, thus removing the need to model the details of the fragmentation, mixing, and heat transfer processes which occur inside the reaction zone.

Although the Board–Hall model was a conceptual breakthrough in the study of vapor explosions, it soon became evident that it was far from a complete picture. As described above, the model assumed that all of the melt transferred all of its energy to the coolant. This gave rise to extremely high shock pressures and propagation velocities that were greatly in excess of anything observed in experiments [39, 40, 41, 42].

The model postulated that behind the shock front the melt was fragmented due to the relative velocity between the melt and coolant. Some workers were skeptical as to whether this process was fast enough or whether it was possible to trigger a detonation [43], but experimental data showed that shock-induced fragmentation of drops of one liquid in another could occur, and some argued that it was fast enough [44]. To explain why the experimentally measured pressures were so much lower than those predicted by their original model, Hall and Board [45] suggested that sideways flow, interphase slip, and the presence of vapor in the interaction zone could all play a role in reducing the predicted pressures. Unfortunately, none of these effects are easy to quantify or support experimentally as yielding the principle cause. Perhaps more importantly, the model then changes from one with no free parameters to one with a large number of parameters.

Condiff [46] pointed out that the analogy with chemical detonations could not be carried too far. The well-known technique [35] of finding the point of tangency of the Rayleigh line with the detonation adiabatic on a P–V diagram only works for a one-component system. It is only a valid procedure when there is complete fragmentation in a system at a point where all the species are always in mechanical equilibrium. Otherwise multiple solutions may exist, because there is a different detonation adiabatic for each degree of fragmentation. Also, it is not possible to construct the point of tangency by drawing a straight line from the point corresponding to the initial conditions unless all of the species have the same velocity. He also questioned the validity of assuming a "thin" shock with zero momentum transfer across it.

Other criticisms were raised and in order to answer these there was a gradual progression from a purely steady-state model in which the behavior in the propagation wave is not considered, to steady-state models in which the detailed variation across the wave was calculated. Following on from this transient models were developed in which the shock was either fitted or captured. An extensive review of these modeling efforts has been produced by Fletcher and Anderson [38]. In addition, other modes of propagation have been proposed and are considered later.

E. OTHER WORK

Research in the 1980s was dominated by the drive in the nuclear industry to develop a better quantification of the likelihood of α-mode failure. A number of very extensive medium-scale experiments were performed at various centers around the world, including Ispra in Italy (KROTOS, FARO), Winfrith in the UK (THERMIR, Rig A and B, and MFTF programmes), and at Argonne National Laboratory (CWTI, CCM), and at Sandia National Laboratories (the EXO-FITS and FITS series) in the USA. These experiments were generally performed to obtain information on energy conversion ratios for systems as close to prototypic as possible for use in reactor safety assessments. They highlighted the importance of the premixing phase, the study of which has become one of the most active of steam explosion research. Summaries of these experimental programes can be found in the reviews by Corradini *et al.* [21], Corradini [47], and Fletcher [16], and the references therein.

An integrated assessment of steam explosions for the reactor geometry by Theofanous and co-workers [48] led to the conclusion that α-mode failure is physically unreasonable, but it also led to rather intense discussions and further work, especially on multifield modeling of premixing, and towards the end of the decade on extensions of the modeling approach to propagation. In the 1990s, the work, in fact, further intensified, but it also started coming together in supporting the major advances needed to address both the basic mechanisms and the prediction of steam explosion energetics. This work is described in the following sections. We use the Premixing–Triggering–Propagation format, and although focusing on recent work, we attempt to provide a coherent evolution of the main ideas all the way from the beginning in each of these three areas. The last two sections are on chemical effects, a newly reinvigorated but still poorly understood subject, and on integral assessments, an older but recently significantly clarified subject.

IV. Premixing

Limits to mixing have become an important concept in the study of large-scale steam explosions. The fact that some initial coarse mixing, or premixing, must take place if ton quantities of melt are to be converted to 100 μm particles was pointed out by Cho *et al.* in 1976 [49]. Based on the consideration of the energy required to mix fluids because of drag forces, they showed that an intermediate step in which the melt was mixed to ~10 mm droplets on the ~1 s time scale was needed prior to the fine fragmentation step. There was much debate about the steps to be assumed in the mixing process, with Baines *et al.* [44] showing that if the large mass of fuel moves and particles are stripped from it then less energy is required than the progressive mixing process proposed by Cho *et al.* [49]. However, it is evident that some form of premixing is needed if a large mass of melt is to be involved in a coherent explosion.

A. PREMIXING MODELS

Attention turned away from the energy requirement to mix the melt and water and has become focused on the hydrodynamic limitations, that is, on the quantity of water that could remain mixed with coarsely broken-up melt. This limitation results from steam production and has become known as the water depletion phenomenon. The first workers to introduce this idea were Henry and Fauske [50], who based their limit to mixing on a critical heat flux (CHF) model. Just as in the flat plate boiling crisis, where liquid can no longer reach the heater surface, they assumed that mixing would stop when the steam flow generated by the melt droplets in contact with water is sufficient to prevent water entering the mixture. If the water is saturated they used the following correlation:

$$(Q/A)_{\text{CHF,SAT}} = 0.14 h_{fg}\, \rho_v^{1/2} [g\sigma(\rho_c - \rho_v)]^{1/4}, \qquad (12)$$

which is modified in subcooled conditions via

$$(Q/A)_{\text{CHF,SUB}} = (Q/A)_{\text{CHF,SAT}}\left[1 + 0.1\left(\frac{\rho_v}{\rho_l}\right)^{1/4}\left(\frac{c_c\,\rho_c\,\Delta T_{\text{sub}}}{\rho_v h_{fg}}\right)\right]. \quad (13)$$

The above correlations are given in, for example, Tong [51] and are applicable to one-dimensional, steady-state situations where radiative heat transfer is negligible. They were obtained from experimental CHF pool-boiling data for flat plate heaters, where return flow of water to the heater is governed by Rayleigh–Taylor instability at the liquid–vapor interface.

In the proposed limit to mixing, the maximum possible removal rate of energy is the product of the cross-sectional area of the mixing vessel and the critical heat flux. The total energy transfer from the droplets is the product of the fuel droplet surface area and their heat flux, assumed to be a sum of contributions from radiation and film boiling. Equating these fluxes gives the following expression for the minimum possible droplet size the fuel can form before the mixture is dispersed by steam flow:

$$D_p = \frac{6m_f(Q/A)_{\text{melt}}}{\rho_f A_{\text{vessel}}(Q/A)_{\text{CHF}}}, \tag{14}$$

where m_f is the mass of fuel and ρ_f is its density. Although this model was used by Henry and Fauske to explain the results of some earlier experiments (notably those of Long [2]), it was not long before it was criticized on the basis of its oversimplification (mixing is an unsteady, multi-dimensional process). Their claim that the mixing of ton quantities of material "can readily be ruled out on the basis of first principle arguments" was rejected.

Although it is fair to say that the above model did not close the coarse mixing issue, it did provide an important input into the modeling area and identified the key physical process, namely, that as melt and water mix vast quantities of steam are produced (at least at low pressure) and the system is then very unstable. Corradini and Moses [52] used data from the FITS experiments [53] to construct limits to mixing based on the idea that either the melt droplets or the water droplets within the mixture would be fluidized if the melt fragmented too much. These limits are reviewed in detail by Fletcher [54].

Bankoff and co-workers then started the modern trend of mixing modeling by writing conservation equations for mass and momentum for the mixture based on the interpenetrating continuum (multiphase) approach. In order to simplify things they assumed that: (i) the melt is in the form of spherical droplets of a single fixed size; (ii) that the melt temperature is fixed; (iii) heat transfer is by radiation and convective film boiling; and (iv) the coolant is saturated (so that there is no condensation).

Initially, Bankoff and Han [55] presented a steady-state model in which they calculated the void distribution above a front of melt droplets falling into a water pool at their terminal velocity. This work showed that both a homogeneous flow model and a separated flow model gave very rapid dispersion of the mixture and they concluded that the simpler homogeneous flow model was adequate. Later calculations extended this approach, using the homogeneous flow assumption, to transient 1D [56] and 2D [57] situations. The transient calculations were scoping in nature, used very coarse numerical grids, and the 2D results were very limited and only for high ambient pressures. This work was not pursued by Bankoff *et al.* because of numerical problems with the code they were using.

At about this time there was a renewed interest in the α-mode failure issue, which together with the availability of more powerful computers led various workers to pick up the work of Bankoff and co-workers. In the UK,

Fletcher [54] reviewed the models and concluded that a two-dimensional, transient, three-phase, separated flow model was needed. Over the following years Fletcher and Thyagaraja developed the CHYMES model, which is presented in detail in [58]. The model has the following features:

1. It is transient two dimensional (planar or cylindrical geometry).
2. It models three phases, namely, melt, liquid water, and steam. Each phase has its own velocity field. Interfacial momentum transfer is modeled via simple expressions that assume that the melt phase is always dispersed and that the region of bubbly flow is small.
3. The coolant and steam are assumed to be saturated. This greatly simplifies the modeling and allows a very efficient incompressible flow algorithm to be used [59]. However, later work showed that this assumption is not valid in certain applications, where the pressure in the system can increase due to the rapid production of steam which induces subcooling [60].
4. The melt is assumed to be in the form of prefragmented spheres, which are allowed to fragment further (based on a simple relative velocity fragmentation model). The melt temperature is calculated from an enthalpy conservation equation. The heat transfer rate is calculated as a sum of contributions from convective film boiling and radiation. The forced convection film boiling model of Witte [61] is used and a factor is included in the usual gray body radiation model to account for the partial absorption of thermal radiation of water when the melt temperature is high. A void fraction dependence is also included to reduce the heat transfer as the water becomes depleted.
5. The current version is now compressible and can treat coolant subcooling.

At about the same time Theofanous and co-workers started to develop a model based on a similar approach. Their early model assumed that the coolant flow was homogeneous [62], but this restriction was dropped in later versions of the model. The final version, called PM-ALPHA, has the following features [63]:

1. It is transient and two dimensional.
2. It models three phases, namely, melt, liquid water, and steam. Each has its own velocity field. Interfacial momentum transfer is modeled using a relatively complex flow regime map.
3. The coolant and steam each have their own temperature fields. The flow is assumed to be compressible, so that pressure-induced subcooling is modeled.

4. The melt is assumed to be in the form of prefragmented spheres of fixed size. The melt temperature is calculated from an enthalpy conservation equation. Heat transfer from the melt to the water is modeled as a sum of forced convection film boiling (again using the Witte correlation) and radiation at low fractions and via radiation and forced convection at high void fractions.

5. The current code version includes an interfacial area transport model for the melt with mechanisms for both breakup (large-scale) and fragmentation (small particles). Also, there are constitutive laws for film boiling in two-phase flow derived from special experiments performed by Liu and Theofanous [64] and a radiation model that allows for nonlocal transport, which is important for melt temperatures exceeding 3000°C.

Models similar to the preceding are also being developed at Sandia (the IFCI code) [65], in France (the MC3D code) [66], and Germany (the IVA3 code) [67]. Corradini and co-workers have used a modified version of the TEXAS code to study premixing in one dimension [68]. Their model differs from the others in that they have used a Lagrangian approach to model the fuel field. However, as this model is only one-dimensional it is unable to treat the essentially two-dimensional phenomenon of premixing adequately. Finally, we should note that there is considerable model development in this area currently taking place in Japan. Workers at the University of Tokyo [69] are developing a model called CHAMP/VE and workers at JAERI are adopting the MISTRAL code to simulate premixing [70]. Both of these models are in the early stages of development and validation. It is fair to say that all of the above models have undergone less validation than the CHYMES and PM-ALPHA models and as yet have not been used in any assessment of steam explosion concerns in reactor accidents.

In addition to the above models of the dynamics of the mixing process, other models to describe the breakup of jets are being developed. These models assume that the dominant breakup mode, at least following penetration of the melt jet into the water pool, is via Kelvin–Helmholtz instability on the surface of the jet. A model called THIRMAL [71, 72] has been developed at Argonne National Laboratory. The model assumes that behind the initial vortex at the head of the jet the upflowing steam causes stripping of particles from the jet that mix with water and generate steam. The change in flow regime with distance from the leading edge is included. This model has been used to examine the likely fraction of melt which could be quenched in a pool of water below the reactor vessel in ex-vessel accident situations.

Workers at the University of Stuttgart [73] have performed extensive studies of Kelvin–Helmholtz instability on jets surrounded by liquid coolant and vapor films. They have determined the fastest growing wavelength for cases in which the velocities are discontinuous in the two layers and for which the velocities are continuous (using the Miles formulation which allows for shear layer profiles). The wave crests are assumed to be stripped when they reach a critical fraction of their wavelength. The stripping rate and fragment size is determined via an energy balance which involves kinetic energy, surface energy, and flow energy. The Miles formulation appears to lead to a model that can explain the breakup of jets of molten Wood's metal at low temperature. The applicability of the model to film boiling jets is still under investigation.

B. PREMIXING EXPERIMENTS

With the vast amount of analytical work came the need for data against which to validate the models. Although there was already a vast amount of existing data on steam explosions, modelers soon found that it was not very helpful in validating their models. This is because the data were not collected to study premixing, so that the initial conditions were not well-specified and details of the premixing process were not recorded. After only a few seconds' thought one realizes that the task of probing inside the mixture is incredibly difficult. The mixing flow is highly transient and model validation requires knowledge of the *local* composition within the mixture itself. If you add to this the complexities of melt fragmentation and high temperatures, one is almost tempted to give up from the start. However, much progress has been made via a combination of the use of simulant systems and/or the development of ingenious diagnostics.

We will start the discussion by looking at simulant experiments. The main reason for using simulants is to allow the use of better diagnostics. Experiments at Oxford University [74, 75] have been performed in which 6-mm steel balls were released into a narrow tank with a known volume fraction and injection velocity. The tank was only slightly wider than the ball diameter so that visibility was very good. These experiments allowed the spreading of the jet to be studied and to be compared with calculations from the CHYMES code. For this very simple system the experimental data and code comparisons were in rather poor agreement, with the code failing to predict the "head" formed at the front of the falling jet. Analysis of the experimental data suggested that for the situation considered the assumption that the "melt" forms a continuum phase is not valid [75]. Subsequently, Fletcher and Witt [76] showed that the "head" formation process could be calculated provided that a high-order numerical scheme was used to minimize numerical diffusion effects. In addition, allowing for the influence of the walls on the particle drag law was found to be important. Thus these experiments showed that, while simulant experiments may be useful, they must be designed so that they represent the model being tested. Hall at Berkeley Nuclear Laboratories [77] performed experiments in a similar geometry in which molten tin was poured into subcooled and saturated water. These experiments provided data on mixing of a molten liquid with water in a 2D geometry. Again the experimental data were compared with CHYMES calculations and agreement was found to be rather poor.

Theofanous and co-workers have performed simulant experiments in which tens of kilograms of mm-size steel balls were heated to typically 1000°C and released into water [78, 79]. The experimental series, known as MAGICO, was designed to demonstrate at 1/8 scale the water-depletion phenomenon expected in the lower head of a PWR. In order for the experiments to provide useful data for code comparison, they were instrumented with high-speed cameras, X-ray imaging, and a novel system, known as FLUTE, which provides data on the local void fraction [80]. The FLUTE system works by adding fluorescein to the water and measuring the intensity of the light emitted from a local fluid region using small optical fiber probes located in the water pool. This technique allows the local void fraction to be estimated and compared with calculated values. Note that the melt volume fraction is very low, typically 2%, so that the instrumentation is not affected by the presence of the "melt." Void fractions of up to ∼60% were measured.

The model calculations performed using PM-ALPHA are in good agreement with both the global void fraction (obtained from visual records of level swell) and the local data obtained using the FLUTE system. As well as providing confirmation of the code's ability to reproduce the void generation aspect of premixing, the experiments showed that once the water has been expelled from the central core of the mixture a static pressure imbalance can develop which forces water located around the edges of the mixture to reenter the mixing zone and cause a burst of vapor generation. This effect was christened ETHICCA, for Energetic Transfer of Heat in a Counter-Current Ambient [79].

Extensions to conditions covering higher particle temperatures (up to 1800 K already, and 2270 K eventually) so as to include significant radiation effects (see Fig. 7, color plate), moderate and high coolant subcoolings, and much higher inlet velocities (up to ∼5 m/s) have been made in the MAGICO-2000 facility [81]. Also, cold particle pours were investigated, with particles of different densities (steel, ZrO_2, Al_2O_3, SiC). Flash X-ray instrumentation was used to obtain qualitative information on

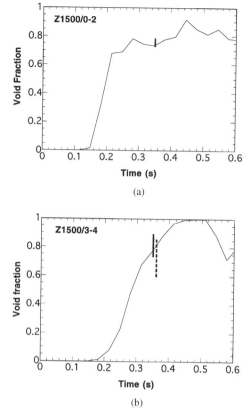

FIG. 9. (a) Comparison of predicted local (chordal average) void fraction in the mixing region in the MAGICO-2000 run shown in Fig. 8a, against the PM-ALPHA prediction. Experimental uncertainty is also shown. (b) Same condition, except 3 K subcooling.

the mixing region (i.e., particle distribution) as well as quantitative information on local void fraction distribution. For saturated and slightly subcooled water *local* void fractions of over 80% were measured, while no measurable (less than ~10% void) could be found for 18 K subcooling (see Fig. 8, color plate). The experimental facility was designed with the help of PM-ALPHA, and the results were found to be in excellent agreement (see Fig. 9). These MAGICO-2000 tests and previous ones in MAGICO provide the first experimental demonstration of the water depletion phenomenon mentioned earlier. Further, the code comparisons indicate that this important effect is quantitatively predictable.

The above approach to the validation of mixing models is also being pursed in France, at CEA Grenoble, in an experimental apparatus known as BILLEAU, and in Germany in an experiment known as QUEOS [82, 83]. In the QUEOS experiments Mo and ZrO_2 shot with diameters in the range 4.2–10 mm and initial temperatures up to 2270 K are released into a pool of water. No published data are available yet from either facility.

A series of medium-scale mixing experiments, called the MIXA experiments, was performed ar Winfrith Technology Centre, UK [84, 85]. The experiments involved the release of ~ 3 kg of molten fuel simulant (81% uranium dioxide and 19% molybdenum metal at a temperature of 3600 K) into a pool of water. A novel droplet former was used that ensured that the melt entered the water as a stream of droplets with a diameter of approximately 6 mm. "Skirts" of varying lengths were attached beneath the droplet former to control radial spreading of the stream of melt droplets. The mixing vessel was of square section with a side of 0.37 m, and a pool depth of 0.6 m was used. The initial pressure was 0.1 MPa in all of the experiments and the water was initially saturated in all but one of them. The mixtures formed were relatively weak, having a melt fraction of typically 1%. The mixing process was recorded using cine photography with intense backlighting. The vessel was left open to the atmosphere via a vent line that contained a flowmeter to measure the steam produced as the melt entered the water.

The above experiments were specifically designed and performed to assist with the task of CHYMES validation. Therefore, the experimental conditions were chosen to meet the validation requirements. Thus, these are the first experiments in which a stream of droplets of prototypic material have been injected into water. In addition, the experimental configuration was such that the assumptions of incompressible flow and saturated water were expected to be valid. One of the experiments was performed with the water initially subcooled by 20 K to provide a means of benchmarking the effect of subcooling.

Even with these experimental features, analysis of the data was not straightforward. After some experimentation, a well-defined two-dimensional inflow of melt was generated, but even then the rapid steam generation and level swell that occurred when the melt entered the water obscured the movie, so that the initial inflow rate could only be observed for a fraction of the inflow period. The movies gave a very clear picture of the mixing process, and features such as melt holdup and the existence of an extensive steam void in the mixing zone were observed. No steam explosions occurred.

There were five successful experiments in the series. Tests 01, 04, and 06 used increasing lengths of "skirt" around the droplet former and resulted in a range of melt pours from one-dimensional initially (i.e., a pour of melt droplets over most of the vessel width) in 01 to a central pour (i.e., a pour over a small region in the center of the vessel) in 06 [84]. Experiment 05 was similar to 06, except that the water was initially subcooled by 20 K in 05. Experiment 07 was similar to 06, except that a modified droplet former was used which resulted in larger droplets (10–30 mm instead of 6 mm).

An analysis of the MIXA01 experiment, in which the melt pour was initially one-dimensional, is given in Denham *et al.* [84], and an analysis of the MIXA06 experiment, in which the melt pour was two-dimensional, is given in Fletcher and Denham [85]. The validation exercise carried out using data from these experiments has led to the following conclusions [85]:

- CHYMES predicts the mixing behavior when the flow is one-dimensional reasonably well. For two-dimensional pours the code underpredicts the melt spreading rate significantly. In the calculations the first melt arriving at the surface falls to the base of the vessel with virtually no spreading; subsequent spreading occurs because of steam levitation of the remaining melt rather than via progressive radial mixing.
- The initial slowing of the melt front observed in the experiments is not reproduced by CHYMES but the melt front speed through the lower half of the mixing vessel is better predicted.
- In general the steam production rate is calculated to rise too slowly and to persist for too long. It is believed that this is a consequence of the failure to reproduce the observed melt dynamics. However, the calculated steam flow rate was generally within a factor of two of the measured values during mixing.
- The fragmentation model produces final particle sizes consistent with those observed in the experiment.

Thus, while the code was able to reproduce the gross features of the experiments, it was not able to predict the detail. This failure could be due to the fact that the mixtures were very weak, so that the continuum assumption is not valid, and/or because of a need for improved constitutive models for drag and heat transfer. It is the authors' view that it is most likely the former, since a comparison of calculations of the CHYMES and PM-ALPHA codes has shown that the melt dynamics are not very sensitive to the chosen constitutive physics for heat transfer and interphase momentum transfer [86].

A series of large-scale (20–50 kg melt mass) mixing experiments is being performed in the FARO facility at JRC Ispra to investigate the quenching of large masses of corium in water [87]. The melt is composed of 80% UO_2

Fɪɢ. 7.   A particle pour in MAGICO-2000. ZrO_2 at 1800 K.

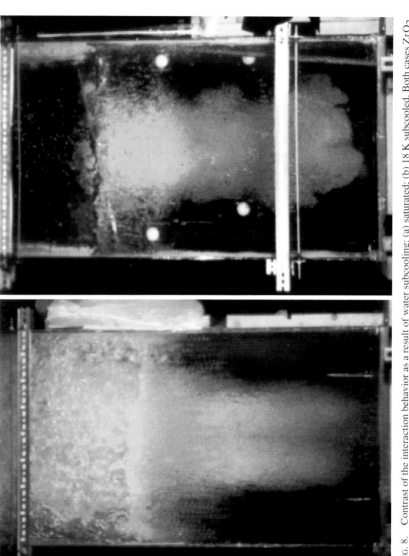

Fig. 8. Contrast of the interaction behavior as a result of water subcooling: (a) saturated; (b) 18 K subcooled. Both cases ZrO_2 at 1800 K.

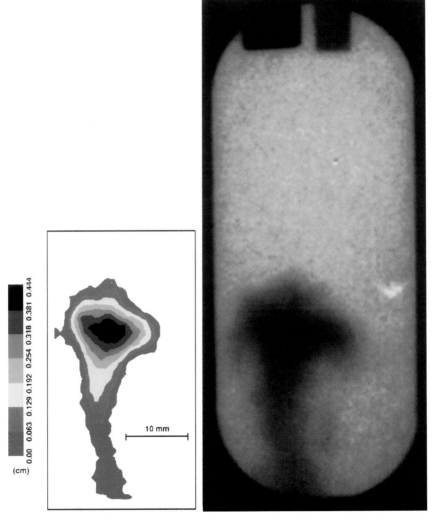

FIG. 10.   Examples of the X-ray-based droplet fragmentation data of Yuen *et al*. Taken from [108].

RUN 307M (Drop Temperature: 1300°C, Shock Strength: 20 MPa)

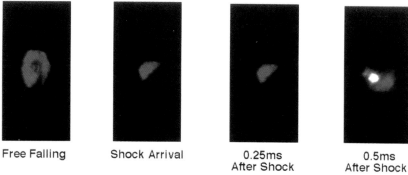

| Free Falling | Shock Arrival | 0.25ms After Shock | 0.5ms After Shock |

RUN 306M (Drop Temperature: 1500°C, Shock Strength: 20 MPa)

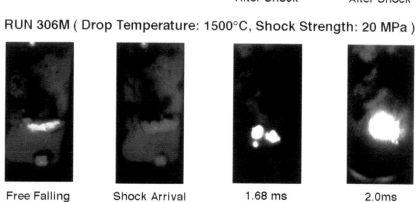

| Free Falling | Shock Arrival | 1.68 ms After Shock | 2.0ms After Shock |

FIG. 14.   High-speed movie records illustrative of the light flash phenomena during the interactions.

(by weight) and 20% ZrO_2. The apparatus consists of a melt generator and an interaction vessel with a volume of 1.5 m^3 that can withstand a pressure of 10 MPa at a temperature of 673 K. The water pool can be up to 2.5 m deep and the vessel diameter is 0.71 m. To date two experiments have been reported, the main features of which are given in Table I.

The experiments are difficult to instrument because of the large scale and the high pressure. So far the data have comprised thermocouple readings within the vessel, pressurization data, and resistance probes above the initial water level to record the level swell. The data show that approximately 2/3 of the melt was fragmented to a median size of 4 mm, with the remaining 1/3 forming a pool on the base of the vessel. These experiments have proved relatively difficult to model conclusively because of uncertainties in the initial conditions. The melt starts as a jet and vessel pressurization data show that there must have been significant heat transfer and fragmentation prior to melt water contact. This uncertainty means that various modeling groups have been able to reproduce the experimental data based on a variety of assumptions about the initial conditions at melt–water contact (see [66, 67, 79]). Further experiments in this series are planned using increased melt masses and melts containing zirconium.

C. Summary

The area of premixing remains one of the most active in steam explosion research. This is because of its central role in the α-mode failure issue and in accident management studies, including quenching in deep-water pools. Experimental investigations using simulants and prototypic melts are continuing. As always, the closer the experiment is to prototypic conditions the harder it is to instrument and the more difficult it is to analyze.

TABLE I

Test Conditions for the FARO Experiments

	Scoping test (ST)	Quenching test 2 (QT2)
Melt mass (kg)	18	44
Melt temperature (K)	2923	3023
Melt flow rate (kg/s)	64	119
Ambient pressure[a] (MPa)	5.4	6.1
Water subcooling (top) (K)	2	12
Water subcooling (bottom) (K)	38	20

[a] At melt–water contact.

However, the FARO program is providing good data and presenting a challenge to modelers.

On the modeling front there appears to be convergence on the form of model being used. Essentially, all of the important models are transient, 2D, compressible, three-phase flow codes that allow transfer of momentum and energy between all of the phases. A wide variety of constitutive physics appears to be in use but the underlying formulation is consistent. The main limitation is that the models do not adequately treat fragmentation of a melt jet. This physics is either deliberately excluded or a framework is included within the code but is not validated. Model comparison exercises and experimental interpretation are likely to lead to increased convergence of the modeling approach.

V. Triggering

As was discussed in the introduction, the transition from the relatively benign process of premixing to an explosive interaction requires a triggering event. It is widely acknowledged that this event involves the local collapse of the insulating vapor layer around a melt droplet (or a number of melt droplets) somewhere within the mixture. It may occur spontaneously in the bulk of the mixture, at the base of the mixing vessel, or be induced by some external means, such as the firing of a detonator.

In *spontaneous explosions*, which occur as melt is falling through water, the melt surface temperature falls below the minimum film boiling temperature, causing the vapor film to become unstable and to collapse. This mode of collapse often occurs in experiments using low temperature melts, for example, tin–water droplet experiments [88], and occurs either because a droplet cools or as it falls into a region of colder water. Triggering can also occur in the LNG–water system when the composition of the LNG changes due to evaporation of a particular component that causes the effective minimum film boiling temperature to change [1].

In *triggered explosions*, water is forced into contact with the melt due to some external event. This may be caused by an applied pressure pulse (the usual experimental means of triggering an explosion), the bulk flow of water, or local coolant entrapment. In the first mechanism, the pressure pulse induces a velocity in the water, toward the melt, at the liquid–vapor interface. If this motion is sufficient to drive the water into contact with the melt, triggering occurs. In the second mechanism, the bulk flow of water past a droplet (without a pressure wave being present) causes the vapor layer to be convected away from the melt, causing film collapse. In the third mechanism, water is entrapped within the melt or against the vessel wall by the melt, and is superheated until its temperature rises to the homogeneous nucleation temperature, at which point it flashes into steam, throwing the melt surrounding it into contact with water, and triggering occurs.

Board and Hall [89] were among the first to demonstrate that explosions could be triggered by either a mechanical disturbance or the addition of cold water. Schins [90] has described some of the means (e.g., exploding bridgewire or release of compressed gas) which have been used to trigger an explosion in experiments. Detonators have also been widely used to trigger explosions.

The importance of using realistic triggers in experiments if they are to be relevant to reactor safety application has recently been discussed by Henry [91]. He has developed a simple empirical criterion, based on the mixing energy required to cause more melt droplets to participate, to decide when an explosion that occurs following external triggering is a propagating event or when it is simply energy released from the metastable mixture because of the fragmentation caused by the trigger. Thus care has to be taken to distinguish between experiments that use artificial triggers and say something about the fundamental physics of steam explosions and those that use realistic triggers and say something relevant to reactor safety.

The available modeling and experimental evidence on triggering has recently been reviewed by Fletcher [16, 92]. He examined the very limited amount of modeling work, looked at evidence from small-scale experiments, and summarized the data from most of the large-scale experimental test series. This material is summarized below; details are given in the review papers cited above and the references contained therein.

A. Vapor Film Collapse Experiments

In order to study the vapor film collapse process many workers have used a solid heater. This has the advantages that the geometry is well-defined, it is possible to control and measure the surface temperature of the "melt," and explosions cannot occur. Thus this type of experiment decouples the processes of vapor film collapse and droplet fragmentation.

Inoue and Bankoff [93] investigated the triggered collapse of film boiling of Freon 113 or ethanol on an electrically heated nickel tube, using a pressure step. The magnitude of the pressure rise varied between 0.1 and 0.5 MPa, and the rise time of the pulse was varied between 80 μs and 344 ms. The instrumentation consisted of measurements of the temperature and pressure at the heater surface. Vapor film collapse was observed to occur when the pressure step had a magnitude greater than three times the ambient pressure and a rise time of less than 150 ms. Continuing this work, Inoue *et al.* [94] investigated the triggered collapse of film boiling for the system of an electrically heated platinum foil immersed in water. The trigger used was a pressure step with a magnitude of between 0.1 and 1.5 MPa with a rise time between 0.1 and 7.5 ms. The occurrence of collapse or otherwise was hard to detect in the experiments, but it appeared that partial collapse was triggered by a 0.5 MPa step and was more extensive as the pressure rise time was reduced.

Naylor [95] studied untriggered and triggered film boiling collapse on the surface of a brass rod with a hemispherical end immersed in a pool of water. Metal temperatures up to 770 K and water subcoolings at ambient pressure ranging from 0 to 80 K were considered. They were instrumented using a variety of photographic means (including an IMACON camera), a contact resistance device that allowed the fraction of the film area that had collapsed to be estimated, thermocouples to determine the metal and water temperatures, and pressure transducers. He observed that the film could be collapsed by either a pressure pulse (generated by a shock tube in his experiments) or by the bulk flow of liquid. The collapse of the film was often accompanied by a loud noise that resembled a "rifle shot." Observations suggested that collapse occurred when the average film thickness was less than the sum of the surface roughness plus the amplitude of interfacial waves on the liquid/vapor interface. By using a steady-state film boiling model to predict the vapor film thickness, he was able to confirm the collapse criterion proposed by Knowles [96] (see later) that at low pressure and for low temperature surfaces, vapor film collapse is unresisted.

B. MODELS OF VAPOR FILM COLLAPSE

Most modeling attempts have followed a very similar approach. A one-dimensional model consisting of a melt layer, a vapor layer, and a liquid slug is usually assumed. The idealized geometry represents a section of the vapor film surrounding a melt droplet. Steady-state film boiling is assumed to be established prior to the arrival of a pressure wave at the liquid–vapor interface. Conservation equations for mass, momentum, and energy are then used to determine the transient evolution of the system. The main difference between the various models is in the level of complexity of the equation system used and in the physical processes modeled.

The earliest model appears to be that of Drumheller [97], who considered the symmetric collapse of film boiling around a sphere. The liquid was assumed to be incompressible and energy considerations were used to derive an equation similar to Rayleigh's classical bubble collapse equation, but with phase change terms. The assumption of spherically symmetric collapse would appear to be questionable because of the finite time required for a pressure pulse to pass the sphere. No comparisons with experimental data were made using this model.

The first detailed model to be developed and compared with experiment was that of Inoue et al. [98]. In their model, a full nonequilibrium kinetic theory treatment of evaporation and condensation was used at the vapor–liquid interface and a Knudsen layer was modeled at the melt–vapor interface. The heat conduction equation was solved in the moving liquid slug by assuming a temperature profile that was a quadratic function of the distance from the vapor–liquid interface. A Newtonian model of the slug dynamics was used to determine the motion of the liquid. No vapor flow out of the film was modeled. The results of the calculations for a liquid of Freon 113 highlighted the importance of choosing the evaporation–condensation accommodation coefficient correctly. The paper contains a discussion of the effect of the presence of a permanent gas in reducing the accommodation coefficient because of the increased interfacial mass transfer resistance.

A simplified model was also developed by the above workers in which the heat storage in the vapor film was neglected and a heat balance was applied at the liquid–vapor interface to determine the condensation or evaporation rate, with the liquid–vapor interface temperature set to the local saturation temperature. This resulted in a much simpler set of model equations. The authors concluded that the full model was more reliable because it gave better agreement with measured heat flux data from Freon 113 vapor film collapse experiments. For any applied pressure pulse, the calculated response of the system was either total collapse of the film (judged to have occurred when the film thickness was of the same size as the surface roughness) or oscillation of the film thickness, with the pressure in the film rising sufficiently as collapse started to occur that it pushed the slug away.

This work was extended by Inoue et al. [94] to allow for mass flow out of the film. The study showed that pressure pulses with steep fronts (i.e., shocks) were more effective at collapsing the vapor film than slow pressure rises. It was also noted that the collapse behavior was very sensitive to the ambient pressure, since at higher ambient pressures there is more mass and energy in the vapor layer.

A similar analysis was pursued by Corradini [99], who also examined the effect of different mass transfer assumptions and came to the same conclusions as Inoue et al. [98]. He concluded that the equilibrium model was valid for shock rise times greater than 100 μs. This conclusion was based on a comparison of the computed peak heat fluxes with measured values for experiments performed by Inoue and Bankoff [93]. He noted that neglect of the Rayleigh–Taylor instability at the interface in the modeling would tend to reduce the calculated heat fluxes.

Also, Knowles [96] developed a one-dimensional model based on similar assumptions to those of Inoue *et al.* [98]. However, his model used a more rigorous treatment for slug dynamics and heat transfer into the vapor layer. He solved mass and momentum equations in the liquid slug, so that its compressibility was taken into account and the detailed behavior of the incident pressure pulse could be modeled. Also he solved a finite-difference form of the conduction equation in the liquid layer and the melt, rather than assuming given temperature profiles. Equations from kinetic theory were used to simulate evaporation–condensation processes, with the kinetic theory equations being modified to allow for the net velocity of the interface. The pressure and temperature dependence of thermophysical properties was included.

Observations from simulations of the triggered collapse of low-pressure films in contact with low-temperature surfaces lead him to suggest the following criterion for collapse:

$$\tau_p > \delta \left/ \left(\frac{2 p_{\text{trig}}}{\rho c} \right) \right. , \tag{15}$$

where τ_p is the duration of the trigger pulse, δ is the initial vapor film thickness, p_{trig} is the trigger pressure, ρ is the slug fluid density, and c is the sound speed in the slug. This equation is derived from the equations that govern sound wave transmission at an interface and says that, if the particle velocity of the liquid at the slug–vapor interface multiplied by the duration of the pressure pulse is greater than the thickness of the vapor film, collapse will occur. It applies when film collapse is essentially unresisted. His simulations showed that at higher melt temperatures stability was maintained by evaporation from the advancing slug. He argued that the vapor flux from the slug would mix up any permanent gas present, so that it would not produce a mass transfer barrier. In his simulations the main effect of the gas was to increase the thermal conductivity of the film. At high pressures he was unable to obtain a satisfactory numerical solution and attributed this to the development of a metastable state in the "vapor" phase during collapse.

More recently, Inoue *et al.* [100] have modeled the effect of liquid droplet fragmentation on the film stability. Their new model, which is based on their earlier one, allows for the fact that as fragmentation occurs the surface area of the droplet increases, causing a stretching, and consequent thinning, of both the vapor layer and the layer of heated coolant adjacent to the melt. This results in high heat fluxes and a less stable film. Thus collapse occurs more easily in this situation than the idealized one-dimensional system discussed earlier.

Although a number of models have been developed, it is fair to say that none are validated and they cannot be used with confidence to predict limits on triggerability. They are difficult to apply in any real situation because of uncertainties in the initial vapor film thickness and the geometry, and the poor knowledge of the trigger characteristics. The evidence from model predictions is that triggering becomes more difficult at higher pressure and for higher melt temperature. As the pressure increases the vapor mass and energy densities increase and the latent heat of vaporization decreases so that it becomes more difficult to compress the film, more difficult to condense the vapor, and easier to evaporate the leading edge of the water slug. However, this process has yet to be fully quantified for realistic vapor film geometries.

C. Single Droplet Experiments

There have been a vast number of experimental studies of untriggered and triggered interaction of melt droplets in water. Experiments using tin [88], iron oxide [101], lead [102], copper [103], and a host of other materials have all yielded information on the vapor film collapse and the subsequent fragmentation process. The role of the interface temperature in determining the fragmentation behavior has been examined for a variety of molten metals by Watts et al. [104]. They examined the triggered fragmentation of tin, gallium, bismuth, and Cerrolow in water and noted that more energetic interactions were obtained for materials with low surface tension and low density. This type of experiment yields information not only on the vapor film collapse process but also on the violence of the fragmentation event that follows.

Dullforce and co-workers [88] developed the concept of a Temperature Interaction Zone (TIZ). They performed hundreds of experiments in which molten tin droplets were released into water and determined the region in the melt temperature–water temperature space in which spontaneous explosions occurred. The region in which they occurred was termed the TIZ. It should be noted, however, that the boundary of the TIZ depends on (i) the chosen melt, (ii) the melt mass, and (iii) the time available for film collapse to occur. In addition, it is possible to trigger interactions for a situation in which a spontaneous interaction does not occur. Various theories have been advanced to explain the boundaries of this region, usually based on the need for the interface contact temperature to be above the spontaneous nucleation temperature.

Corradini [105] has analysed the data from over 300 single droplet experiments performed by Nelson at Sandia National Laboratories. The experiments involved the release of small droplets of various melts (stainless steel, metallic corium and oxidic corium) into water. The triggering behavior of the system was studied for various melt compositions, water temperatures and ambient pressures. Two different triggers were used. The first was an exploding bridgewire, giving a peak pressure of 1 MPa at a distance of 40 mm and a rise time of ~ 1 μs. The second was a detonator, giving a pressure of 10 MPa at a distance of 40 mm and a rise time of ~ 20 μs. His analysis of the data from these experiments led him to draw the following conclusions:

- Certain melt compositions did not lead to explosions and this could be explained by the presence of noncondensable gases, particularly hydrogen in the case of metallic melts. This point is also very clearly illustrated in the paper by Akiyoshi et al. [106], who study the different fragmentation behavior of droplets falling through air or steam.
- Explosions were suppressed at higher water temperatures and high ambient pressures because of the increased vapor film stability.
- Increasing the trigger magnitude can result in an explosion for a case where an explosion did not occur for a weaker trigger.

Experiments performed at Sandia National Laboratories have suggested that there is a trigger threshold for efficient fragmentation of a single droplet of melt [107]. A three order of magnitude increase in the volume of the vapor bubble formed following the fragmentation of a single melt droplet was observed when the trigger strength, defined as the product of the trigger pressure and impulse, exceeded a critical value. This result suggests a sudden change in the fragmentation mechanism when the violence of film collapse is increased beyond a certain threshold.

The experiments described in this section have led to the development of a vast number of models of fragmentation. However, it should be borne in mind that these mechanisms are applicable to isolated droplets that either spontaneously explode or are subjected to rather short-duration pressure pulses. Thus the mechanisms identified in this type of experiment may not necessarily apply during the propagation stage of an explosion. Such concerns have motivated Theofanous and co-workers to investigate the fragmentation of droplets following sustained pressure loadings [108] (see later).

D. Triggering in Integral Experiments

The following conclusions were drawn by Fletcher [16] after examination of the data from integral experiments involving typically 5–25 kg of melt released (usually by pouring) into a pool of water:

- Experimental data show very clearly the random nature of the triggering process, that is, two notionally identical experiments may trigger at completely different locations or one may not trigger.
- Explosions can be triggered as the melt enters the water pool, as it is falling through the water pool, upon base contact, or after melt has collected on the base of the mixing vessel. Explosions frequently occur without an applied external trigger.
- The spontaneous explosions which occur when melt contacts the water can be suppressed by a small increase in the ambient pressure (as little as 0.5–1.0 MPa is often sufficient). Data from Sandia (FITS) [109] and Winfrith (SUW) [110] support this view.
- There is no clear evidence for a triggered explosion occurring at pressures above about 3 MPa. An explosion was triggered at 5.8 MPa in the HPTR experiments performed at Winfrith [111], but this involved the injection of a slug of cold water into the mixture. An explosion was triggered at a pressure of 1.0 MPa in the SUW experiments when the end-cap from the charge container hit the base of the mixing vessel [110]. In experiments performed at JRC Ispra, in which molten salt was dumped into water [112], spontaneous explosions occurred at ambient pressure but did not occur at pressures of 0.5, 1, 2, and 4 MPa. In later work various triggers were used, the strongest being a minidetonator charged with black powder [113]. In this case, steam explosions were not observed for pressures above 3 MPa. In the FARO experiments described earlier, tests were performed at pressures of 5.4 and 6.1 MPa and no explosions occurred, even though the water was subcooled by up to 40 K in some regions of the water pool.
- Explosions are much more likely to occur in subcooled conditions compared with saturated conditions. This conclusion is supported by the results from the experimental series performed at Sandia (e.g., CM [114], OM [114], FTS [53, 109]), Winfrith (WUMT) [115], Ispra (KROTOS) [31], and JAERI [116].
- There is considerable evidence that if the melt is predispersed it is much less likely that an explosion will trigger. Again results from experiments performed at Sandia (Open Geometry Tests) [114], Winfirth (MIXA) [84, 85], and JAERI [116] support this statement.
- There is evidence that if the melt has a low superheat partial solidification during the melt–water interaction can inhibit triggering. This is discussed at length in the reports on the early Sandia tests [117]. If the freezing temperature of the melt is above the minimum film boiling temperature, then triggering is unlikely to occur, unless a

physical disturbance forces melt and water into contact before melt freezing occurs.

E. METHODS OF PREVENTING TRIGGERING

Many industries have sought to find practical ways of avoiding explosions without necessarily understanding why or how they work.

The first work in this area was the classic study of aluminum explosions by Long [2], who performed an extensive series of experiments in which molten aluminum was poured into a tank of water. He observed that if the melt was prefragmented (using a wire grill) prior to contact with the water it was less likely to explode than if it was released as a coherent mass. He found that coating the base of the container with lime and gypsum or allowing it to rust led to explosions in situations that would not have led to an explosion if the vessel base was made of uncoated degreased steel. Conversely, explosions did not occur if the vessel base was coated with grease or oil. Explosions were also avoided by painting the vessel base with a bituminous paint (called Tarset). These treatments obviously affect the wetability of the surface and change the likelihood of entrapping water beneath a puddle of melt. This work has been developed further by Nelson et al. [118], who have examined the effect of a vast number of surface finishes and have correlated the explosivity with the wetability of the surface by the coolant. Nelson et al. postulate that not only does good wetting cause a thin film of water to be trapped but it also promotes superheating of the water because nucleation is suppressed [51]. Note that in this mechanism the melt is initially assumed to be separated from the water by a vapor film and that this film must be collapsed by some disturbance before the superheating process can occur.

As early as 1969, when workers were still establishing the nature of the fragmentation process, Flory et al. [119] observed that increasing the viscosity of water by about five times using carboxymethlcellulose caused the fragmentation of lead, tin, and bismuth to be greatly reduced or totally prevented. Since then a number of workers have investigated the use of additives to prevent or mitigate explosions.

Experiments performed at Sandia by Nelson and Guay [120] showed that increasing the viscosity of the coolant could suppress explosions when tin drops were released into water. The viscosity was increased using glycerol or cellulose gum. Also a 50-kg scale test in which iron–alumina melt was dropped into a cellulose gum solution did not explode when similar tests using untreated water did. It was postulated that the increased coolant viscosity prevents microjets of water penetrating the melt droplets following film collapse, so that rapid fragmentation cannot occur (see later for a description of this fragmentation mechanism). However, it was also noted that increasing the viscosity of the water caused more air to be entrained into the vapor film, and, as noted earlier, this has a stabilizing effect on the vapor film.

Single-droplet experiments using an iron oxide melt have also been performed at the University of Wisconsin [121]. An external trigger of varying strength was applied as the droplets fell through the water pool. The water viscosity was varied between 0.04 and 0.24 kg/m · s by adding cellulose gum to the water. Increasing the viscosity was found to considerably reduce the likelihood of an explosion occurring.

Workers at the Georgia Institute of Technology have investigated the effect of surfactants on the spontaneous interaction of tin droplets at 1073 K dropped into water [122, 123]. A variety of surfactants were used and were found to reduce the peak pressure (to about 65% of that observed with pure water) of the spontaneous explosions that occurred. Experiments using dilute solutions of the aqueous polymer poly(ethylene oxide) showed that if the viscosity was increased by between 2 and 13% the observed pressures were actually increased [124]. Increasing the viscosity further was found to reduce the occurrence of spontaneous explosions, until they were completely suppressed when the viscosity was increased by a factor of two. Clearly, there are competing effects at work that are as yet not understood.

Although none of the practices described above are applicable to, say, prevention of explosions in the reactor application, they do highlight the sensitivity of the triggering process to the coolant and vessel properties. There is much anecdotal evidence (see, e.g., [125]) that additives are routinely used in some metal preparation processes to prevent explosions, but the documentation of these procedures is sparse.

F. SUMMARY

There are a variety of ways by which an explosion can be triggered, and these are apparently very sensitive to a wide range of system parameters. Models of the triggering process are not well-advanced and none is validated. There is much evidence from single-droplet and medium-scale tests on the factors that effect triggering. These are very useful when examining the likelihood that a given system will trigger spontaneously but should only be treated as a guide. Of potential industrial importance is the possibility that explosions may be preventable by modifying the properties of the coolant or the vessel. While vessel coatings are used widely in the aluminum industry to prevent base-triggered explosions, there is little detailed evidence on the use of coolant additives. The whole area of triggering is difficult to address mechanistically because of the stochastic nature of the event, but much useful data is available in the literature to the experienced analyst.

VI. Propagation

In order to determine pressures and energy conversion ratios it is necessary to be able to calculate the propagation event that follows triggering, if the premixing conditions are "correct." It is not possible to specify exactly what the "correct" conditions are, but the consensus is that the mixture should not be too lean in melt and there should not be too much steam present. Also, if the propagation process is to be efficient the melt and water must be in close proximity (i.e., premixed with extensive interfacial area). If this is not the case, the evidence is that explosions are still possible but are much weaker. In the following sections we review the available information on droplet fragmentation, describe the experimental database on propagation, and review the modeling. We then discuss stratified explosions, covering the limited amount of modeling and experimental work. Finally, we review recent ideas on deflagration-like modes of propagation.

A. EXPERIMENTAL STUDIES OF FRAGMENTATION

There have been numerous studies of droplet fragmentation, and it appears that there are almost as many proposed fragmentation models as there are studies. They can be split into two broad categories: (i) *hydrodynamic fragmentation* caused by differential velocity effects, and (ii) *thermal fragmentation* caused by phase change effects. These modes of fragmentation have been reviewed in detail by Corradini *et al.* [21] and Fletcher and Anderson [38]. Given the central importance of fragmentation in the propagation process, the main models and results will be summarized here.

1. *Hydrodynamic Fragmentation*

There is a vast literature on relative velocity fragmentation of liquid droplets in both gases and liquids. However, as we shall see, very little of this data is directly applicable to the study of steam explosions. Nowhere is the effect of the vapor blanket and its subsequent collapse taken into account. Good reviews of the available data have been presented by Tan and Bankoff [126] and Pilch and Erdman [127]. There is still a considerable amount of uncertainty in the actual breakup mechanisms, with boundary-layer stripping, Rayleigh–Taylor instability, or a combination of both of them being the most popular [128].

Before presenting a summary of the available data it is worth defining two dimensionless groups that are often used in the study of fragmentation. The first is the Weber number, which is the ratio of the destabilizing inertial force to the stabilizing force due to surface tension, and is defined by

$$We = \frac{\rho_l D v_{rel}^2}{2\sigma}.$$ (16)

A related group is the Bond number, which represents the ratio of accelerational forces to surface tension forces, and is given by

$$Bo = \frac{\rho_d a D^2}{4\sigma}.$$ (17)

For a liquid drop the acceleration, a, can be expressed in terms of the drag force, so that we can obtain the identity

$$Bo = \tfrac{3}{8} c_d We,$$ (18)

and since $c_d \approx 2.5$ [127, 128, 129] for fragmenting drops we see that the Bond and Weber numbers are virtually the same. (Note that some authors replace $D/2$ by D, giving rise to slightly different definitions for the Weber and Bond numbers.)

a. Gas–liquid Systems Initially, vapor explosion workers looked to the existing data on water droplet breakup by shock waves in air, as there were data available from the field of aerodynamics. Ranger and Nicholls [130] studied the breakup of raindrops in air and found that the breakup was due essentially to boundary-layer stripping. They correlated their data by introducing the following dimensionless time:

$$T = \frac{tu}{D_0} \sqrt{\frac{\rho_l}{\rho_d}},$$ (19)

where u us the velocity behind the shock front and D_0 is the initial drop diameter. Their total breakup time data was then correlated by

$$T_b \sim 5.$$ (20)

They also extended the work of Taylor [131] to produce the following stripping correlation:

$$\frac{dm}{dt} = (12\pi^3)^{1/2} \left(\frac{\rho_l}{\rho_d}\right)^{1/3} \left(\frac{v_l}{v_d}\right)^{1/6} \left(\frac{D}{D_0}\right)^{3/2} \rho_d v_d^{1/2} v_{rel}^{1/2} r_d^{3/2}.$$ (21)

Predications obtained from this model were found to be in reasonable agreement with their experimental data.

Waldman *et al.* [132] pointed out that boundary-layer stripping alone cannot account for the experimentally observed breakup times. In their experiments, boundary-layer stripping was followed by catastrophic breakup of the remainder of the drop. Their data were correlated by

$$T_b \sim 3.5. \tag{22}$$

The droplet mass was found to vary with time in the following manner:

$$m = \frac{m_0}{2}(1 + \cos(\pi T/T_b)), \tag{23}$$

so that the stripping rate is given by

$$\frac{dm}{dt} = m_0 \frac{\pi u}{2T_b D_0} \sqrt{\frac{\rho_d}{\rho_l}} \sin\left(\frac{\pi T}{T_b}\right), 0 \le T \le T_b. \tag{24}$$

One criticism leveled against this model was that it does not allow for the fact that the relative velocity between the droplet and the fluid decreases with time. Scott and Berthoud [133] and Sharon and Bankoff [134] modified the dimensionless time so that

$$T = \int_0^t \frac{v_{rel}}{D_0} \sqrt{\frac{\rho_d}{\rho_l}} \, dt' \tag{25}$$

in order to account for this. (Note that this modification changes the stripping rate given in Eq. (24). See [134] for details.) Although this modification makes the breakup rate a function of the local relative velocity, it is not clear that the practice has any physical justification. This expression was widely used in the early steady-state models [133, 134].

b. Liquid–Liquid Systems   Experiments have also been performed in liquid–liquid systems. Patel and Theofanous [128, 135] studied the fragmentation of mercury, gallium, and acetylene tetrabromide drops in water, using a shock tube to produce a pressure pulse to initiate fragmentation. They observed a symmetrical blowup of the drops without apparent deformation or stripping. They also found that there was no Bond number threshold for the onset of fragmentation, as was observed in earlier experiments. Their results suggested that breakup was much faster than observed in previous experiments, and that

$$T_b \sim 0.4, \tag{26}$$

or, more specifically,

$$T_b = 1.66 \text{Bo}^{-1/4}. \tag{27}$$

The above equation predicts that breakup times are much faster in liquid–liquid systems, since the gas–liquid data of Simpkins and Bales [37] was fitted by the functional form except that the coefficient was 22 for the commencement of Rayleigh–Taylor instability and 65 for complete fragmentation of the drop. They considered that a coefficient of 22 represented a lower bound for the estimate of the breakup time and was consistent with the data of Reinecke and Waldman. The original estimates of fragmentation times for the Board–Hall model were made using this correlation, so the work of Patel and Theofanous suggested that fragmentation was indeed fast enough to support a detonation. Based on both their experimental results and theoretical analysis, Patel and Theofanous postulated that Rayleigh–Taylor instability was the dominant fragmentation process.

In a further set of experiments Theofanous *et al.* [136] studied the fragmentation of *stationary* mercury drops using flash X-ray diagnostics and produced the following correlation:

$$T_b = 10.3 \text{Bo}^{-1/4}, \tag{28}$$

which is somewhat slower than that observed in the studies by Patel and Theofanous [128, 135]. The reason the observed breakup rate in this set of experiments was slower than for the earlier study was attributed to the fact that in this case the droplets were fragmented without initial surface instabilities being present (stationary drops), so that the time for instabilities to shatter the drop was somewhat longer. Still, for a Bond number of 10^4 this yields a $T_b \sim 1$; that is, 4 to 5 times faster than given in Eq. (20). Both the X-ray photos and the quantitative analysis of them can be found in [137]. These indicate clearly the existance of Rayleigh–Taylor instabilities on the forward face of the drop, its cross-stream deformation (flattening), and the stripped fine particles convected downstream by the flow. The superposition of the convected fine particles and of the drop deformation produced a "round" overall envelope that explained the nearly symmetrical "blowup" phenomenon observed in the high-speed movies of Patel and Theofanous, as noted earlier.

In contrast to this, Baines [138] and Baines and Buttery [139] found that in their studies of the fragmentation of mercury drops in water both capillary wave growth and boundary-layer stripping contributed to fragmentation. For Weber numbers in the range 100–2000 the stripping process was dominant. For Weber numbers of the order of 2000 Rayleigh–Taylor instability was observed on the windward surface of the drop. They found that over the Weber number range 100–2000 the breakup time was approximately constant and given by

$$T_b \sim 4.0. \tag{29}$$

There has been considerable discussion as to the reason why breakup appeared to occur so much faster in the experiments of Theofanous et al. [44]. The answer seems to lie in a difference in interpretation of the experimental data. In the experiments of Baines, breakup was considered to have occurred when the droplet was observed to be completely fragmented and the fragments had been dispersed. High-speed movies were employed for these observations, and the experimental technique (a horizontal tube full of water containing the mercury drop was suddenly accelerated by a hammer action) did not lend itself to high resolution in the photographs obtained. On the other hand, Theofanous et al. determined fragmentation by high-resolution quantitative X-ray radiography, which could reveal very precisely when the drop actually had become a cloud, and well before this cloud was dispersed by the flow. Moreover, it should be noted that in the Baines experiments the drops were "sitting" on the tube side and thus were flattened in the direction of the flow, that is, a direction orthogonal to that at which flow-induced flattening was supposed to occur.

Tan and Bankoff [126] examined the fragmentation of mercury drops in a bubbly aqueous liquid. They concluded that Rayleigh–Taylor instability was the dominant fragmentation mechanism for $5 < We < 117$ and that the Taylor waves occurred on the lee side of the drop due to backflow in that region. They also found that there was a critical Weber number of 17 below which fragmentation did not occur.

Bankoff and Yang [140] have continued this work by studying the fragmentation of droplets of Wood's metal and mercury. These experiments covered a Weber number range of 5–644. Their experimental data agree with the analysis of Harper et al. [141], who showed that for this Weber number range Rayleigh–Taylor instability at the upstream surface of the drop does not occur. Instead they observed a continuous increase in the drop size. They suggested that this was due to vorticity production at the trailing edge of the drop. In a heated system they postulate that internal vaporization of the entrained fluid could enhance fragmentation and melt freezing could retard it [142]. Thus at low Weber numbers fragmentation may be due to a combination of relative velocity and thermal effects.

Kim et al. [143] have studied the hydrodynamic fragmentation of gallium drops in water for Weber numbers in the range 30–3519. They observed a gradual increase in the efficiency of boundary-layer stripping as the Weber number was increased. The fragmentation process was observed to change from the stripping of a coherent skin that subsequently broke up, for $30 < We < 300$, to the direct stripping of fragments for $We > 1300$, with a transition region existing for intermediate Weber numbers. The fragmentation time was given by

$$T_b \sim 3.5 \rightarrow 6.6, \tag{30}$$

with no clear dependence on the Weber number. Again, as in the case of Baines, fragmentation was deduced from photographic images and considered complete when the fragments were completely dispersed.

Crachalios *et al.* [144] used these data to produce a dynamic fragmentation model. In their model the stripping rate from a single particle is given by

$$\frac{\mathrm{d}m}{\mathrm{d}t} = c_{\mathrm{frag}} v_{\mathrm{rel}} \pi D_d^2 \sqrt{\rho_l \rho_d}, \tag{31}$$

where $c_{\mathrm{frag}} \sim 1/6$. Assuming spherical droplets, this equation can be written in the equivalent form:

$$\frac{\mathrm{d}D}{\mathrm{d}t} = -\tfrac{1}{3} v_{\mathrm{rel}} \sqrt{\frac{\rho_l}{\rho_d}}. \tag{32}$$

Carachalios *et al.* [144] replaced the fluid density by an effective fluid density that allows for the fact that fragments have been added to the fluid stream.

Subsequently, the data of Kim *et al.* have been further analyzed and two detailed fragmentation models have been produced [145]. The first is based on deformation breakup and Rayleigh–Taylor instability and the second is based on crest stripping of capillary waves due to shear flow instabilities. These models were constructed by assuming an ellipsoidal droplet and calculating the growth rate of the appropriate instability. In the case of Rayleigh–Taylor instability, breakup was assumed to occur when waves penetrated the drop. In the stripping model, droplets were assumed to be stripped when the wave amplitude exceeded a specified value. In both cases the fragment size was calculated by the model. It is clear that a large number of assumptions are required to construct such a model. For example, the initial perturbation of the surface must be specified following stripping and the fastest growing wavelengths have to be changed as the relative velocity between the drop and the surrounding fluid decays. The various parameters needed in the models were chosen so that they predicted breakup times, and fragment sizes are in agreement with the data of Kim *et al.* [143].

The above model is the only relative velocity fragmentation model (known to us) which predicts the fragment size mechanistically. Some workers, for example, Jacobs [146], have obtained the fragment size by assuming that fragments are formed with a size that gives them the instantaneous critical Weber number, that is,

$$d = \frac{2\sigma \, We_{crit}}{\rho_l v_{rel}^2}. \tag{33}$$

Pilch and Erdman [127] have shown that for a constant relative velocity this procedure can produce particle sizes that are too small by almost two orders of magnitude. In their review they present an empirical method for determining the fragment size. However, they note that there are *no* fragment size data for the liquid–liquid system and that their empirical method is based on gas–liquid data.

More recently, workers at UCSB [108] have investigated fragmentation of mercury and molten tin droplets under sustained pressure loadings using a hydrodynamic shock tube and quantitative X-ray radiography. Their apparatus is known as the SIGMA facility. Their work is specifically targeted at producing a fragmentation correlation for their propagation model, and they therefore correlated the data using instantaneous properties. For mercury droplet fragmentation they obtained the following correlation:

$$\frac{dm}{dt} = \frac{\pi D_f^2(t) \, |v_{rel}(t)|}{6T_b} (\rho_l \rho_f)^{1/2}, \tag{34}$$

where

$$T_b = 13.7 Bo_i^{-1/4}, \tag{35}$$

and Bo_i is the instantaneous Bond number. This equation provided an excellent fit to their data for driver pressures in their shock tube of 20, 34, and 47.6 MPa. They also showed that no constant value for T_b could be found to match theory data over the entire Bond number and pressure ranges.

The picture for tin droplet fragmentation was much less straightforward. At low tin temperature (360°C), with water at 85°C to prevent spontaneous explosions, the fragmentation caused by a 6.6-MPa shock was negligible after 2 ms. At intermediate tin temperatures (670°C) fragmentation was again negligible for the 6.6-MPa shock but was catastrophic before 1 ms for a 20-MPa shock. The observed fragmentation in the later case was far faster then expected from hydrodynamic fragmentation and resulted in very fine fragments. At the highest tin temperature (1000°C) thermally driven fragmentation was present even in the 6.6-MPa shock case and was again faster at higher pressures. They observed that there was a delay time before thermally driven fragmentation occurred and that this delay decreased as the shock pressure and/or tin temperature increased.

They were also able to produce maps of the metal density from their X-ray data that show how the fragments are dispersed within the water. A number of different fragmentation morphologies were observed, as shown in Figure 10 (color plate). No simple correlation could be extracted from the molten tin data. The micromixing data obtained in these experiments allowed the authors to make estimates of the volume of water into which the debris mixes, and they coined the word *microinteractions* to describe this mixing phenomenon. More details are given in the modeling section below.

2. Thermal Fragmentation

Although most of the detonation models assume relative velocity fragmentation, some assume thermal fragmentation or assume that thermal fragmentation is important during escalation. These can be lumped under the title thermal mechanisms because they are all based in some manner on heat transfer and coolant vaporization effects. These models have been reviewed by Cronenberg and Grolmes [147]. Aside from the more popular models based on vaporization effects, they include fragmentation due to solidification (i.e., mechanical stresses set up in a thin crust as the melt freezes) [148] or due to gas release [149].

A set of related models assume that vapor film collapse causes jetting of coolant into the melt, a fragmentation mechanism which has been observed by a number of experimenters (e.g., [15]) Ochiai and Bankoff [150] proposed that coolant tongues that cross the vapor film will vaporize upon contact with the melt if the conditions for spontaneous nucleation are met. They treated the fragmentation event as being similar to the splash crater caused by impact of a cylinder and a liquid surface. Other workers have postulated that coolant jets may be formed either by Rayleigh–Taylor instability as the film collapses or by the collapse of vapor bubbles which form following melt–coolant contact. The injected coolant vaporizes and rips the drop apart. Detailed models of this process have been developed by Buchanan [151] and Kim and Corradini [152]. This fragmentation mechanism is discussed at length in the paper by Nelson *et al.* [153], which presents photographic evidence for the proposed mechanism. Tang and Corradini [154] have produced a simplified version of Kim and Corradini's model for use in their propagation code. Based on the assumption that it is the jet penetration stage that provides the rate-controlling step, they produced the following correlation for the fragmentation rate:

$$\frac{\mathrm{d}m}{\mathrm{d}t} = -C\,\frac{m_d}{D_d}\left(\frac{p - p_0}{\rho_l}\right)^{1/2},\qquad(36)$$

where p is the local pressure, p_0 is the ambient pressure, and C is an empirical constant (set to 0.01). While this model may be appropriate during escalation, it is hard to see how it can apply to a propagating event, where the "memory" of the initial pressure is lost and pressures may exceed the critical value, so that the phase change effect is lost.

Ciccarelli and Frost [129, 155, 156] have recently used simultaneous high-speed photography and flash X-ray radiography to study fragmenting drops. They used 0.5-g drops of Cerrolow (an alloy that is liquid below 100°C) and tin dropped into water. These droplets were triggered using either a bridgewire (peak pressure 10 Mpa) or a blasting cap (peak pressure 40 MPa), both having a duration of about 20 μs. They found that for low-flow velocities (<5 m/s, We ~ 150) cold drops fragmented via stripping and long wavelength penetration of the drop, whereas hot drops were fragmented by the growth and collapse of a vapor bubble. Their X-ray pictures showed that during bubble growth the surface of the drop became very convoluted and fine filaments of melt protruded from the surface (giving it the appearance of a thistle). These filaments were then torn off and mixed with water during bubble collapse, resulting in very rapid energy transfer and the formation of a larger bubble. Later experiments, using a variety of materials, showed that the first bubble growth was largely independent of the melt (tin, Cerrolow, bismuth, or gallium) but that there was more spread in the growth rate and ultimate size of the second bubble [104]. For higher flow velocities (>45 m/s, We $\sim 15,000$) the vapor bubble growth was diminished, but the high-speed vapor flow caused the filaments to be dispersed into the wake of the droplet by the flow.

The fragmentation mechanism observed by Ciccarelli and Frost is very different from that proposed by Nelson *et al.* and is only apparent because they were able to image both the bubble boundary and the location of the fragments [129]. It is also clear from their work that the mechanism is very dependent on the flow velocity, which is a function of the escalation distance in a large-scale explosion. It should also be noted that they used a very-short-duration trigger, which is very different from the conditions in an explosion where a sustained pressure may prevent any bubbles from growing (i.e., as discussed at the end of the previous section). In addition, the observed fragmentation process was relatively slow, with the initial growth and collapse cycle taking ~ 1.5 ms [129].

B. PROPAGATION EXPERIMENTS

Although there have been many studies of intermediate-scale steam explosions, most of these have been designed for purposes other than validation of propagation models. As discussed earlier, many experiments have shown propagation behavior (observed on high-speed films or via an array of pressure transducers) they have generally been poorly characterized in terms of the conditions prior to the explosion and the trigger characteristics. The few experiments performed to study propagation are discussed below.

The first reasonably well-characterized experiments were performed by Baines [157]. He set up the system shown in Fig. 11 in which 1.5 kg of molten tin at a temperature of about 800°C was poured into nearly saturated water. The water was contained in a vertical tube about 1 m tall and 30 mm in diameter. After release the melt fell through the water and generated steam. Explosions were triggered when molten tin reached the base of the vessel and encountered cold water. Values were closed so that the explosion products had to push a magnetic piston along a tube. The motion of the magnet through a series of sensing coils allowed the work done to be calculated. In addition, an array of pressure transducers along the tube allowed the propagation behavior to be observed.

He was able to measure *average* values for the melt and vapor fractions. Typically, the average melt fraction was in the range 0.08–0.14 and the vapor fraction was in the range 0.07–0.24. Observation showed that the mixture was nonuniform because of the tendency for vapor to rise and melt to sink. Also, the water temperature became stratified during mixing, so that more vapor was produced in the warmer top region. When triggered, these mixtures gave rise to propagating interactions with peak pressures of the order of 4–10 MPa and plateau pressures behind the von Neumann spike of typically 2 MPa. Propagation velocities were in the range 50–250 m/s. The efficiency of conversion of thermal to mechanical energy was very low, typically $\leq 0.4\%$.

Baines performed very careful analysis of the experiments and drew attention to the fact that, even in the apparent 1D geometry that he chose, nonequilibrium effects were important in the water phase. He calculated that typical thermal boundary layer thicknesses around fragments would be of the order of 10–100 μm. Thus, unless the mixture is very rich in melt, most of the water will remain unheated during the propagation event. He calculated that if *all* of the water was involved the Chapman–Jouguet pressure would be ~3 bar and the propagation pressure would be ~35 m/s. Thus, his experiments provide clear evidence for the need to allow for the nonuniform heating of the water.

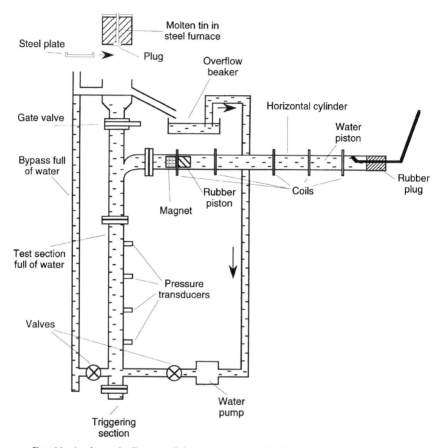

FIG. 11. A schematic diagram of the apparatus used by Baines to study propagation.

The KROTOS facility located at JRC Ispra has been used to examine steam explosion propagation behavior for a number of years. Initially, the experiments involved molten tin and water [158]. These gave results consistent with those reported by Baines [157]. More recently, experiments have been performed using aluminium oxide and water [31]. In these experiments ~1.5 kg of melt at an initial temperature of ~2600 K was poured into a test section with a diameter of 0.4 m and a height of 2.2 m. The test section was instrumented with pressure transducers, thermocouples (to determine the melt location), and a level swell meter. So far, all of the experiments have been performed at ambient pressure and the main

TABLE II
TEST CONDITIONS AND RESULTS FOR THE RECENT KROTOS EXPERIMENTS

Experiment	27	28	26	29	30
Melt Mass (kg)	1.0	1.22	<1.0	1.5	1.5
Water subcooling (K)	10	13	40	80	80
Event[a]	B	TSE	TSE	SSE	SSE

[a] TSE ≡ Triggered steam explosion. SSE ≡ Spontaneous steam explosion.
B ≡ Benign interaction.

variable has been the water subcooling. A water depth of 1.1 m was used in all of the tests. A triggering device that works by releasing high-pressure argon is located at the base of the test section. The main features of the tests are given in Table II.

The experiments are described below in order of increasing water subcooling. (A detailed description is warranted as these experiments are the most widely used for propagation model validation.) In experiment 27, with "nearly saturated water," there was a long period of steaming lasting several minutes but no explosion. In experiment 28, a steam explosion was triggered for almost the same conditions as experiment 27 using the strong gas trigger (15 cm^3 of argon at 8.5 MPa released at the base of the vessel) activated a preset time after melt release. A propagating interaction was observed, with pressure levels in excess of 50 MPa (the limit of the transducer range) being recorded. In experiment 26, the trigger was activated when the melt had penetrated only a small way into the water pool and an explosion was triggered that also gave rise to supercritical pressures.

In experiment 29, a spontaneous trigger occurred when the melt was still 150 mm from the base of the interaction vessel. The explosion was very strong and gave pressures of the order of 100 MPa. Experiment 30 was a repeat of experiment 29 but with some modifications to try to avoid the occurrence of a spontaneous interaction. (A tin membrane used to slow the melt at entry was removed to eliminate the possibility of a trigger from a tin–water interaction and a Plexiglass liner was inserted in the interaction vessel.) An explosion was again spontaneously triggered, this time when the melt front was about halfway through the water pool. Again pressures in excess of 100 MPa were recorded and significant damage was done to the apparatus.

The experimenter estimated, based on the assumption of homogeneous 1D mixing, that the melt volume fraction was approximately 0.08 in experiments 28 and 29, and about 0.18 in experiment 30. The corresponding estimates of vapor fraction are 0.04 and 0.23. In experiments 28–30 the propagation velocity was in the range 650–1000 m/s. Thus, these experiments provide direct evidence of supercritical propagating pressures in 1D system.

Corradini and co-workers at the University of Wisconsin have developed an experimental apparatus in which they aimed to make use of the best features of the above two systems [159, 160]. The experiment is referred to as WFCI. The apparatus is very similar to that of Baines discussed previously. The interaction tube has a height of 1.5 m and a diameter of 90 mm. As a triggering device they developed an electromagnetic hammer driven by a capacitor discharge. It produces a trigger pressure of ~4 MPa with a duration of 170 μs. Experimental series have been run to examine the trigger characteristics, to check for experimental consistency and to study spontaneous and triggered explosions. Typical experiments involve a few kg of tin mixing with water and then a weak explosion, which involves only a small fraction (~2%) of the metal. In all cases the explosion pressures have been subcritical (typically ~5 MPa) and conversion ratios have been about 0.2%. Also it was observed that spontaneously triggered explosions tended to result in multiple propagations but that triggered explosions produced an escalating wave which was not fully developed at the end of the tube.

C. PROPAGATION MODELING

Most propagation models are based on the ideas introduced by the Board–Hall model, with additional effects of finite rates of heat and momentum transfer being included and the escalation stage being studied. Fletcher and Anderson [38] have provided a detailed review of the model developments that took place up until 1990. In that review they note the gradual progression from the steady-state modeling approach of the 1970s to the formulation of transient models in the 1980s. Until recently most models have adopted a common approach, with the differences between the CULDESAC model [161, 162, 163], the ESPROSE model [164, 165, 108, 166, 167, 168], and the IDEMO model [169] being limited essentially to the choice of constitutive physics, apart from the important distinction that ESPROSE is two-dimensional and the others are one-dimensional.

The models adopt a common framework, very similar to that described for the premixing models. A multiphase approach is used, with the species being the melt droplets, the melt fragments, and the coolant phases (vapor and liquid). The models differ in their assumptions regarding phase equilibrium, but typically the fragments are assumed to equilibrate their velocity and temperature with the surrounding fluid as soon as they are formed. The fragmentation rate is based on a correlation of experimental data for hydrodynamic breakup, with differences in the chosen para-materization.

Corradini and co-workers [154, 170] have adopted a different approach and assumed that fragmentation is caused by the high pressures generated during vapor blanket collapse can lead to coherent propagations (see Sect. VI.B.2 for details). This approach seems to be at variance with the evidence, as it known that at high relative velocities hydrodynamic frag-mentation plays an important role [129].

As soon as attempts were made to compare these models with experi-mental data, severe problems were identified. Fletcher [162] tried to simulate the tin–water propagation experiments performed by Baines [157] and found that he could not produce a sufficient pressure rise if he assumed that all of the water was involved. Medhekar et al. [164] analyzed the same experiments and noted that accounting for local variations in the void fraction was very important. At about the same time, in studies of molten salt water detonations, Bürger et al. [171] concluded that neither hydrodynamic fragmentation nor the thermal fragmentation model of Corradini [170] gave the required degree of fragmentation to explain the experimental data. They also had a similar problem in their analysis of the KROTOS tin–water data [158] and found that they had to add significant thermal fragmentation to the hydrodynamic fragmentation to match the data. Clearly, all of the preceding were pointing to the need for the inclusion of new physics.

The new physics required was that of "microinteractions" first proposed by Theofanous and co-workers. Recall that in the discussion of fragmenta-tion presented earlier we described how they observed the mixing of the fragments into the coolant phase. Based on this observation they decided that the coolant phase must be split into two parts: (i) that which mixes with the fragments, gets rapidly heated, and hence pressurizes the system, and (ii) that which is heated on a much slower time scale but serves as a compressible medium and hence influences the pressure level. They made a sequence of steps before they reached their final formulation of the "microinteractions" concept. Initially, when examining a 1D experiment they used two radial nodes in the simulation, mixed the melt with water in only one of them, and *specified* that a fraction of the heat from the fragments produced vapor directly [108]. This calculation produced a propagating explosion for a situation where the original model failed to predict one. Clearly, it was having the effect of simulating the nonequilib-rium in the coolant phase but in a nonpredictive manner.

In order to perform simulations for genuinely 2D situations they re-tained the assumption of some of the fragment energy producing vapor directly and multiplied the fragmentation rate by a factor to account for thermal fragmentation during the escalation phase [166]. This model was called ESPROSE.a (the "a" being for augmented). Subsequently, data from the SIGMA facility (see Sect. VI.A.1) was used to determine the rate of mixing with the coolant and a fourth phase was added to their model (i.e., the phases comprised melt droplets, unheated coolant, heated coolant, and melt fragments). In this case the predicted behavior depends on the specified rate of mixing, which can be determined from experiments. This model is known as ESPROSE.m. A full description of this model is given in the appendix to this paper. This modified model was able to reproduce the transient pressure development observed in two KROTOS tests: one that produced a mild propagation and the other that produced a supercritical detonation.

ESPROSE.m has been used to perform calculations for ex-vessel steam explosions [166, 167]. A number of important points have emerged from these simulations:

- The premixing stage must be simulated so that the presence of mixture inhomogeneities is reproduced. Regions of high melt fraction and high void fraction can change the predicted pressures significantly. This means that 2D simulations are essential for reactor simulations.
- The inclusion of a model for "microinteractions" is essential for realistic predictions of propagation, and hence for the determination of pressure loads.
- The "steam chimney" that forms above the mixing zone plays an important role in venting the explosion pressure. The propagation of pressure waves into these voided regions determines the duration of pressure loads on the surrounding structures.
- Consideration of reflections off the pool boundaries (both free, if any, and rigid) is essential to properly predict the pressure wave dynamics and resulting loads.

Work is continuing to refine the constitutive description of microinteractions (simultaneous flash X-ray radiography and high-speed movies), as well as to examine various melts with key property differences (iron, various oxides, etc.) [168].

D. Stratified Explosions

There has been little theoretical work on the study of propagation in a stratified geometry. There are no developed models and no clear mechanistic picture exists of how stratified explosions propagate, what causes the mixing or what controls the amount of material involved in the explosion.

An extensive program of experiments using a variety of materials and experimental geometries have been performed at Argonne National Laboratory [172]. The best-characterized of these experiments were performed in a vertical geometry using freon and water separated by a freon vapor layer. The interface between each species was preserved by separating them with thin mylar diaphragms. These diaphragms were subsequently removed, causing the liquid columns to contact at the base of the experimental vessel, so that propagation occurred in a well-defined geometry. Propagating events with a speed of 100–150 m/s and a peak pressure of 0.5–1.0 MPa were observed. The propagation velocities were considerably slower than the speed of sound in either of the liquids present. Analysis of the experimental data showed that a total mixing depth of ∼6 mm would be required to explain the propagation behavior if the two liquids came into complete thermodynamic equilibrium at the front, with this figure being increased if thermodynamic equilibrium was not achieved. Thus, even though the experiments gave rise to very low pressures, a significant depth of mixing occurred.

Extensive experimental investigations have been performed by Frost and co-workers at McGill University using tin and water [173, 174]. They have identified a minimum water depth (∼70 mm) overlying the tin below which propagation does not occur. Propagations with a velocity of ∼40 m/s and peak pressures of 2–10 bars were observed. The explosions were weak, with only ∼0.7% of the thermal energy being converted to mechanical work. These authors also performed an interesting analysis showing that the general shape of the measured pressure field in the overlying water could be explained using a potential flow model in which the interaction zone is represented by a moving wedge (with an angle of 10°) that displaces the water above it. Increasing the confinement, by restricting the vertical expansion of the system, the impulse was increased by a factor of about three, but the propagation speed was increased by less than 10 m/s. In addition, they performed experiments in a cylindrical system, so that the pressure wave was able to spread as a circular wavefront. The work yield per unit surface area was reduced by 30% and the depth of mixing was decreased significantly.

Bang and Corradini [175] have investigated stratified explosions between water and freon-12, and water and liquid nitrogen. In both cases the explosions were weak, giving pressures of only a few bars and low efficiencies. The depth of premixing was also small. The proposed a model of the mixing process which is based on a Rayleigh–Taylor instability mechanism, but the physical mechanism in unclear and use of the model is not recommended.

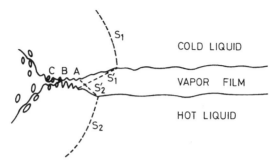

FIG. 12. The stratified propagation process as envisaged by Harlow and Ruppel [176].

Colgate and Sigurgeirsson [9] performed a number of scoping calcula-tions for the situation of molten lava submerged in water. They suggested that mixing could be caused by Rayleigh–Taylor instability, Kelvin–Helmholtz instability, or cracking of the lava surface upon contact with cold water. Simple order-of-magnitude calculations showed that all of these mechanisms could lead to fragmentation. They envisaged a geometry where an initial pressure release at the lava–water interface was large enough to "crater" the crust and to start a pressure wave running out radially into the lava. The pressure wave would travel faster in the water than in the lava. Hence an annular high pressure region would push down on the lava, pushing it back towards the rarefaction at the point of "cratering." This downward and radially inward acceleration of the lava interface would lead to Rayleigh–Taylor and Kelvin–Helmholtz instability and subsequent mixing of the lava and water.

Harlow and Ruppel [176] have performed some scoping calculations of propagation in a stratified system. Their work is based on the conceptual picture shown in Fig. 12. In this picture the traveling-wave configuration moves to the right with a constant speed. The region C denotes the zone of rapid heat transfer and boiling that supports shock waves S_1 and S_2. These travel at velocity much greater than the speed of sound in the vapor film. Thus, the shocks implode the film, deflecting the interfaces inward, pre-ceded by transmitted shocks S_1' and S_2'. Until the shocks reach the opposite surface (point A), the interfaces are stable. After this, the interface in the region A–B is Rayleigh–Taylor unstable, causing the two fluids to mix. Heat is transferred in the zone B–C. At point C vapor forms, driving the fluids apart and supplying the energy required to sustain shock S_1 and S_2.

By assuming that the hot and cold fluids had the same physical proper-ties, they were able to perform numerical simulations for a pair of symmetrical shocks and demonstrated that these shocks could collapse the vapor film. Heat transfer was not modeled directly, so they could not simulate the complete process. They worked in a frame of reference moving with the shocks and modeled heat transfer by applying a boundary condition of a fixed high pressure at the left-hand end of the solution domain. However, since the interaction zone velocity is much lower than the two different shock velocities, it is not possible to transform to a coordinate system in which steady-state conditions exist. Indeed, experimental data [173] suggest that the shock waves associated with triggering have passed through the system and decayed long before the explosion propagates a significant distance.

Thus the current picture so far as stratified explosions is concerned is that experiments show that they are relatively weak and that the depth of mixing in small-scale experiments is limited to the order of 10 mm. There are no developed models of this propagation process, largely because the physical mechanism governing the mixing process has not yet been identified.

E. OTHER MODES OF PROPAGATION

Over the past five years there have been repeated suggestions that the mode of propagation is not necessarily analogous to a detonation but may be more like a deflagration. Recall that in the Board–Hall model discussed in Sect. III.D the reaction front propagated supersonically compared with the sound speed ahead of the front and the combustion products follow the reaction front. In a deflagration the reaction front moves subsonically with respect to the gas ahead of it and combustion products move in the opposite direction to the front. The pressure falls across the front. However, because the front is moving subsonically the conditions ahead of the front are changed by the generated flow. Thus, the pressure ahead of the combustion zone rises if combustion is occurring in a closed vessel and the conditions at the start of the reaction front do not correspond to point O in Fig. 6. This explains why the net pressure rise in the system can be large even if the pressure is falling across the reaction front. The application of the above ideas to multiphase systems has been discussed by a number of workers, including Sharon and Bankoff [133, 177], Condiff [46], and more recently by Frost *et al.* [178, 179] and McCahan and Shepherd [180]. It should be noted that the fact that the sound speed is not well-defined in multiphase systems and that the various components can have different temperatures and velocities complicates the analysis.

When Board and Hall developed their model of steady-state propagation they noted that rapid fragmentation of the melt behind the shock front would lead to a detonation-like structure [26]. They were, however, quick to point out that this steady-state wave would only be possible in very special circumstances and that in reality the propagation behavior would be more complex and pressures would be lower due to effects such as vapor formation in the reaction zone [45]. They concluded that the deflagration mode of propagation had no stable analogue in the vapor explosion context. Their argument was based on the following observations. First, the equivalent subsonic initiation process would be heat conduction ahead of the front, which would enhance the stability of the vapor blanket around the melt droplets. Also, the fall in pressure through the deflagration wave would enhance the film boiling stability and impede energy transfer. Second, they felt that the available experimental evidence suggested that explosions were triggered by relatively small pressure or velocity changes. Thus, they expected that the deflagration-to-detonation transition that often occurs in gaseous combustion would happen more easily in the multiphase situation.

Berman and Beck [181] have suggested that the complete range of propagation modes observed in gaseous combustion may also occur in vapor explosions. They envisage that, as the coupling between the passage of the shock front and fragmentation becomes stronger, the wave changes from being similar to an accelerated flame to a detonation. They noted the importance of the fragmentation mechanism in determining which type of propagation mode occurs and performed experiments to show that the droplet fragmentation time could vary significantly with the trigger strength.

Hugoniot analysis for the tin-water system has been performed independently by Frost *et al.* [178, 179] and McCahan and Shepherd [180]. For equal-volume mixtures of saturated water, steam, and tin at a temperature of 1500°C and an initial pressure of 0.1 MPa, the detonation solution gives a C–J pressure of 180 MPa and a propagation velocity of 650 m/s. The deflagration solution gives a pressure of approximately 0.05 MPa and a velocity of 0.1 m/s [178]. Both sets of workers have concluded that the observed propagations in this system result in pressures much closer to those corresponding to the C–J deflagration solution than to the C–J detonation solution. However, they state that there are many reasons why the steady-state analysis does not apply and note that there is no conclusive experimental evidence for deflagration-type solutions. McCahan and Shepherd [180] went one step further and put together a simple model for a deflagration-type wave in a tube closed at one end. However, as this is a thermodynamic model, all the melt and water are assumed to be in thermal equilibrium, an assumption now known to be wrong.

Fletcher [182] has used the CULDESAC model to investigate what conditions must be satisfied by the fragmentation process for a deflagration-like solution to be obtained. He compared calculations using a hydrodynamic fragmentation model with those obtained by assuming a simple model of thermal fragmentation in which a fixed time elapsed between vapor film collapse and fragmentation. His calculations showed that the two different propagation modes were possible but that the hydrodynamic fragmentation model gave predicted pressure transients much closer in shape to those observed experimentally.

It is also noteworthy that various workers have postulated that a complete spectrum of modes is possible in the case of evaporation of a superheated liquid. The process considered is one of rapid evaporation of a liquid in a metastable state following the release of an applied pressure. Shepherd *et al.* [183] have modeled the process as one of a deflagration-type wave, while Fowles [184] has postulated that the process can be a weak detonation in some circumstances.

It is important to realize that most of the above ideas have come about in an attempt to explain the low pressures observed in most experiments. While the deflagration mode has some appealing features in this respect, there is very little evidence beyond its prediction of low pressures to support it.

F. Summary

There has been considerable progress in the development of propagation models for the pre-mixed system over the last five years. With the introduction of the "microinteractions" concept it is now possible to analyze experiments and perform reactor-scale calculations in a consistent manner. At present the key need is for a more extensive database from which to refine the constitutive laws for the fragmentation and micro-mixing rates. Predictive modeling of the initial thermal-fragmentation-dominated stage of a propagation event is also needed if prediction of the escalation stage is required.

The situation for stratified explosions is much less satisfactory. There is a growing database to suggest that they are weak in laboratory-scale experiments. Their role in industrial accidents is thought to be important, but because of the unknown state of the melt and water at the time explosions occur in such accidents, it is not possible to draw any firm conclusions.

VII. Combined Thermal–Chemical Explosions

There are a number of important situations in which a chemical explosion can accompany a steam explosion. For this to occur the melt must be highly reactive with steam, as is the case for aluminum and zirconium. For these materials it is possible that the fine fragmentation associated with propagation can bring a large surface area of the metal in contact with steam, leading to rapid oxidation of the melt. As this reaction is highly exothermic, it causes a rapid rise in melt temperature and a very energetic interaction. If the reaction rate is fast enough, the heat supplied by the chemical reaction exceeds that lost by convection and the particle temperature increases, causing the reaction rate to increase. This very rapid oxidation event is known as *ignition*, and the explosion is accompanied by a bright flash due to the very high temperatures generated. The hydrogen produced in the interaction can subsequently explode in air. This gaseous explosion is quite separate from the vapor explosion and, although caused by the hydrogen produced from the intense interaction between the melt and water, it should not be confused with it.

In order to illustrate the importance of the chemical reaction, we consider the case of molten aluminum interacting with water. As a useful scale it is worth noting that the heat of detonation of TNT is 4.6 MJ/kg [185]. If aluminum at a temperature of 700°C is cooled to 100°C (the temperature at which it can no longer produce steam) the heat release is 1.04 MJ/kg [186]. If it reacts chemically the reaction taking place is

$$2Al + 3H_2O \rightarrow Al_2O_3 + 3H_2, \qquad (37)$$

and the heat released if the aluminum reacts completely with steam at 700°C to produce Al_2O_3 is 17.6 MJ/kg [186]. (Note that experimental data suggest that not all of the aluminum forms Al_2O_3 but some will form AlO. However, in experiments reported by Reid [1, p. 170], 98.3% of the oxide was Al_2O_3.) Using the above values one obtains an energy per kilogram of aluminum reacted that is equivalent to that from 4 kg of TNT. This, however, represents a gross upper bound, as the steam explosion efficiency cannot be 100% and the chemical reaction will not be complete.

The issue of combined chemical–thermal explosions is of interest to the aluminum industry, where steam explosions have often been accompanied by a flash of light, indicating chemical reaction [187]. In addition, it became a significant issue when the feasibility of building a New Production Reactor (NPR) to produce tritium was considered in the United States, and it was proposed to use aluminum-clad fuel [188]. In addition, the same effect can occur in the lithium–water system. Finally, the possibility that zirconium and, to a much lesser extent structural steel, can lead to significant hydrogen production during a severe accident in a PWR has been considered [189].

A. THEORETICAL MODELING

There has been a relatively limited amount of modeling of combined thermal–chemical explosions. The modeling that has been performed has been mainly concerned with determination of hydrogen production during the premixing stage. Modeling work in this area has recently been reviewed by Fletcher et al. [190]. Essentially, there are three possible limits on the hydrogen production rate [191]:

1. If steam is not available the reaction cannot proceed. This limit is unlikely to occur in most melt–water interactions but could be important in industrial incidents when water is entrapped in a body of melt.

2. The reaction is limited by the rate at which oxygen can diffuse through the oxide layer formed on the surface of the melt and interact with unreacted metal beneath it. This diffusion rate is relatively well-known for pure materials, such as zirconium, but there is relatively little data for mixed metal–oxide melts [190].

3. Steam must diffuse towards the surface of the melt and the hydrogen produced must diffuse away. This limiting process is complicated by advection effects if there is significant slip between the melt particle and the surrounding steam.

The various models based on the above limits that have been applied to premixing and the propagation stage are described in detail in the review article by Fletcher et al. [190] and the references therein. Of particular note is the paper by Epstein et al. [192], which provides a detailed analysis of reaction process at the stagnation point of a sphere in film boiling. In addition, Young et al. [193] have summarized the hydrogen data from the FITS experiments performed at Sandia and analysed them using relatively simple models.

More recently, a number of more sophisticated mechanisms have been proposed to explain the apparent temperature threshold below which an ignition type interaction is not observed. Epstein and Fauske [194] have suggested that the delay in nucleation of the aluminum oxide within the aluminum may play an important role in the process. They suggest that in the tiny fragments of aluminum, referred to as microspheres, the process proceeds in three distinct stages, viz.: (i) chemical reaction between a dissolved species of oxygen and molten aluminum to metastable oxide; (ii) nucleation of solid aluminum oxide crystals within the metastable melt, and (iii) growth of crystals into a continuous scale of solid aluminum oxide. They point out that the oxidation rate during stages (i) and (ii) will be very high and then fall rapidly as stage (iii) occurs. Thus, in their picture the competing rates are the chemical reaction rate and the oxide crystallization rate. (They discuss other mechanisms and discount the possibility of vapor phase burning on the basis that the metal evaporation rate is too slow.) Their theory is presented as one that requires further development but that may be useful in guiding future work.

Fauske and Epstein [195] have built upon the above model in an attempt to explain the difference between experimental reactor incidents in which little oxidation occurred and some pouring mode experiments in which significant oxidation occurred. They propose that to have an efficient oxidation reaction there must be a significant amount of steam present prior to the explosion. Their model, based on oxidation kinetics and fragmentation rates, requires the fundamental assumption that the interaction zone length is approximately the same in all explosions. This assumption needs further examination.

More recently, Cho [196, 197] has proposed that the collapse of steam voids behind the shock can lead to very high temperatures because the process is almost adiabatic and that these temperatures are sufficiently high to melt the oxide layer, causing rapid oxidation of the aluminum. He draws the analogy between this process and the hot-spot mechanism for the initiation of explosions in liquids. Again this hypothesis needs further investigation.

B. EXPERIMENTAL DATA

The available database from which to estimate the likely extent of the chemical reaction is very sparse. A recent review by Cho [196, 197] contains information on experiments with a mass scale of up to 23 kg. There is no consistent picture as to when a chemical reaction is triggered. Some experimental data appear to suggest that there is a temperature threshold below which explosions do not occur; other data contradict this hypothesis. It is clear that such reactions progress more rapidly at higher temperatures, so that once a reaction is initiated it can "run away," as discussed previously.

Recent tests performed at Sandia National Laboratory as part of the NPR program [198] have provided information on intermediate-mass-scale (2–10 kg) molten aluminum poured into water at a temperature of 290 K. Metal temperatures in the range 1000–1540 K were examined. At an initial temperature of 1540 K an untriggered explosion occurred that caused 40% of the chemical energy in 7 kg of melt to be released. Combined thermal–chemical interactions were triggered using a detonator for melt temperatures of 1440, 1320, and 1150 K. At a melt temperature of 1000 K only a very weak thermal explosion was triggered, and in the absence of a trigger no explosion was observed. The investigators postulate that the temperature threshold for a combined thermal–chemical reaction lies in the range 1000–1150 K, which corresponds to metal superheats in the range 67–227 K. Nelson [199] has performed a large number of single-droplet (2–10 g) melt–water interaction experiments using pure aluminum and aluminum–lithium alloys. Spontaneous explosions were never observed during free fall or at base contact. Moderate thermal explosions could be triggered using an exploding bridgewire if the melt temperature was in the range 1273–1673 K. Paradoxically, when the strength of the trigger was increased by exploding the bridgewire at a focus of an elliptic reflector, the bubble growth rate was reduced significantly. However, for an initial melt temperature of 1773 K he observed a combined thermal–chemical explosion characterized by a bright flash of light and bubble growth an order of magnitude faster than normal.

Nelson discusses the effect of the trigger on film collapse and proposes that the explosion mechanism is similar to that observed by Frost *et al.* when molten tin droplets explode in water [155]. It is important to distinguish between the very-short-duration trigger used here and the sustained pressure used in the experiments discussed below.

Theofanous *et al.* [200] have investigated the interaction of gram quantities of molten aluminum at droplet temperatures of up to 1973 K with water. Explosions were triggered by applying a sustained pressure loading (using a hydrodynamic shock tube) of up to 40.8 MPa and the interaction was recorded using high-speed photography. The results of the experiments are given in Fig. 13. Essentially, it was observed that for metal temperatures above ~1500°C ignition occurred irrespective of the size of the pressure applied (in the range 5–40 MPa). The investigators observed three different types of behavior:

1. At temperatures below 1400°C the debris was highly convoluted and in some cases fragmented into millimeter-size particles but there was no chemical reaction.

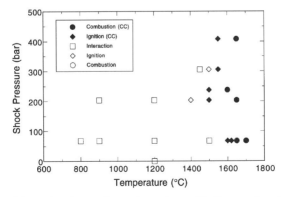

FIG. 13. Map of the various interaction regimes observed in the experiments of Theofanous *et al.* [200].

2. For the temperature range 1400–1600°C the debris was composed of both oxidic fragments (10–40%) and metal fragments in the size range 100–1000 μm.

3. At temperatures above about 1600°C the debris consisted of a fine powder in the 1–10 μm size range with essentially complete oxidation.

The high-speed photography revealed very bright sources of white light (see Fig. 14, color plate) at temperatures as low as 1300°C, but these did not "run away" and consume the complete drop. It is postulated that this is because the chemical heating rate is too slow to keep up with the cooling rate at the droplet surface. Various oxidation mechanisms are discussed in the paper, including vapor phase burning, as well as oxidation of suboxides (Al_2O and AlO) as a means of initially supporting a flame. In addition to the discussion of the oxidation behavior, the authors also note the lack of fine fragmentation present at low temperatures, even though it would be expected on the basis of the usual hydrodynamic fragmentation correlations. This raises questions about the surface properties of the droplet that may be of significance in understanding oxidation behavior.

C. Discussion

It is clear that combined thermal–chemical reactions are of significant importance in the study of aluminum–water explosions. Their relevance to other materials, such as molten core material, has yet to be established. While there is a growing body of experimental data on such explosions, it is hard to define the conditions under which they can occur. There is clearly a temperature effect, but it seems that there is not a simple temperature threshold. Nelson [199] complied the available data and suggested that the ignition temperature falls with mass scale. However, this picture is complicated by the fact that almost every experiment uses either a different contact mode or a different trigger magnitude or duration.

The various experimental studies have led to the postulation of a variety of theories to explain the observed data. However, it is fair to say that none of these has been sufficiently well-validated to be used with confidence.

VIII. Integral Explosion Assessments

Although this paper is principally concerned with the physics of steam explosions, it would be a grave omission not to outline how this information has been used in steam explosion assessments. To understand the background it is important to realize that the needs of different industries regarding the assessment of steam explosions have been very different. As we have seen, the process industry is concerned with finding methods of preventing explosions. These methods vary from the simple expedient of isolating melt from water to the coating of casting pits with special paints. During everyday operation in many casting plants the possibility of a steam explosion is present and has to be guarded against very carefully.

The situation in the nuclear industry is very different. During normal operation there is no possibility of a steam explosion, and it is only in the highly unlikely event of a severe accident that molten material can be formed. And then, only if a significant amount of melt forms, relocates, mixes with water, and explodes efficiently is there the possibility that the event will fail the reactor vessel and the containment, thus having off-site consequences. The nuclear industry is almost unique in having to perform assessments of these highly unlikely events (the core melt frequency for a large modern PWR is typically 10^{-4} to 10^{-5} per reactor year). Attention has been focused on the sequence of events in which there is significant core melting, the melt relocates en masse into the lower head of the reactor, explodes in a coherent manner, drives a slug of water, melt, and debris against the upper head, fails the upper head producing a missile, and the missile subsequently fails the containment. This sequence of events was first put forward in the WASH-1400 study [201] and has become known as α-mode failure. In the above study, an energetic large-scale steam explosion was judged to occur one out of ten times that a large mass of core material contacted water. Clearly, such a high probability represents an unacceptably high risk.

The obsession in the nuclear industry with addressing the α-mode failure issue led to much controversy and initially diverted much funding and effort away from the task of understanding the fundamental issues. However, in the last five years the emphasis has changed from using "expert judgment" to producing an assessment based on sound physical understanding using validated models. It is this drive that has led to the development and validation of the multiphase models of mixing and propagation discussed in the earlier sections. Before giving a brief history, it is important to realize that we are not dealing with everyday probabilities of the "coin tossing" kind, as there is no statistical data upon which to base the argument. Instead, they should be seen as a "degree of belief" based on the results from a combination of physical models, experimental studies, and expert knowledge. The approach to performing such assessments has been the subject of much controversy. However, there seems to be convergence on an accepted methodology, called Risk-Orientated Accident Analysis Methodology (ROAAM), developed by Theofanous [202].

The key feature of this approach is the production of a well-argued and documented initial assessment, which is then subjected to an open peer review process and is modified as necessary as a result of this procedure.

Earlier studies split the event up into a number of discrete events (e.g., mass of melt relocating, likelihood of triggering, energy conversion ratio) and assigned probabilities or probability distributions to these events. These studies were based largely on expert judgment and were either performed by a small team (e.g. the assessment by Corradini and Swenson [203]) or by a pooling of expert knowledge. The meeting of the Steam Explosion Review Group (SERG) resulted in an assignment of a probability of 0.01 based on the typical response of the members [204]. The above meeting was called partly in response to a Monte Carlo study performed by Berman, Swenson, and Wickett that concluded that the α-mode failure probability lies somewhere between zero and one [205]. This conclusion was reached by ascribing uniform probability distributions to all events and therefore taking no account of physical limitations, that is, all masses of melt participating were assumed to be equally likely, when all the available evidence says that it is harder to mix large masses (even if we cannot quantify exactly how difficult it is to mix large masses we must acknowledge that the limitation exists when performing such an assessment).

The first rigorous assessment was presented by Theofanous and co-workers in 1987 in a suite of four papers which dealt with probabilistic aspects [48], premixing limits [62], expansion and energy partition [206], and impact mechanics, dissipation, and vessel head failure [207]. This study treated the assessment process in a much more mechanistic manner and made use of limits to mixing arguments by quantifying the mass mixed as a function of the pour rate. In addition, Hicks–Menzies calculations were performed to estimate explosion pressures and work potentials, the effect of failure of the lower head in mitigating the upward explosion force was quantified, and account was taken of the energy dissipated as the slug crushed internal structures, deformed the core barrel, failed the upper head, and generated a missile. Although the papers created much controversy regarding the models used and the assumptions made, they provide the framework for all subsequent analyses.

In the United Kingdom Turland et al. [208] have adapted the above approach and applied it to the quantification of the α-mode failure probability of the Sizewell B PWR for the complete range of system pressures. They made use of the assessment described above but split the process up into rather different events by prescribing 12 probability distributions. The difference is that Theofanous et al. placed more emphasis on "limits to mixing" and lower head failure while the UK study highlighted the importance of slow melt pours (because of the design of the core support structure), a range of triggering times, and low conversion ratios (which is equivalent to limits to mixing). Again, in this study it was found that careful account had to be taken of the melt relocation behavior, the structural dynamics of the system, and the absence of a coherent slug to transfer explosion energy to the upper head. The failure probability given a core melt was assessed to be a few parts in 10,000, which is negligible.

Theofanous and Yuen [60] have reexamined the original study of Theofanous et al. described earlier in the light of the developments of their premixing model (PM–ALPHA) and their propagation model (ESPROSE.m) over the past five years. Their conclusion is that the mixing arguments are much stronger because of both intermodel comparison and experimental validation, and that propagation calculations are now sufficiently well-developed to be useful in the assessment process (in place of thermodynamic arguments). Examples of ex-vessel propagation calculations which provide pressure–time histories on the cavity walls have already been performed [167]. The main conclusion is that there is a large degree of conservatism in the earlier assessment. This view also dominated at the second meeting of the Steam Explosion Review Group held in June 1995 [209].

It is appropriate that we should end this section by referring to the summary of the latest CSNI specialist meeting on fuel–coolant interactions [82]. This meeting concluded that significant progress had been made in confirming and quantifying the water depletion phenomenon (i.e., limit to mixing) and analyzing propagation behavior. These advances used in the α-mode failure assessments presented at the meeting confirmed the very low likelihood of α-mode failure. In addition, a number of papers focused on the need to address other melt–water contact modes and issues, such as accident management. Such research is likely to be beneficial to all who are interested in the steam explosion phenomenon.

IX. Appendix: Formulation of the ESPROSE.m Field Model

This appendix contains the conservation equations and constitutive physics models that make up the ESPROSE.m model, as described in Theofanous and Yuen [167], and including some minor refinements made since.

A. CONSERVATION EQUATIONS

There are four phases, namely, "microinteractions" fluid, coolant liquid, fuel (melt) drops, and fuel debris. They will be referred to as m-fluid, liquid, fuel, and debris, respectively. Each phase is represented by one flow field with its own local concentration and temperature. The debris is assumed to be part of the m-fluid in thermal and hydrodynamic equilibrium. Thus, we have four continuity equations, three momentum equations, and three energy equations. In the usual manner, the fields are allowed to exchange energy, momentum, and mass with each other. With the definition of the macroscopic density ρ_i' of phase i defined via

$$\rho_i' = \theta_i \rho_i \quad \text{for } i = m, l, f, \text{ and } db, \tag{A1}$$

and the compatibility condition

$$\theta_m + \theta_l + \theta_f + \theta_{db} = 1, \tag{A2}$$

these equations can be written rather directly [210].

1. *Continuity Equations*

Mass conservation equations are readily formulated based on the defined transfer rates:

m-Fluid:
$$\frac{\partial \rho_m'}{\partial t} + \nabla \cdot (\rho_m' u_m) = E + J. \tag{A3}$$

Liquid:
$$\frac{\partial \rho_l'}{\partial t} + \nabla \cdot (\rho_l' u_l) = -E - J. \tag{A4}$$

Fuel:
$$\frac{\partial \rho_f'}{\partial t} + \nabla \cdot (\rho_f' u_f) = -F_r. \tag{A5}$$

Debris:
$$\frac{\partial \rho_{db}'}{\partial t} + \nabla \cdot (\rho_{db}' u_m) = F_r. \tag{A6}$$

2. *Momentum Equations*

The momentum equations are formulated by balancing convective transport by the pressure gradient, drag forces, momentum exchange due to mass transfer, and gravity:

m-Fluid:

$$\frac{\partial}{\partial t} \left((\rho'_m + \rho'_{db})u_m \right) + \nabla \cdot \left((\rho'_m + \rho'_{db})u_m u_m \right)$$

$$= -(\theta_m + \theta_{db})\nabla p - F_{ml}(u_m - u_l)$$

$$- F_{mf}(u_m - u_f) + Eu_l + F_r u_f$$

$$+ J(H[J]u_l + H[-J]u_m) + (\rho'_m + \rho'_{db})g + \nabla \cdot [\alpha \hat{\sigma}_m].$$

$$(A7)$$

Liquid:

$$\frac{\partial}{\partial t} (\rho'_l u_l) + \nabla \cdot (\rho'_l u_l u_l)$$

$$= -\theta_l \nabla p + F_{ml}(u_m - u_l) - F_{lf}(u_l - u_f)$$

$$- Eu_l - J(H[J]u_l + H[-J]u_m) + \rho'_l g + \nabla \cdot [(1 - \alpha)\partial_l].$$

$$(A8)$$

Fuel:

$$\frac{\partial}{\partial t} (\rho'_f u_f) + \nabla \cdot (\rho'_f u_f u_f)$$

$$= -\theta_f \nabla_p + F_{mf}(u_m - u_f) + F_{lf}(u_l - u_f) + \rho'_f g - F_r u_f. \quad (A9)$$

3. *Energy Equations*

The energy equations are readily formulated by balancing convective energy transfer by the pressure work, enthalpy transfer due to mass transfer, and intercomponent energy transfer:

m-Fluid:

$$\frac{\partial}{\partial t} (\rho'_m I_m + \rho'_{db} I_{db}(T_m)) + \nabla \cdot [(\rho'_m I_m + \rho'_{db} I_{db}(T_m))u_m]$$

$$= EI_l + Jh_m + F_r I_f - p\left[\frac{\partial}{\partial t} (\theta_m + \theta_{db}) + \nabla \cdot [(\theta_m + \theta_{db})u_m] \right]$$

$$- R_{ms}(T_m - T_s) + \dot{q}_{fm}. \quad (A10)$$

Liquid: $\dfrac{\partial}{\partial t}\left(\rho_l' I_l\right) + \nabla \cdot \left(\rho_l' I_l u_l\right)$

$$= -p\left[\frac{\partial}{\partial t}\left(\theta_l\right) + \nabla \cdot \left(\theta_l u_l\right)\right] - EI_l - Jh_l$$

$$- R_{ls}(T_l - T_s) + \dot{q}_{fl}. \tag{A11}$$

Fuel: $\dfrac{\partial}{\partial t}\left(\rho_f' I_f\right) + \nabla \cdot \left(\rho_f' I_f u_f\right) = -\dot{q}_{fm} - \dot{q}_{fl} - F_r I_f.$ (A12)

In the above equations, $H(J)$ is the Heaviside step function that becomes unity for positive values of the argument and zero otherwise. When $T_m < T_s$, the m-fluid is liquid and J is set to be zero, and T_s is an "equivalent" interface temperature given by

$$T_s = \frac{R_{ms}T_m + R_{ls}T_l}{R_{ms} + R_{ls}}. \tag{A13}$$

When $T_m > T_s$, J is an evaporation rate given by

$$J = \frac{1}{h_m - h_l}\left[R_{ms}(T_m - T_s) + R_{ls}(T_l - T_s)\right]. \tag{A14}$$

It should be pointed out that diffusive energy transport *within* each field has been ignored in the above formulation, as it is insignificant.

4. Fuel Interfacial Area Transport Equation

Finally, the fragmentation of the fuel drop is simulated by an interfacial area transport equation:

$$\frac{\partial}{\partial t}\left(\frac{\theta_f}{D_f}\right) + \nabla \cdot \left(\frac{\theta_f}{D_f}u\right) = \dot{S}_f, \tag{A15}$$

where D_f represents the drop diameter and $\dot{S}_f$ the source term due to fragmentation.

B. THE CONSTITUTIVE LAWS

The interfacial exchanges of momentum and heat are clearly regime-dependent, and uncertainties remain even for simple two-phase flows. Only experiments specifically oriented to this problem and detailed *local* measurements provide the basis for appropriate assessment, particularly if one of the phases is the "microinteractions" fluid. The ESPROSE.m approach is to treat the m-fluid as a "pseudo"-gas and to utilize a similar set of constitutive laws as in the previous work of Medhekar *et al.* [164]. Only the constitutive laws for fuel fragmentation, which was not presented in the previous work, are included in this appendix.

The processes of fragmentation are responsible for the source term that appears on the r.h.s. of the fuel length scale transport [Eq. (A15)]. In order to relate physically to this source term, and to obtain its general form, it is best to begin with the interfacial area transport equation, written in conservative form, per unit volume of the total flow field, as

$$\frac{\partial A_f}{\partial t} + \nabla \cdot (A_f u_f) = \dot{S}'_f. \tag{A16}$$

In this equation $\dot{S}'_f$ corresponds to interfacial area source–sink terms due to fragmentation per unit volume of mixture. The above equation can be derived in the usual manner by using the Reynolds transport theorem and Green's theorem for a "material" volume in the fuel field, including the source term in the statement of conservation, and letting the volume shrink to infinitesimally small dimensions.

Now, assuming that the interfacial area of the fuel can be characterized by that of a cloud of spherical particles with a single effective length scale, A_f can be written as

$$A_f = n_f \pi D_f^2, \tag{A17}$$

where n_f is the number density of particles and therefore relates to the particular volume fraction and length scale by

$$\theta_f = n_f \frac{1}{6} \pi D_f^3. \tag{A18}$$

From these equations we obtain

$$A_f = \frac{6\theta_f}{D_f}. \tag{A19}$$

In the material volume mentioned above, the changes in A_f (and hence the source terms) can be obtained by simple differentiation for fragmentation as

$$\dot{S}'_f \equiv 6\dot{S}_f \equiv \frac{dA_f}{dt} = \frac{6}{D_f} \frac{d\theta_f}{dt} - \frac{6\theta_f}{D_f^2} \frac{dD_f}{dt}. \tag{A20}$$

Since the fragmentation is assumed to be occurring without affecting the particle number density, but only their size, the first term on the r.h.s. of

Eq. (A20) can be written [using Eq. (A18)] as

$$\frac{6}{D_f}\frac{d\theta_f}{dt} = \frac{18\theta_f}{D_f^2}\frac{dD_f}{dt},$$ (A21)

and collecting Eqs. (A19) to (A21) into Eq. (A16), we finally obtain:

$$\frac{\partial}{\partial t}\left(\frac{\theta_f}{D_f}\right) + \nabla \cdot \left(\frac{\theta_f}{D_f}\cdot u_f\right) = \frac{2\theta_f}{D_f^2}\frac{dD_f}{dt}.$$ (A22)

The physics of fragmentation has been discussed in previous publications (Yuen et al. [108]; Chen et al. [168]). We follow the instantaneous Bond number formulation of Yuen et al. [108]. The formulation consists of defining a total fragmentation time, t_f^*, correlating it to the instantaneous Bond number, and assuming that the instantaneous fragmentation rate is given by the ratio of the current droplet volume to the instantaneous fragmentation time, namely,

$$t_{fi}^* \equiv \frac{|u_f - u_i|\,t_{fi}}{D_f}\,\epsilon^{-1/2} = \beta_f \text{Bo}^{-1/4},$$ (A23)

with

$$\text{Bo} \equiv \frac{3C_d\,\rho_c D_f\,|u_f - u_i|^2}{16\sigma},\ \epsilon = \frac{\rho_f}{\rho_i},\ i = l, m,$$ (A24)

and

$$\left[\frac{d}{dt}\left(\frac{\pi}{6}D_f^3\right)\right]_i = \gamma_t\,\frac{\pi D_f^3}{6t_{fi}},$$ (A25)

which can be reduced to

$$\left(\frac{dD_f}{dt}\right)_i =: \frac{\gamma_t}{3}\frac{D_f}{t_{fi}}.$$ (A26)

In Eq. (A23) β_f is an empirical constant that needs to be determined from experiments (see [168]). A factor γ_t is introduced in Eqs. (A25) and (A26) to simulate the possible thermal effect not accounted for by the "instantaneous Bond number" model (which considers Taylor instability as the dominant mechanism leading to fuel fragmentation [168]). The two-phase character of the coolant is then approximately taken into account by weighing the above result by the m-fluid and liquid volume fraction to obtain the derivative we are looking for:

$$\left(\frac{dD_f}{dt}\right)_f = \tfrac{1}{3}D_f\left\{\frac{\alpha}{t_{fm}} + \frac{1-\alpha}{t_{fl}}\right\}. \tag{A27}$$

The fragmentation rate, F_r, required for the mass conservation equations for fuel and debris, is given by

$$F_r = \rho'_f\left\{\frac{\alpha}{t_{fm}} + \frac{1-\alpha}{t_{fl}}\right\}. \tag{A28}$$

In addition to the fragmentation rate, the entrainment rate, E, is a key parameter in the microinteraction concept. Based on the limited data generated so far by the SIGMA experiment (the only experiment in which the growth of an m-fluid–debris mixture under a sustained pressure wave can be quantitatively observed), the following empirical expression for the entrainment rate is adopted for the initial period of fragmentation:

$$E = f_e F_r \frac{\rho_l}{\rho_f}, \tag{A29}$$

where f_e is an entrainment factor (see [168]).

Nomenclature

		J	evaporation rate
A_f	interfacial area of fuel	j	mass flux in shock frame
a	acceleration	k	thermal conductivity
Bo	Bond number (defined in Eq. 17)	m	mass of droplet
c	sound speed	$\dot{m}$	fragmentation rate for a fuel droplet
c_d	drag coefficient	n_f	number density of fuel particles
c_p	specific heat at constant pressure	p	pressure
D	diameter of droplet	Q/A	heat flux per unit area
d	diameter of fragments	$\dot{q}_{fl}$	heat transfer rate between fuel and liquid
E	entrainment rate		
e	internal energy	$\dot{q}_{fm}$	heat transfer rate between fuel and m-fluid
F	factor for interfacial momentum exchange		
		R_{ls}	heat transfer coefficient between liquid and the liquid/m-fluid interface
F_r	fragmentation rate		
f_e	entrainment factor		
f_f	enhancement factor of fragmentation rate	R_{ms}	heat transfer coefficient between the m-fluid and the liquid/m-fluid interface
g	acceleration due to gravity		
g	gravity	r	drop radius
I	specific internal energy	$\dot{S}_f$	source term for length scale
h	heat transfer coefficient or enthalpy	$\dot{S}'_f$	source term of interfacial area
h_{fg}	latent heat of vaporization		

T	temperature or dimensionless time	τ	characteristic time
t	time	ν	kinematic viscosity
t_f	fragmentation time		
t_f''	dimensionless fragmentation time		
u	velocity behind the shock front in fragmentation expts.		
			SUBSCRIPTS
u	velocity		
V	specific volume		
v	velocity	b	breakup
We	Weber number (defined in Eq. 16)	c	coolant
		crit	value at critical point
		d	droplet

GREEK SYMBOLS

		db	debris
		f	fuel
α	thermal diffusivity or void fraction	hn	homogeneous nucleation
α_m	"void fraction" of microinteractions fluid	l	coolant (m-external fluid)
		m	melt droplets or microinteractions fluid
γ_t	thermal enhancement factor		
ΔT_{sub}	subcooling	rel	relative
θ	volume fraction	s	saturation value
ρ	microscopic density	sn	spontaneous nucleation
ρ'	macroscopic density ($\equiv \rho\theta$)	v	vapor
σ	surface tension	0	original value
$\hat{\sigma}$	viscous stress tensor	1	ahead of front
		2	behind front

Acknowledgments

The authors learned the subject of this review while researching it as a long-term commitment. They are grateful to their respective organizations (UKAEA, UCSB), funding agencies (Nuclear Electric, USNRC, and USDOE), to their colleagues and collaborators (A. Thyagaraja, W W. Yuen, S. Angelini, X. Chen, S. Medhekar, and W. H. Amarasooriya), and friends who stimulated their interest in the subject (S. J. Board, R. W. Hall, H. K. Fauske, and T. Speis). We thank Eileen Horton for her considerable assistance in the production of this manuscript. Finally, David would like to acknowledge his therapist, Evi Perlmutter, without whose skill and support this contribution would not have been possible.

References

1. Reid, R. C. (1983). Rapid phase transitions from liquid to vapor. *Advances in Chemical Engineering* **12**, 105–208.
2. Long, G. (1957). Explosions of molten metal in water—causes and prevention. *Metal Progress* **71**, 107–112.
3. Page, F. M., Chamberlain, A. T., and Grimes, R. (1987). The safety of molten aluminium–lithium alloys in the presence of coolants. *J. de Phys. Colloq. C3* (Suppl. 9), **48**, 63–73.

4. Epstein, S. G. (1992). Molten aluminum-water explosions: An update. In *Light Metals 1993* (S. K. Das, ed.) pp. 845-853. Metallurgical Society of AIME.
5. Lipsett, S. G. (1966). Explosions from molten metals and water. *Fire Tech.* **2**, 118-126.
6. Anonymous (1976). *The Explosion at Appleby-Frodingham Steelworks, Scunthorpe, 4 November 1975.* Health and Safety Executive Report, HMSO, ISBN 0-11-880331-X.
7. Katz, D. L., and Sliepcevich, C. M. (1971). Liquefield natural gas/water explosions: Cause and effect. *Hydrocarbon Process.* **50**, 240-244.
8. Arakeri, V. H., Catton, I., and Kastenberg, W. E. (1978). An experimental study of the molten glass/water thermal interaction under free and forced conditions. *Nucl. Sci. Eng.* **66**, 153-166.
9. Colgate, S. A., and Sigurgeirsson, T. (1973). Dynamic mixing of water and lava. *Nature* **244**, 552-555.
10. Francis, P., and Self, S. (1983). The eruption of Krakatau. *Sci. Am.* **249**, 146-159.
11. Theofanous, T. G. (1995). The study of steam explosions in nuclear systems. *Nucl. Eng. Des.* **155**, 1-26.
12. Theofanous, T. G., and Saito, M. (1981). An assessment of Class-9 (core melt) accidents for PWR dry-containment systems. *Nucl. Eng. Des.* **66**, 301-332.
13. Stanmore, B. R., and Desai, M. (1993). Steam explosions in boiler ash hoppers. *Proc. Instn. Mech. Engnrs., Part A: Journal of Power and Energy* **207**, 133-142.
14. Akiyoshi, R., Nishio, S., and Tanasawa, I. (1989). Production of rapidly solidified particles by thermal interaction between molten metal and water. In *Heat Transfer Manufacturing and Materials Processing.* Vol. HTD-V113, pp. 71-76. American Society of Mechanical Engineers.
15. Board, S. J., Farmer, C. L., and Poole, D. H. (1974). Fragmentation in thermal explosions. *Int. J. Heat Mass Transfer* **17**, 331-339.
16. Fletcher, D. F. (1994). A review of the available information on the triggering stage of a steam explosion. *Nucl. Safety* **35**, 36-57.
17. Katz, D. L. (1972). Superheat-limit explosions. *Chem. Eng. Prog.* **58**, 68-69.
18. Henry, R. E., and Fauske, H. K. (1979). Nucleation processes in large scale vapor explosions. *ASME J. Heat Tranfer* **101**, 280-287.
19. Cronenberg, A. W. (1980). Recent developments in the understanding of energetic molten fuel-coolant interactions. *Nucl. Safety* **21**, 319-337.
20. Hicks, E., and Menzies, D. C. (1965). Theoretical study on the fast reactor maximum accident. In *Proc. Conf. on Safety, Fuels, and Core Design in Large Fast Power Reactors*, 11-14 October, Argonne, Il. Argonne National Laboratory Report ANL-7120, pp. 654-670.
21. Corradini, M. L., Kim, B. J., and Oh, M. D. (1988). Vapor explosions in light water reactors: A review of theory and modelling. *Prog. Nucl. Energy* **22**, 1-117.
22. Anderson, R. P., and Armstrong, D. R. (1974). Comparison between vapor explosion models and recent experimental results. *AIChE Symp. Ser.* **70**, 31-47.
23. Cho, D. H., Ivins, R. O., and Wright, R. W. (1971). *A Rate-Limited Model of Molten Fuel/Coolant Interactions: Model Developments and Preliminary Calculations.* Argonne National Laboratory Report ANL-7919.
24. Berthoud, G., and Newman, W. H. (1984). A description of a fuel-coolant thermal interaction model with application in the interpretation of experimental results. *Nucl. Eng. Des.* **82**, 381-391.
25. Fauske, H. K. (1973). On the mechanism of uranium dioxide-sodium explosive interactions. *Nucl. Sci. Eng.* **51**, 95-101.
26. Board, S. J., Hall, R. W., and Hall, R. S. (1975). Detonation of fuel coolant explosions. *Nature* **254**, 319-321.

27. Carslaw, H. S., and Jaeger, J. C. (1959). *Conduction of Heat in Solids*. Clarendon Press, Oxford.
28. Skripov, V. P. (1974). *Metastable Liquids*. Wiley, New York.
29. Fauske, H. K. (1974). Some aspects of liquid–liquid heat transfer and explosive boiling. In *Proc. First Conf. on Fast Reactor Safety*, 2–4 April, Beverly Hills, CA, pp. 992–1005. ANS.
30. Henry, R. E., and Miyazaki, K. (1978). Effects of system pressure on the bubble growth from highly superheated water. In *Topics in Two-Phase Heat Transfer and Flow* (S. G. Bankoff, Ed.), pp. 1–10. ASME, New York.
31. Hohmann, H., Magallon, D., Schins, H., and Yerkess, A. (1995). FCI experiments in the aluminium oxide/water system. *Nucl. Eng. Des.* **155**, 391–403.
32. Bankoff, S. G. (1978). Vapor explosion: A critical review. In *Proc. 6th Int. Heat Transfer Conf.*, 7–11 August, Toronto, Canada, Vol. 6, pp. 355–360. Hemisphere Publishing Corporation, Washington, D.C.
33. Board, S. J., and Hall, R. W. (1974). *Propagation of Thermal Explosions, Part 1: Tin–Water Experiments*. Central Electricity Generation Board Report RD/B/N2850.
34. Board, S. J., and Hall, R. W. (1974). *Propagation of Thermal Explosions, Part 2: Theoretical Model*. Central Electricity Generation Board Report RD/B/N3249.
35. Courant, R., and Friedrichs, K. O. (1948). *Supersonic Flow and Shock Waves*. Interscience, London.
36. Zeldovich, I. B., and Kompaneets, A. S. (1960). *Theory of Detonations*. Academic Press, London.
37. Simpkins, P. G., and Bales, E. L. (1972). Water drop response to sudden accelerations. *J. Fluid Mech.* **55**, 629–639.
38. Fletcher, D. F., and Anderson, R. P. (1990). A review of pressure-induced propagation models of the vapour explosion process. *Prog. Nucl. Energy* **23**, 137–179.
39. Hall, R. W., Board, S. J., and Baines, M. (1979). Observations of tin/water thermal explosions in a long-tube geometry; their interpretation and consequences for the detonation model. In *Proc. 4th CSNI Specialists Meeting on Fuel–Coolant Interactions in Nuclear Reactor Safety*, 2–5 April, Bournemouth, England, Vol. 2, pp. 450–476.
40. Briggs, A. J. (1976). Experimental studies of thermal interactions at AEE Winfirth. In *Proc. Third Specialists Meeting on Sodium Fuel Interactions*, 23–25 March, Tokyo, Japan, Vol. 1, pp. 75–96.
41. Fry, C. J., and Robinson, C. H. (1980). *Propagating Thermal Interactions between Molten Aluminium and Water*. AEE Winfrith Report AEEW-R1309.
42. Mitchell, D. E., and Evans, N. A. (1983). The effect of water to fuel mass ratio and geometry on the behavior of molten core-coolant interaction at intermediate scale. In *Proc. Int. Meeting on Thermal Nuclear Reactor Safety*, 29 August–2 September, 1982, Chicago, NUREG/CP-0027, Vol. 2, pp. 1011–1025.
43. Williams, D. C. (1976). A critique of the Board–Hall model for thermal detonations in UO_2–Na systems. *Proc. Int. Meeting on Fast Reactor Safety and Related Physics*, 5–8 October, Chicago, Vol. 4, pp. 1821–1832.
44. Baines, M., Board, S. J., Buttery, N. E., and Hall, R. W. (1980). The hydrodynamics of large-scale fuel–coolant interactions. *Nucl. Tech.* **49**, 27–39.
45. Hall, R. W., and Board, S. J. (1979). The propagation of large scale thermal explosions. *Int. J. Heat Mass Transfer* **22**, 1083–1093.
46. Condiff, D. W. (1982). Contributions concerning quasi-steady propagation of thermal detonations through dispersions of hot liquid fuel in cooler volatile liquid coolants. *Int. J. Heat Mass Transfer* **25**, 87–98.

47. Corradini, M. L. (1991). Vapor explosions: A review of experiments for accident analysis. *Nucl. Safety* **32**, 337–362.
48. Theofanous, T. G., Najafi, B., and Rumble, E. (1987). An assessment of steam-explosion-induced containment failure. Part I: Probablistic aspects. *Nucl. Sci. Eng.* **97**, 259–281.
49. Cho, D. H., Fauske, H. K., and Grolmes, M. (1976). Some aspects of mixing in large-mass, energetic fuel–coolant interactions. In *Proc. Int. Meeting on Fast Reactor Safety and Related Physics*, 5–8 October, Chicago, Vol. 4, pp. 1852–1861.
50. Henry, R. E., and Fauske, H. K. (1981). Required initial conditions for energetic steam explosions. *Fuel Coolant Interactions, HTD* [*Publ.*] (*Am. Soc. Mech. Eng.*) **19**, 99–108.
51. Tong, L. S. (1965). *Boiling Heat Transfer in Two-Phase Flow*. Wiley, New York.
52. Corradini, M. L., and Moses, G. A. (1985). Limits to fuel/coolant mixing. *Nucl. Sci. Eng.* **90**, 19–27.
53. Corradini, M. L. (1984). Molten fuel/coolant interactions: recent analysis of experiments. *Nucl. Sci. Eng.* **86**, 372–387.
54. Fletcher, D. F. (1985). *A Review of Coarse Mixing Models*. Culham Laboratory Report CLM-R251, HMSO.
55. Bankoff, S. G., and Han, S. H. (1983). Mixing of molten core material and water. *Nucl. Sci. Eng.* **85**, 387–395.
56. Han, S. H., and Bankoff, S. G. (1986). An unsteady one-dimensional two-fluid model for fuel–coolant mixing. *Nucl. Eng. Des.* **95**, 285–295.
57. Hadid, A., and Bankoff, S. G. (1985). Fuel–coolant mixing in the lower plenum of a PWR in a severe accident. In *Proc. Third Int. Topical Meeting on Reactor Thermal Hydraulics*, 15–18 October, Newport, RI, Vol. 2, Paper 19K.
58. Fletcher, D. F., and Thyagaraja, A. (1991). The CHYMES coarse mixing model. *Prog. Nucl. Energy* **26**, 31–61.
59. Thyagaraja, A., and Fletcher, D. F. (1988). Buoyancy-driven, transient, two-dimensional thermo-hydrodynamics of a melt–water–steam mixture. *Comput. Fluids* **16**, 59–80.
60. Theofanous, T. G., and Yuen, W. W. (1995). The probability of alpha-mode containment updated. *Nucl. Eng. Des.* **155**, 459–474.
61. Witte, L. C. (1968). Film boiling from a sphere, *I & EC Fundamentals*, **7**, 517–518.
62. Abolfadl, M. A., and Theofanous, T. G. (1987). An assessment of steam-explosion-induced containment failure. Part II: Premixing limits. *Nucl. Sci. Eng.* **97**, 282–295.
63. Amarasooriya, W. H., and Theofanous, T. G. (1991). Premixing of steam explosions: A three-fluid model. *Nucl. Eng. Des.* **126**, 23–39.
64. Liu, C., and Theofanous, T. G. (1995). Film boiling from spheres in single- and two-phase flow. Part I: Experimental studies; Part II: A theoretical study. In *Proc. National Heat Transfer Conf.*, 5–9 August, Portland, Oregon, pp. 34–61.
65. Young, M. F. (1991). Application of the integrated fuel–coolant interaction code to FITS-type pouring mode experiments. *Prog. Astronaut. Aeronaut.* **134**, 356–386.
66. Berthoud, G., and Valette, M. (1994). Development of a multidimensional model for the premixing phase of a fuel–coolant interaction. *Nucl. Eng. Des.* **149**, 409–418.
67. Jacobs, H. (1994). Analysis of large-scale melt–water mixing events. In *Proc. of the CSNI Specialist Meeting on Fuel Coolant Interactions*, 5–8 January, 1993, Santa Barbara, CA, NUREG/CP-0127, pp. 14–26.
68. Chu, C. C., and Corradini, M. L. (1989). One-dimensional transient fluid model for fuel/coolant interaction analysis. *Nucl. Sci. Eng.* **101**, 48–71.
69. Ohashi, H., Takano, T., Yang, Y., and Akiyama, M. (1995). Simulations of the coarse mixing process of the vapor explosion using a multi-component model. Presented at *NSF/JSPS Seminar on the Physics of Vapor Explosions*, 9–13 June, Santa Barbara, CA.

70. Yamano, N., Koriyama, K., Maruyama, Y., Kudo, T., and Sugimoto, J. (1995). Study of premixing phase of steam explosion in ALPHA program. Presented at *NSF/JSPS Seminar on the Physics of Vapor Explosions*, 9–13 June, Santa Barbara, CA.
71. Wang, S. K., Blomquist, C. A., and Spencer, B. W. (1989). Modeling of thermal and hydrodynamic aspects of molten jet/water interactions. In *ANS Proc. 26th Nat. Heat Transfer Conf.*, 6–10 August, Philadelphia, Vol. 4, pp. 225–232. ANS, La Grange Park.
72. Chu, C. C., Sienicki, J. J., Spencer, B. W., Frid, W., and Löwenhielm, G. (1995). Ex-vessel melt–coolant interactions in deep water pools: studies and accident management for Swedish BWRs. *Nucl. Eng. Des.* **155**, 159–214.
73. Berg, E. v., Bürger, M., Cho, S.-H., and Schatz, A. (1994). Modeling of the breakup of melt jets in liquids for LWR safety analysis. *Nucl. Eng. Des.* **149**, 419–429.
74. Gilbertson, M. A., Fletcher, D. F., Hall, R. W., and Kenning, D. B. R. (1992). Isothermal coarse mixing: Experimental and CFD modelling. *IChemE Symp. Ser.* **129**, 547–556.
75. Gilbertson, M. A., Kenning, D. B. R., and Fletcher, D. F. (1994). Small-scale coarse mixing experiments: problems of comparison with multiphase flow models. In *Proc. International Conference on New Trends in Nuclear System Thermohydraulics*, 30 May–2 June, Pisa, Italy, Vol. 2, pp. 247–255.
76. Fletcher, D. F., and Witt, P. J. (1996). Numerical studies of multiphase mixing with application to some small-scale experiments. *Nucl. Eng. Des.* (in press).
77. Hall, R. W., and Fletcher, D. F. (1995). Validation of CHYMES: simulant studies. *Nucl. Eng. Des.* **155**, 97–114.
78. Angelini, S., Takara, E., Yuen, W. W., and Theofanous, T. G. (1994). Multiphase transients in the premixing of steam explosions. *Nucl. Eng. Des.* **146**, 83–95.
79. Angelini, S., Yuen, W. W., and Theofanous, T. G. (1995). Premixing-related behaviour of steam explosions. *Nucl. Eng. Des.* **155**, 115–157.
80. Yan, H., Yuen, W. W., and Theofanous, T. G. (1992). The use of fluorescence in the measurement of local liquid content in transient multiphase flows. In *Proc. NURETH-5*, 21–24 September, Salt Lake City, UT, Vol. 5, pp. 1271–1278.
81. Angelini, S., Theofanous, T. G., and Yuen, W. W. (1995). The mixing of particle clouds plunging into water. In *Proc. NURETH-7*, 10–15 September, Saratoga Springs, NUREG/CP-0142, Vol. 1, pp. 1754–1778.
82. Anonymous (1994). Summary of the CSNI Specialist Meeting on Fuel Coolant Interactions. 5–8 January, 1993, Santa Barbara, CA, NUREG/CP-0127.
83. Jacobs, H., Lummer, M., Meyer, L., Steehle, B., Thurnay, K., and Väth, L. (1995). Multifield simulations of premixing experiments. Presented at *NSF/JSPS Seminar on the Physics of Vapor Explosions*, 9–13 June, Santa Barbara, CA.
84. Denham, M. K., Tyler, A. P., and Fletcher, D. F. (1994). Experiments on the mixing of molten uranium dioxide with water and initial comparison with CHYMES code calculations. *Nucl. Eng. Des.* **146**, 97–108.
85. Fletcher, D. F., and Denham, M. K. (1995). Validation of the CHYMES mixing model. *Nucl. Eng. Des.* **155**, 85–96.
86. Fletcher, D. F. (1992). A comparison of the coarse mixing predictions obtained from the CHYMES and PM-ALPHA models. *Nucl. Eng. Des.* **135**, 419–425.
87. Magallon, D., and Hohmann, H. (1995). High pressure corium melt quenching tests in FARO. *Nucl. Eng. Des.* **155**, 253–270.
88. Dullforce, T. A., Buchanan, D. J., and Peckover, R. S. (1976). Self-triggering of small-scale fuel-coolant interactions: 1. Experiments. *J. Phys. D: Appl. Phys.* **9**, 1295–1303.
89. Board, S. J., and Hall, R. W. (1975). Thermal explosions at molten tin/water interfaces. In *Moving Boundary Problems in Heat Flow and Diffusion* (J. R. Ockenden and W. R. Hodgkins, eds.), Oxford Univ. Press, London.

90. Schins, H. (1986). Characterization of shock triggers used in thermal detonation experiments. *Nucl. Eng. Des.* **94**, 93-98.

91. Henry, R. E. (1995). Externally triggered steam explosion experiments: Amplification or propagation? *Nucl. Eng. Des.* **155**, 37-44.

92. Fletcher, D. F. (1995). Steam explosion triggering: A review of theoretical and experimental investigations. *Nucl. Eng. Des.* **155**, 27-36.

93. Inoue, A., and Bankoff, S. G. (1981). Destabilisation of film boiling due to arrival of a pressure shock. Part I: Experimental. *ASME J. Heat Transfer* **103**, 459-464.

94. Inoue, A., Aoki, S., Aritomi, M., Kataoka, H., and Matsunaga, A. (1983). Study on transient heat transfer of film boiling due to arrival of pressure shock. In *Proc. 7th Int. Heat Transfer Conf.*, 6-10 September, 1982, Munich, Germany, Vol. 4, pp. 403-408. Hemisphere Publishing Corporation.

95. Naylor, P. (1985). Film boiling destabilisation. Ph.D. Thesis, University of Exeter, UK.

96. Knowles, J. B. (1985). *A Mathematical Model of Vapour Film Destabilisation.* AEE Winfrith Report AEEW-R1933, HMSO.

97. Drumheller, D. S. (1979). The initiation of melt fragmentation in fuel-coolant interactions. *Nucl. Sci. Eng.* **72**, 347-356.

98. Inoue, A., Ganguli, A., and Bankoff, S. G. (1981). Destabilization of film boiling due to arrival of a pressure shock. Part II: Analytical. *ASME J. Heat Transfer* **103**, 465-471.

99. Corradini, M. L. (1983). Modeling film boiling destabilization due to a pressure shock arrival. *Nucl. Sci. Eng.* **84**, 196-205.

100. Inoue, A., Takahashi, K., Takahashi, M., and Matsuzaki, M. (1995). Transient film boiling under conditions related to vapor explosions (effects of transient flow and fragmentation under a shock pressure). *Nucl. Eng. Des.* **155**, 55-66.

101. Nelson, L. S., and Duda, P. M. (1976). Steam explosion experiments with single drops of iron oxide melted with CO_2 laser. *High Temp.-High Pressures* **14**, 259-281.

102. Fröhlich, G., Müller, G., and Unger, H. (1976). Experiments with water and hot melts of lead. *J. Non-Equilib. Thermodyn.* **1**, 91-103.

103. Zyszkowski, W. (1976). Study of the thermal explosion phenomenon in molten copper-water system. *Int. J. Heat Mass Transfer* **19**, 849-868.

104. Watts, P., Frost, D. L., and Barbone, R. (1993). Effect of fluid properties on the vapor explosion of single molten metal drops. In *ANS Proc. Nat. Heat Transfer Conf.*, 8-11 August, Atlanta, GA, Vol. 7, pp. 275-286.

105. Corradini, M. L. (1981). Phenomenological modeling of the triggering phase of small-scale steam explosion experiments. *Nucl. Sci. Eng.* **78**, 154-170.

106. Akiyoshi, R., Hishio, S., and Tanasawa, I. (1990). A study of the effect of noncondensible gas in the vapor film on vapor explosion. *Int. J. Heat Mass Transfer* **33**, 603-609.

107. Beck, D. F., Berman, M., and Nelson, L. S. (1991). Steam explosion studies with molten iron-alumina generated by thermite reactions. *Prog. Astronaut. Aeronaut.* **134**, 326-355.

108. Yuen, W. W., Chen, X., and Theofanous, T. G. (1994). On the fundamental microinteractions that support the propagation of steam explosions. *Nucl. Eng. Des.* **146**, 133-146.

109. Marshall, B. W. (1988). Recent fuel-coolant interaction experiments conducted in the FITS vessel. In *ANS Proc. Nat. Heat Transfer Conf.*, 24-27 July, Houston, Vol. 3, pp. 265-275.

110. Bird, M. J. (1984). An experimental study of scaling in core melt/water interactions. Presented at *22nd Nat. Heat Transfer Conf.*, 5-8 August, Niagara Falls, New York.

111. Tattersall, R. B., and Maddison, R. J. (1989). *Molten Fuel Coolant Interactions in Water at High Pressures: Final report on Experiments Carried Out in the High Pressure Thermite Rig.* AEE Winfrith Internal Report AEEW-R2448.

112. Henry, R. E., Hohmann, H., and Kottowski, H. (1979). The effect of pressure on NaCl–H₂O explosions. In *Proc. Fourth CSNI Specialist Meeting on Fuel–Coolant Interactions in Nuclear Reactor Safety*, 2–5 April, Bournemouth, UK, Vol. 1, pp. 308–323.

113. Hohmann, H., Kottowski, H., Schins, H., and Henry, R. E. (1982). Experimental investigations of spontaneous and triggered vapor explosions in the molten salt/water system. In *Proc. Int. Meeting on Thermal Nuclear Reactor Safety*, August 29–September 2, Chicago, Vol. 2, pp. 962–971.

114. Marshall, B. W., Berman, M., and Krein, M. S. (1986). Recent intermediate-scale experiments on fuel–coolant interactions in an open geometry (EXO-FITS). In *Proc. Int. ANS/ENS Topical Meeting on Thermal Reactor Safety*, 2–6 February, San Diego, Paper II.5.

115. Fletcher, D. F. (1988). The particle size distribution of solidified melt debris from molten fuel–coolant interaction experiments. *Nucl. Eng. Des.* **105**, 313–319.

116. Yamano, N., Maruyama, Y., Kudo, T., Hidaka, A., and Sugimoto, J. (1995). Phenomenological studies on melt–coolant interactions in the ALPHA program. *Nucl. Eng. Des.* **155**, 369–389.

117. Buxton, L. D., Benedick, W. B., and Corradini, M. L. (1980). *Steam Explosion Efficiency Studies. Part II: Corium melts* U.S. Nuclear Regulatory Commission Report NUREG/CR-1746.

118. Nelson, L. S., Eatough, M. J., and Guay, K. P. (1988). Initiation of explosive molten aluminum–water interactions at wet or underwater surfaces. In *ANS Proc. Nat. Heat Transfer Conf.*, 24–27 July, Houston, Vol. 3, pp. 219–227.

119. Flory, K., Paoli, R., and Mesler, R. (1969). Molten metal–water explosions. *Chem. Eng. Prog.* **65**, 50–54.

120. Nelson, L. S., and Guay, K. P. (1986). Suppression of steam explosions in tin and Fe–Al₂O₃ melts by increasing the viscosity of the coolant. *High Temp.–High Pressures* **18**, 107–111.

121. Kim, H., Krueger, J., and Corradini, M. L. (1989). Single droplet vapor explosions: Effect of coolant viscosity. In *Proc. Fourth Int. Topical Meeting on Nuclear Reactor Thermal-Hydraulics*, 10–13 October, Karlsruhe, Germany, Vol. 1, pp. 261–267.

122. Kowal, M. G., Dowling, M. F., and Abdel-Khalik, S. I. (1993). An experimental investigation of the effects of surfactants on the severity of vapor explosions. *Nucl. Sci. Eng.* **115**, 185–192.

123. Skelton, W. T. W., Kowal, M. G., and Abdel-Khalik, S. I. (1995). Effect of boric acid on the severity of vapor explosions in pure water and surfactant solutions. *Nucl. Eng. Des.* **155**, 359–368.

124. Dowling, M. F., Ip, B. M., and Abdel-Khalik, S. I. (1993). Suppression of vapor explosions by dilute aqueous polymer solutions. *Nucl. Sci. Eng.* **113**, 300–313.

125. Becker, K. M., and Lindland, K. P. (1991). *The Effect of Surfactants on Hydrodynamic Fragmentation and Steam Explosions*. Department of Nuclear Reactor Engineering, Royal Institute of Technology, Stockholm, Sweden, Report KTH-NEL-50.

126. Tan, M. J., and Bankoff, S. G. (1986). On the fragmentation of drops. *ASME J. Fluids Eng.* **108**, 109–114.

127. Pilch, M., and Erdman, C. A. (1987). Ude of breakup time data and velocity history data to predict the maximum size of stable fragments for acceleration-induced breakup of a liquid drop. *Int. J. Multiphase Flow* **13**, 741–757.

128. Patel, P. D., and Theofanous, T. G. (1981). Hydrodynamic fragmentation of drops. *J. Fluid Mech.* **103**, 207–223.

129. Ciccarelli, G., and Frost, D. L. (1993). Effect of fluid flow velocity on the fragmentation mechanism of a hot melt drop. *Prog. Astronaut. Aeronaut.* **154**, 334–361.

130. Ranger, A. A., and Nicholls, J. A. (1969). Aerodynamic shattering of liquid drops. *AIAA J.* **7**, 285–290.

131. Taylor, G. I. (1963). The shape and acceleration of a drop in a high-speed air stream. In *The Scientific Papers of G. I. Taylor* (G. K. Batchelor, ed.), Vol. 3. Cambridge Univ. Press, Cambridge.

132. Waldman, G. D., Reinecke, W. G., and Glenn, D. C. (1972). Raindrop breakup in the shock layer of a high speed vehicle. *AIAA J.* **10**, 1200–1204.

133. Scott, E. and Berthoud, G. J. (1978). Multiphase thermal detonation. In *Topics in Two-Phase Heat Transfer and Flow* (S. G. Bankoff, ed.). American Society of Mechanical Engineers, New York.

134. Sharon, A., and Bankoff, S. G. (1981). On the existence of steady supercritical plane thermal explosions. *Int. J. Heat Mass Transfer* **24**, 1561–1572.

135. Patel, P. D., and Theofanous, T. G. (1978). Fragmentation requirements for detonating vapor explosions. *Trans. Am. Nucl. Soc.* **28**, 451–452.

136. Theofanous, T. G., Saito, M., and Efthimiadis, T. (1979). The role of hydrodynamic fragmentation in fuel coolant interactions. In *Proc. 4th CSNI Specialists Meeting on Fuel–Coolant Interactions in Nuclear Reactor Safety*, 2–5 April, Bournemouth, UK, Vol. 1, pp. 112–128.

137. Theofanous, T. G., Gherson, P., Nourbakhsh, H. P., Hu, K., Iyer, K., Viskanta, R. and Lommers, L. (1983). *LWR and HTGR Coolant Dynamics: The Containment of Severe Accidents.* U.S. Nuclear Regulatory Commission Report NUREG/CR-3306.

138. Baines, M. (1979). Hydrodynamic fragmentation in a dense dispersion. In *Proc. 4th CSNI Specialists Meeting on Fuel–Coolant Interactions in Nuclear Reactor Safety*, 2–5 April, Bournemouth, UK, Vol. 1, pp. 90–111.

139. Baines, M., and Buttery, N. E. (1979). *Differential Velocity Fragmentation in Liquid–Liquid Systems.* Central Electricity Generating Board Report RD/B/N4643.

140. Bankoff, S. G., and Yang, J. W. (1989). Studies relevant to in-vessel steam explosions. In *Proc. 4th Int. Topical Meeting on Nuclear Reactor Thermal Hydraulics*, 10–13 October, Karlsruhe, Germany, Vol. 1, pp. 312–318.

141. Harper, E. Y., Grube, G. W., and Chang, I. D. (1972). On the breakup of accelerating liquid drops. *J. Fluid Mech.* **52**, 565–591.

142. Yang, J. W., and Bankoff, S. G. (1987). Solidification effects on the fragmentation of molten metal drops behind a pressure shock wave. *ASME J. Heat Transfer* **109**, 226–231.

143. Kim, D. S., Bürger, M., Fröhlich, G., and Unger, H. (1983). Experimental investigation of hydrodynamic fragmentation of gallium drops in water flows. In *Proc. Int. Meeting on LWR Severe Accident Evaluation*, 28 August–1 September, Cambridge, MA, Vol. 1, Paper 6.4.

144. Carachalios, C., Bürger, M., and Unger, H. (1983). A transient two-phase model to describe thermal detonations based on hydrodynamic fragmentation. In *Proc. Int Meeting on LWR Severe Accident Evaluation*, 28 August–1 September, Cambridge, MA, Vol. 1, Paper 6.8.

145. Bürger, M., Kim, D. S., Schwalbe, W., Unger, H., Hohmann, H., and Schins, H. (1984). Two-phase description of hydrodynamic fragmentation processes within thermal detonation waves. *ASME J. Heat Transfer* **106**, 728–734.

146. Jacobs, H. (1976). Computational analysis of fuel–sodium interactions with an improved method. In *Proc. Int. Meeting on Fast Reactor Safety and Related Physics*, 5–8 October, Chicago, Vol. 3, pp. 926–935.

147. Cronenberg, A. W., and Grolmes, M. A. (1975). Fragmentation modelling relative to molten UO_2 in sodium. *Nucl. Safety* **16**, 683–700.

148. Cronenberg, A. W., Chawla, T. C., and Fauske, H. K. (1974). A thermal stress mechanism for the fragmentation of molten UO_2 upon contact with sodium coolant. *Nucl. Eng. Des.* **30**, 434–443.

149. Epstein, M. (1974). Thermal fragmentation—a gas release phenomenon. *Nucl. Sci. Eng.* **55**, 462–467.

150. Ochiai, M., and Bankoff, S. G. (1976). Liquid–liquid contact in vapor explosions. In *Proc. ANS/ENS Int. Meeting on Fast Reactor Safety*, 5–8 October, Chicago, Vol. 4, pp. 1843–1851.

151. Buchanan, D. J. (1974). A model for fuel–coolant interactions. *J. Phys. D: Appl. Phys.* **7**, 1441–1457.

152. Kim, B., and Corradini, M. L. (1988). Modeling of small-scale single droplet fuel/coolant interactions. *Nucl. Sci. Eng.* **98**, 16–28.

153. Nelson, L. S., Duda, P. M., Fröhlich, G., and Anderle, M. (1988). Photographic evidence for the mechanism of fragmentation of a single drop of melt in triggered steam explosion experiments. *J. Non-Equilib. Thermodyn.* **13**, 27–55.

154. Tang, J., and Corradini, M. L. (1994). Modelling of the complete process of one-dimensional vapor explosions. In *Proc. of the CSNI Specialist Meeting on Fuel Coolant Interactions*, 5–8 January 1993, Santa Barbara, CA, NUREG/CP-0127, pp. 204–217.

155. Frost, D. L., Ciccarelli, G., and Watts, P. (1993). Flash X-ray visualization of the steam explosion of a molten metal drop. *Prog. Astronaut. and Aeronaut.* **154**, 388–420.

156. Ciccarelli, G., and Frost, D. L. (1994). Fragmentation mechanisms based on single drop steam explosion experiments using flash X-ray radiography. *Nucl. Eng. Des.* **146**, 109–132.

157. Baines, M. (1984). Preliminary measurements of steam explosion work yields in a constrained system. *Inst. Chem. Eng. Symp. Ser.* **86**, 97–108.

158. Bürger, M., Müller, K., Buck, M., Cho, S.-H., Schatz, A., Schins, H., Zeyen, R., and Hohmann, H. (1991). Examination of thermal detonation codes and included fragmentation models by means of triggered propagation experiments in a tim/water mixture. *Nucl. Eng. Des.* **131**, 61–70.

159. Park, H. S., Yoon, C., Corradini, M. L., and Bang, K. H. (1993). Vapor explosion escalation/propagation experiments and possible fragmentation mechanisms. In *Proc. Int. Seminar on the Physics of Vapor Explosions*, 25–29 October, Tomakomai, Japan, pp. 187–196.

160. Park, H. S., Yoon, C., Corradini, M. L., and Bang, K. H. (1993). Experiments on the trigger effect for 1-D large scale vapor explosions. In *Proc. International Conference on New Trends in Nuclear System Thermohydraulics*, 30 May–2 June 1994, Pisa, Italy, Vol. 2, pp. 271–280.

161. Fletcher, D. F. (1991). An improved mathematical model of melt/water detonations. I: Model formulation and example results. *Int. J. Heat Mass Transfer* **34**, 2435–2448.

162. Fletcher, D. F. (1991). An improved mathematical model of melt/water detonations. II: A study of escalation. *Int. J. Heat Mass Transfer* **34**, 2449–2459.

163. Fletcher, D. F. (1995). Propagation investigations using the CULDESAC model. *Nucl. Eng. Des.* **155**, 271–287.

164. Medhekar, S., Abolfadl, M., and Theofanous, T. G. (1991). Triggering and propagation of steam explosions. *Nucl. Eng. Des.* **126**, 41–46.

165. Medhekar, S., Amarasooriya, W. H., and Theofanous, T. G. (1989). Integrated analysis of steam explosions. In *Proc. Fourth Int. Top. Meet. on Nuclear Reactor Thermal-Hydraulics*, 10–13 October, Karlsruhe, Germany, Vol. 1, pp. 319–326.

166. Yuen, W. W., and Theofanous, T. G. (1995). The prediction of 2D thermal detonations and resulting damage potential. *Nucl. Eng. Des.* **155**, 289–309.

167. Theofanous, T. G., and Yuen, W. W. (1994). The prediction of dynamic loads from ex-vessel steam explosions. In *Proc. International Conference on New Trends in Nuclear System Thermohydraulics*, 30 May–2 June, Pisa, Italy, pp. 257–270.

168. Chen, X., Yuen, W. W., and Theofanous, T. G. (1995). On the constitutive description of the microinteraction concept in steam explosions. In *Proc. NURETH-7*, 10–15 September, Saratoga Springs, NY, NUREG/CP-0142, Vol. 1, pp. 1586–1606.

169. Bürger, M., Buck, M., Müller, K., and Schatz, A. (1994). Stepwise verification of thermal detonation models: Examination by means of the KROTOS experiments. In *Proc. of the CSNI Specialist Meeting on Fuel Coolant Interactions*, 5–8 January 1993, Santa Barbara, CA, NUREG/CP-0127, pp. 218–232.

170. Corradini, M. L. (1982). Analysis and modelling of large-scale steam explosion experiments. *Nucl. Sci. Eng.* **82**, 429–447.

171. Bürger, M., Schwalbe, W., and Unger, H. (1983). Application of hydrodynamic and thermal fragmentation models and a steady state thermal detonation model to molten salt–water explosions. In *Proc. Int. Meeting on Thermal Nuclear Reactor Safety*, 29 August–2 September 1982, Chicago, NUREG/CP-0027, Vol. 2, pp. 1378–1387.

172. Anderson, R. P., Armstrong, D., Cho, D., and Kras, A. (1988). Experimental and analytical study of vapor explosions in stratified geometries. In *ANS Proc. Natl. Heat Transfer Conf.*, 24–27 July, Houston, Vol. 3, pp. 236–243.

173. Ciccarelli, G., Frost, D. L., and Zarafonitis, C. (1991). Dynamics of explosive interactions between molten tin and water in stratified geometry. *Prog. Astronaut. Aeronaut.* **134**, 307–325.

174. Frost, D. L., Bruckert, B., and Ciccarelli, G. (1995). Effect of boundary conditions on the propagation of a vapor explosion in stratified molten tim/water systems. *Nucl. Eng. Des.* **155**, 311–333.

175. Bang, K. H., and Corradini, M. L. (1991). Vapor explosions in a stratified geometry. *Nucl. Sci. Eng.* **108**, 88–108.

176. Harlow, F. H., and Ruppel, H. M. (1981). *Propagation of a Liquid–Liquid Explosion*. Los Alamos National Laboratory Report LA-8971-MS.

177. Sharon, A., and Bankoff, S. G. (1978). Propagation of shock waves in a fuel-coolant mixture. In *Topics in Two-Phase Heat Transfer and Flow* (S. G. Bankoff, ed.). American Society of Mechanical Engineers, New York.

178. Frost, D. L., Lee, J. H. S., and Ciccarelli, G. (1991). The use of Hugoniot analysis for the propagation of vapor explosion waves. *Shock Waves* **1**, 99–110.

179. Frost, D. L., and Ciccarelli, G. (1993). Implications for the existence or thermal detonations for the equilibrium Hugoniot analysis. *Prog. Astronaut. Aeronaut.* **154**, 362–387.

180. McCahan, S., and Shepherd, J. E. (1993). Models of rapid evaporation in nonequilibrium mixtures of tin and water. *Prog. Astronaut. Aeronaut.* **154**, 432–448.

181. Berman, M., and Beck, D. (1989). Steam explosion triggering and propagation: Hypothesis and evidence. In *Proc. 3rd International Seminar on Containment of Nuclear Reactors*, 10–11 August, University of California, Los Angeles. (Also available as Sandia National Laboratory Report SAND89-1878C.)

182. Fletcher, D. F. (1994). Vapour explosions: multiphase detonations or deflagrations? *Shock Waves* **3**, 181–192.

183. Shepherd, J. E., McCahan, S., and Cho, J. (1989). Evaporation wave model for superheated liquids. In *Proc. IUTAM Symposium on Adiabatic Waves in Liquid–Vapour Systems*, 28 August–1 September, Göttingen, Germany, pp. 3–12.

184. Fowles, G. R. (1989). Vapor detonations in superheated fluids. In *Proc. IUTAM Symposium on Adiabatic Waves in Liquid-Vapour Systems*, 28 August–1 September, Göttingen, Germany, pp. 407–416.
185. Kinney, G. F., and Graham, K. J. (1962). *Explosive Shocks in Air*. Spinger-Verlag, Berlin.
186. Chase, Jr., M. W., Davies, C. A., Downey, Jr., J. R., Frurie, D. J., McDonald, R. A., and Syverud, A. N. (1985). *JANAF Thermochemical Tables*, 3rd ed. American Chemical Society and the American Institute of Physics for the National Bureau of Standards, New York.
187. Hess, P. D., and Brondyke, K. J. (1969. Causes of molten aluminum–water explosions and their prevention. *Metal Progress* **95**, 93–100.
188. Bergeron, K, ed. (1993). *Ex-Vessel Severe Accident Review for the Heavy Water New Production Reactor*. Sandia National Laboratories Report SAND90-234.
189. Baker, L., Jr. (1983). Hydrogen-generating reactions in LWR severe accidents. In *Proc. Int. Meeting on Light Water Reactor Severe Accident Evaluation*, 28 August–1 September, Cambridge, MA, Vol. 2, Paper 16.
190. Fletcher, D. F., Turland, B. D., and Lawrence, S. P. A. (1992). A review of hydrogen production during melt/water interaction in LWRs. *Nucl. Safety* **33**, 514–534.
191. Baker, L, Jr. (1965). Metal–water interactions. *Nucl. Safety* **7**, 25–34.
192. Epstein, M., Leung, J. C., Hauser, G. M., Henry, R. E., and Baker, L., Jr. (1984). Film boiling on a reactive surface. *Int. J. Heat Mass Transfer* **27**, 1365–1378.
193. Young, M. F., Berman, M., and Pong, L. T. (1988). Hydrogen generation during fuel/coolant interactions. *Nucl. Sci. Eng.* **98**, 1–15.
194. Epstein, E., and Fauske, H. K. (1994). A crystallization theory of underwater aluminum ignition. *Nucl. Eng. Des.* **146**, 147–164.
195. Fauske, H. K., and Epstein, M. (1995). On the requirements for energetic molten aluminum–water chemical reactions. In *Proc. of the CSNI Specialist Meeting on Fuel Coolant Interactions*, 5–8 January, 1993, Santa Barbara, CA, NUREG/CP-0127, pp. 142–147.
196. Cho, D. H. (1993). Combined vapor and chemical explosions of metals and water. In *Proc. Int. Seminar on the Physics of Vapor Explosions*, 25–29 October, Tomakomai, Japan, pp. 157–164.
197. Cho, D. H., Armstrong, D. R., and Anderson, R. P. (1995). Combined vapor and chemical explosions of metal and water. *Nucl. Eng. Des.* **155**, 405–412.
198. Rightly, M. J., Beck, D. F., and Berman, M. (1993). *NPR/FCI EXO-FITS Experiment Series Report*. Sandia National Laboratories Report SAND91-1544.
199. Nelson, L. S. (1994). Steam explosions of single drops of pure and alloyed aluminum. *Nucl. Eng. Des.* **155**, 413–425.
200. Theofanous, T. G., Chen, X., Di Piazza, P., Epstein, M., and Fauske, H. K. (1994). Ignition of aluminum droplets behind shock waves in water. *Phys. Fluids* **6**, 3513–3515.
201. Nuclear Regulatory Commision (NRC) (1975). *Reactor Safety Study, WASH-1400*. U.S. Nuclear Regulatory Commission Report NUREG/75-0114.
202. Theofanous, T. G. (1993). Dealing with phenomenological uncertainty in risk analysis. Paper presented at *Workshop I in Advances Topics in Reliability and Risk Analysis*, 20–22 October, Annapolis, MD.
203. Corradini, M. L., and Swenson, D. V. (1981). *Probability of Containment Failure Due to Steam Explosions Following a Postulated Core Meltdown in an LWR*. U.S. Nuclear Regulatory Commission Report NUREG/CR-2214.
204. Steam Explosion Review Group (1985). *A Review of the Current Understanding of the Potential for Containment Failure Arising from In-Vessel Steam Explosions*. U.S. Nuclear Regulatory Commission Report NUREG/1116.

205. Berman, M., Swenson, D. V., and Wickett, A. J. (1984). *An Uncertainty Study of PWR Steam Explosions*. U.S. Nuclear Regulatory Commission Report NUREG/CR-3369.
206. Amarasooriya, W. H., and Theofanous, T. G. (1987). An assessment of steam explosion-induced containment failure. Part III: Expansion and energy partition. *Nucl. Sci. Eng.* **97**, 296–315.
207. Lucas, G. E., Amarasooriya, W. H., and Theofanous, T. G. (1987). An assessment of steam-explosion-induced containment failure. Part IV: Impact mechanics, dissipation, and vessel head failure. *Nucl. Sci. Eng.* **97**, 316–326.
208. Turland, B. D., Fletcher, D. F., Hodges, K. I., and Attwood, G. J. (1995). Quantification of the probability of containment failure caused by an in-vessel steam explosion for the Sizewell B PWR. *Nucl. Eng. Des.* **155**, 445–458.
209. Basu, S., and Ginsberg, T. (1996). *A Reassessment of the Potential for an Alpha-Mode Containment Failure and a Review of the Current Understanding of Broader Fuel–Coolant Interaction (FCI) Issues*. U.S. Nuclear Regulatory Commission Report NUREG-1529.
210. Ishii, M. (1975). *Thermo-Fluid Dynamic Theory of Two-Phase Flow*. Eyrolles, France.

Heat Transfer During Direct
Containment Heating

MARTIN M. PILCH, MICHAEL D. ALLEN,
AND DAVID C. WILLIAMS

Sandia National Laboratories, Albuquerque, New Mexico

I. Introduction

Direct containment heating (DCH) can occur in a nuclear power plant (NPP) if, as the result of a core melt accident, molten core material accumulates on the lower head of the reactor pressure vessel (RPV) causing it to fail by thermally induced rupture, by expulsion of an incore instrument guide tube, or by the thermally induced rupture of an incore guide tube or other penetration outside the RPV boundary [1]. DCH is only of interest if the RPV failure occurs while the reactor coolant system (RCS) is still at elevated pressure. Failure of the lower head of the RPV by one of these mechanisms initiates forcible ejection of molten core material into the reactor cavity located beneath the RPV (Fig. 1). These processes have been termed high-pressure melt ejection (HPME).

The subsequent blowdown of the RCS adds both mass and energy to the containment atmosphere. Some portions of the molten core material that ejected into the reactor cavity can be entrained, fragmented, and dispersed into subcompartments and the upper dome of the containment; the remainder is typically, but not always, ignored in DCH analyses. Fragmented debris dispersed from the reactor cavity can rapidly liberate its thermal energy to blowdown gases and the containment atmosphere. The metallic components of the dispersed material can be oxidized by blowdown steam (predominantly), liberating both energy and hydrogen (which can subsequently burn). These processes will heat the containment atmosphere, possibly to the point at which combustion of hydrogen (preexisting

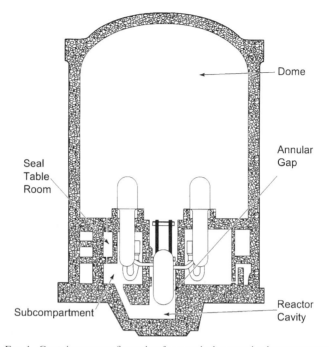

FIG. 1. Containment configuration for a typical pressurized water reactor.

in the atmosphere) can occur under conditions where the atmosphere might otherwise be inerted due to high steam concentrations. Collectively, these processes are termed DCH, which might lead to early failure of the reactor containment building (RCB) by overpressurization.

DCH found its genesis in the Zion Probabilistic Safety Study [2], where it was argued that failure of the RPV would occur in a large number of accident sequences while the RCS was still at elevated pressure. The study also argued that the resulting HPME and blowdown of the RCS would forcibly sweep melt from the cavity into the containment, where the melt would settle into a coolable geometry on the containment floor. Thus, HPME and debris dispersal were thought to lead to a benign or even beneficial termination of the accident. These predictions led to Electric Power Research Institute (EPRI) sponsored programs [3, 4] at Argonne National Laboratory (ANL) and Nuclear Regulatory Commission (NRC) sponsored programs at Sandia National Laboratories (SNL) [5] to study debris dispersal from the cavity. The SNL program also placed heavy emphasis on characterizing the aerosol source term associated with the HPME and debris dispersal processes [6, 7, 8, 9, 10, 11]. Towards this end,

SNL conducted the SPIT-18,19 experiments [11] in a closed chamber to facilitate aerosol measurements. Heating of the chamber atmosphere by dispersed debris damaged the facility in both tests. These tests raised concerns that HPME/DCH processes, overlooked until that time, might be a significant contributor to short-term containment pressurization in an NPP accident.

Coincident with these initial activities, the NRC sponsored the Containment Loads Working Group [12] to examine possible overpressurization threats to the containment. Bounding calculations assuming thermodynamic equilibrium between dispersed debris and the entire containment atmosphere indicated that some containments might be threatened with less than half the core participating in DCH [13, 14]. These early estimates of containment-threatening pressures were counterbalanced by a recognition that actual containments possessed a number of inherent mitigative features by virtue of their design and that certain physical processes could limit DCH loads. The more important of the potential mitigators are: (1) compartmentalized geometry to limit debris mixing with the atmosphere, (2) water in the reactor cavity or in the RPV to quench DCH interactions, and (3) heat transfer to structures to limit peak atmosphere temperatures. At the time, however, there was no convincing way to quantify the magnitude of potential mitigation. Since that time, experimental and analytical programs have extensively addressed HPME/DCH issues.

NUREG-1150 [15, 16] was the first attempt to treat DCH from a probabilistic risk analysis (PRA) perspective that integrates sequence probabilities with uncertainties associated with the initial and boundary conditions and phenomenological uncertainties associated with predicting containment loads. NUREG-1150 addressed only a small number of reference plants and the DCH database was limited at the time, so validation of these early attempts to predict DCH loads was correspondingly limited. The risk perspective of DCH decreased significantly between the draft [14] and final [16] versions of NUREG-1150. Important reasons for this change includes recognition of spontaneous depressurization resulting from creep rupture of the hot leg nozzles, taking credit for power recovery prior to large-scale core melting, more refined assessments of containment loads, and more responsible treatment of uncertainties in general. The individual plant examinations (IPEs) have more recently addressed the DCH issue from a PRA perspective. Their strength is that plant-specific sequence information is fully integrated into the assessment for every plant. On the other hand, the approaches taken in the IPEs to assess containment loads are not completely tied to the existing database.

The low containment failure probabilities resulting from DCH in these level two PRA studies are largely attributable to the relatively low probability of high RCS pressures existing at vessel breach. More specifically, creep rupture of the hot leg nozzles, leading to spontaneous depressurization of the RCS, is expected with high probability unless operator intervention disrupts the buoyancy driven redistribution of core energy to the hot legs.

More recently, system-level code calculations with SCDAP/RELAP5 (see [17], [18, App. C] and [19, App. E]) suggest that spontaneous depressurization is even more likely than previously believed for station blackout accidents. All reactor vendors have now implemented severe-accident guidelines instructing the operator to depressurize the RCS at the first signs of core uncovery. The current perception is that sequences relevant to DCH (i.e., high pressure at vessel breach) are far less likely than envisioned at the time of the Zion Probabilistic Safety Study.

The NRC has also pursued a loads-based approach to resolution of the DCH issue in parallel with their research on intentional and unintentional RCS depressurization. Here assessments of containment pressurization (in relation to the containment strength) are made under the assumption that an HPME event might actually occur. The NRC-sponsored experimental program has played a major role in developing an understanding of the key physical processes in DCH. The technical basis for these scaled experiments was developed by the Severe Accident Scaling Methodology Technical Program Group (SASM-TPG) [20]. Zuber et al. [20] made four recommendations for continued NRC-sponsored experimentation: (1) counterpart tests should be conducted at two different physical scales, (2) integral effects tests should be conducted in plant-specific geometries, (3) initial and boundary conditions should be closely tied to relevant accident sequences, and (4) the experiment programs should be monitored by a technical review group. These recommendations were implemented as part of the SNL/IET and ANL/IET test programs.

The experiment program led to independent and contemporaneous development of two simple physics models, the Two-Cell Equilibrium (TCE) and the Convection Limited Containment Heating (CLCH) model, that attempt to represent the few dominant processes that contribute to DCH loads. In addition, this database was used to assess DCH models in the NRC sponsored MELCOR and CONTAIN systems codes [21, 22].

The first step in the NRC's DCH issue-resolution process was writing NUREG/CR-6075 [23], "The Probability of Containment Failure by Direct Containment Heating in Zion." NUREG/CR-6075 assesses the probability of containment failure by DCH for the Zion NPP and establishes the basic methodology used to address DCH for other NPPs. The methodology involved the definition of four splinter scenarios that were intended to bound all possible sequences under the assumption that the RCS could fail at high RCS pressures while uncertainties in key initial conditions were quantified. Loads distributions were computed with the TCE and CLCH models. The loads distributions were convoluted with a distribution for containment strength to compute the containment failure probability for each splinter. The containment failure probability was sufficiently low for Zion that DCH was considered resolved with wide margins.

Supplement 1 of NUREG/CR-6075 [18] was written in response to the peer review process to close the DCH issue for the Zion plant. It contains a redefinition of some of the scenarios and the additional analyses that working groups indicated were necessary to strengthen the original conclusions. Core-melt-progression analyses were performed using SCDAP/RELAP5 (see App. C in [18]) to better establish the melt mass and composition in a consistent manner. The melt composition was found to be more oxidic than previous estimates (see App. G in [20]). NUREG/CR-6109 [19] used the methodology and scenarios described in NUREG/CR-6075 and NUREG/CR-6075 Supplement 1 to address the DCH issue for the Surry plant. The TCE model alone was used to compute the loads distributions, but point comparisons with CONTAIN (a system-level computer code for evaluating containment response to severe reactor accidents) were performed. NUREG/CR-6338 [24] addresses the DCH issue for all Westinghouse plants with dry containments, which include 34 plants with large dry containments and 7 plants with subatmospheric containments. The containment failure probability was again sufficiently low that DCH was considered resolved for all the plants. Ongoing NRC efforts will examine Combustion Engineering plants, Babcock & Wilcox plants, and Westinghouse plants with ice condenser containments.

This article addresses the hydrodynamic, heat transfer, and mass transfer processes that control DCH in pressurized water reactors (PWRs) with large dry or subatmospheric containments. DCH issues for PWRs with ice condenser containments and for boiling water reactors (BWRs) are only briefly addressed. Models for individual processes and their experimental validation will be reviewed. Integral models suitable for performing experiments and NPP analyses will also be reviewed. Lessons learned from the extensive DCH integral experiments will be discussed. This article will not address the likelihood of DCH events or uncertainties in conditions at the onset of DCH that are established during core melt progression.

II. DCH Integral Database

DCH experiments have been conducted at Sandia National Laboratories (SNL), Argonne National Laboratory (ANL), and Fauske and Associates, Inc. (FAI). Of interest here are experiments employing high-temperature chemically reactive melts, driven under pressure into a simulated reactor cavity, with the whole system confined in a vessel so that containment pressure can be measured. A brief survey of these experiments is presented in Table I along with all relevant references. Experiments have been conducted at five different physical scales, five different cavity designs, and with and without subcompartment structures, reactive and nonreactive blowdown gases, reactive and nonreactive atmospheres, and water in the cavity and basement.

The earliest DCH tests were the ANL/CWTI tests (1 : 30 scale). Spencer *et al.* [42, 43], Blomquist *et al.* [44], and Binder and Spencer [45] concluded that a combination of plant-specific subcompartment structures and cavity water showed significant mitigation of loads. Some researchers felt that the observed containment pressurizations were substantially lower than would be expected at full scale because the time scale available for heat and mass transfer is compressed in small-scale experiments.

The early Sandia experiments (SNL/DCH, SNL/TDS, SNL/WC) were conducted at much larger scale (1 : 10) and without any attempt to simulate the compartmentalized nature of real containments. In this way, separate effects information on heat and mass transfer rates, debris velocity, aerosol production, and other separate effects information could be obtained for model development. The effect of containment compartmentalization was crudely simulated in the SNL/LFP test series by placing a simple concrete slab at an adjustable height above the cavity exit. Henry *et al.* [55] also conducted DCH experiments (FAI/DCH) and included simulations of Zion subcompartment structures. These experiments produced DCH loads significantly less than would be predicted by simple bounding models. However, questions persisted on the effects of physical scale.

These early experiments were reviewed as part of an NRC-sponsored effort known as the Severe Accident Scaling Methodology (SASM) program [20]. As a result, experiment programs were redirected toward performing counterpart experiments at two different physical scales (SNL/IET at 1 : 10 scale and ANL/IET at 1 : 40 scale). These experiments, which benefited from a more focused scaling analysis [20], included detailed simulations of the Zion subcompartment structures and initial conditions closely tied to postulated accident scenarios. This testing program in Zion geometry is now complete. No strong effect of physical scale on containment loads or hydrogen production was found. Other efforts, which are complete, include large-scale (1 : 5.75 and 1 : 10) tests conducted at SNL in Surry geometry [41] and small-scale (1 : 40) tests conducted at ANL using fully prototypic melts [52]. Experiments are currently being

TABLE I

SURVEY OF DCH RELEVANT EXPERIMENTS

Experimental series	Number of tests	Nominal scale	Cavity type	Water location
SNL/DCH Tarbell et al. [25, 26] Allen et al. [27]	4	1 : 10	Zion	None
SNL/TDS Allen et al. [28]	7	1 : 10	Surry	None
SNL/LFP Allen et al. [25]	6	1 : 10	Surry	None
SNL/WC Allen et al. [30, 31]	3	1 : 10	Zion	2 Dry 1 w/shallow pool
SNL/IET-Zion Allen et al. [32, 33, 34, 35, 36, 37, 38, 39, 40]	9	1 : 10	Zion	Cavity Cavity/basement
SNL/IET-Surry Blanchat et al. [41]	4	1 : 5.75	Surry	None Cavity/basement
ANL/CWTI Spencer et al. [42, 43] Blomquist et al. [44] Binder and Spencer [45]	10	1 : 30	Zion-like	None Cavity/basement
ANL/IET Binder et al. [46, 47, 48, 49, 50, 51, 52]	6	1 : 40	Zion	None Cavity
ANL/U Binder et al. [53, 54, 52]	3	1 : 40	Zion	None
FAI/DCH Henry et al. [55]	4	1 : 20	Zion	Basement Cavity/basement
FAI/DCH Spain Hammersley et al. [56]	2	1 : 20	ASCo	Wet basement Dry/wet cavities
SNL/CE Blanchat [57]	6	1 : 10	Calvert Cliffs	None Coejected

Experiment series	Driving gas	Driving pressure (MPa)	Melt mass (kg)	Melt composition	Hole size (m)
SNL/DCH	N_2	2.6–6.7	20, 80	$Fe-Al_2O_3$	0.06
SNL/TDS	H_2O	3.7–4.0	80	$Fe-Al_2O_3-Cr$	0.065
SNL/LFP	H_2O	2.5–3.6	50, 80	$Fe-Al_2O_3-Cr$	0.04–0.09
SNL/WC	H_2O	3.8–4.6	50	$Fe-Al_2O_3-Cr$	0.04–0.10
SNL/IET-Zion	H_2O	5.9–7.1	43	$Fe-Al_2O_3-Cr$	0.04
SNL/IET-Surry	H_2O	12	158	$Fe-Al_2O_3-Cr$	0.07–0.098
ANL/CWTI	N_2	4.7–5.0	2.8–5.1	UO_2-ZrO_2-SS	0.13–0.25
ANL/IET	H_2O	5.7–6.7	0.72, 0.82	$Fe-Al_2O_3-Cr$	0.011
ANL/U	H_2O	3.0–6.0	1.13	$UO_2/Zr-ZrO_2-SS$	0.011
FAI/DCH	N_2, H_2O	2.4–3.2	20	$Fe-Al_2O_3$	0.025
FAI/DCH Spain	H_2O	1.9–2.94	10	$Fe-Al_2O_3$	
SNL/CE	N_2, H_2O	8.3	33	$Fe-Al_2O_3$	0.04–0.057

TABLE I

SURVEY OF DCH RELEVANT EXPERIMENTS (CONT.)

Experiment series	Containment pressure (MPa)	Annular gap around RPV	Atmosphere composition	Containment structures
SNL/DCH	0.08	No	Air, Ar	Open containment
SNL/TDS	0.09–0.23	No	Ar	Open containment
SNL/LFP	0.16	No	Ar	Compartmentalized by slab
SNL/WC	0.16	No	Ar	Essentially open
SNL/IET Zion	0.2	No	N_2, N_2–Air, N_2–Air–H_2, CO_2–Air–H_2	Zion subcompartment structures
SNL/IET Surry	0.13–0.19	None, partial, gap w/insulation	Air–H_2O–H_2	Surry subcompartment structures
ANL/CWTI	0.1	No	Ar	Open and compartmentalized by baffle
ANL/IET	0.2	No	N_2, N_2–Air, N_2–Air–H_2, H_2O–Air–H_2	Zion subcompartment structures
ANL/U	0.2	No	N_2–Air–H_2	Zion subcompartment structures
FAI/DCH	0.1	No	Air–N_2	Zion-like subcompartment structures
FAI/DCH Spain	0.1	No	N_2	ASCo or Vandellos subcompartment structure
SNL/CE	0.2	Yes	N_2 Air–steam–H_2	Calvert Cliffs

performed at SNL in annular cavity geometries found in some Combustion Engineering plants [57].

The integral DCH database has revealed only two dominant sensitivities. First, DCH loads were significantly higher in open-geometry experiments that did not simulate the subcompartment structures that will trap debris in many plants. The total steam supply (i.e., RCS pressure) is an important parameter in predicting DCH loads in those experiments that simulated subcompartment structures. The second dominant sensitivity exhibited in the database was hydrogen combustion. Experiments in which DCH-produced hydrogen could burn exhibited significantly higher DCH loads.

The experimental technique employed in the various tests and facilities is basically the same, and this merits a brief summary. Figure 2 illustrates the basic arrangement for the SNL/IET tests, which is representative of all the other tests. The key components are: (1) an accumulator representing the RCS, (2) a melt generator representing the lower head of the RPV, (3) a mockup of the reactor cavity, and (4) an expansion vessel representing the reactor containment building.

For the SNL/IET tests the accumulator was pressurized with steam to values representative of relevant accident sequences. Melt was created with a thermite reaction contained in a crucible placed within a melt generator vessel. The accumulator and melt generator were initially isolated from each other by a set of rupture disks. After ignition of the thermite, the melt generator was pressurized by failing the rupture disks. Upon completion of the thermite reaction, the melt contacted a fusible brass plug in the bottom of the melt generator, causing it to fail. This initiated the high-pressure melt ejection event into the cavity.

The thermite mixture (including chromium) used in these experiments was the same as that used in the large majority of all the subsequent thermite-driven experiments. The prereaction mixture was analyzed chemically and corresponds to a postreaction composition of Al_2O_3–Fe–Cr–Al equal to $0.373 : 0.505 : 0.108 : 0.014$ by weight, assuming complete reaction. Note that the Fe–Cr ratio is about equal to that of the stainless steel comprising the reactor internals. Hence, the chemical reactivity of the

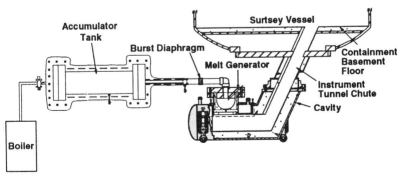

FIG. 2. Representative systems for simulating HPME events in typical DCH experiments.

metal fraction of the melt is comparable to that of molten core debris unless the latter contains significant unoxidized zirconium metal in which case the core debris reactivity would be greater.

The geometry of the reactor cavity varied between tests and facilities depending on the NPP being studied. Figure 2 shows that the cavity was sometimes placed outside the expansion vessel while in other tests (Fig. 3) the cavity was placed inside the containment mockup. The containment mockup sometimes contained very detailed representations of the rooms and compartments existing in the plant. This is illustrated in Fig. 3 for the SNL/IET-Surry tests conducted in the containment technology test facility (CTTF). The CTTF was the largest facility (1 : 5.75 scale) used for DCH testing. To gain perspective on the scale of DCH testing, Fig. 3 compares

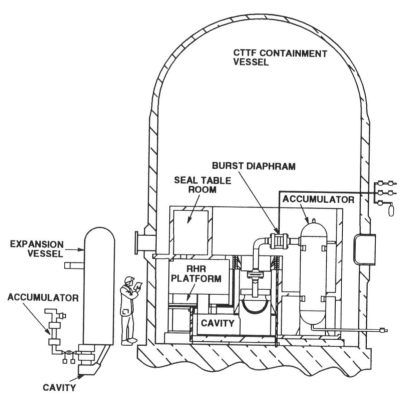

FIG. 3. Comparison of the largest (SNL CTTF facility, 1 : 5.75 scale) and the smallest (ANL COREXIT facility, 1:40 scale) facilities used in DCH integral testing.

CTTF to the smallest facility, the COREXIT facility (1 : 40 scale) used in the ANL/IET tests.

The basic measurements taken in most DCH tests are: (1) containment pressurization, (2) gas composition, and (3) gas temperatures. Consistent and reliable pressurization measurements have been demonstrated in many tests with multiple measurement locations, even on the DCH time scale of one second or less. This is because pressure equilibrates very rapidly (speed of sound), so that any given measurement is representative of the entire compartment volume at any given instant.

This is not the case with gas composition measurements or gas temperatures. These measurements are necessarily a small number of point measurements in space while atmosphere conditions are not well mixed on the DCH time scale (< 1 s). In some of the latter DCH tests, thermocouple arrays were employed to facilitate mole-averaged estimates of the atmosphere (bulk) temperature. It will be noted in the following discussions that different models have adopted different interpretations of the published hydrogen data.

Modeling uncertainties generally dominate uncertainties associated with variations or uncertainties in material properties. In resolution applications, additional consideration of uncertainties in initial and boundary conditions even more clearly dominate material property variations in load models.

A. Summary of DCH Experiments

The overview presented above provides a historical perspective on DCH experimentation. The summary that follows provides a qualitative description of each test series and some of the key conclusions and observations. Many details are omitted that may be found in the cited experiment reports.

1. *Early Exploratory Experiments*

Early experimental investigations of DCH included four experiments performed at Sandia National Laboratories (SNL/DCH series) in the Surtsey facility, five performed at Argonne National Laboratory (ANL/CWTI series) in the COREXIT facility, and four experiments performed at Fauske and Associates, Inc. (FAI/DCH series). The SNL/DCH tests, which were conducted without any representation of subcompartment structures, were intended to provide basic data on heat and mass transfer rates and to quantify aerosol production for source-term analyses. It was found that DCH processes could enhance the radiological source term through vaporization–condensation and mechanical fragmentation processes. Even in open geometry, energy exchange processes did not go to completion. The ANL/CWTI and FAI/DCH tests showed that the compartmentalized nature of containments had a first-order effect in limiting debris mixing with the atmosphere and in reducing containment loads. In the ANL/CWTI-5 and 12 tests, cavity water virtually eliminated heating of the atmosphere, reducing any pressurization to a steam pike. With the exception of one FAI/DCH test, none of these tests employed steam as a driving gas; instead, a chemically inert driving gas (N_2 or Ar) was used. Consequently, hydrogen production resulting from DCH processes did not occur. These experiments provide much useful information which helped to guide subsequent experimental and analytical studies.

2. SNL Technology Development Series (*SNL/TDS*)

The basic purpose of these experiments was to develop the technology for performing experiments using steam-driven thermite melts in the Surtsey facility. In addition, techniques were developed for enhancing melt chemical reactivity by adding chromium metal to the melt in order to better simulate the higher chemical reactivity of molten core debris containing zirconium and steel. The emphasis in these experiments was on technology development, and they were all quite similar in terms of parameters thought to be important to DCH. Realistic plant geometry outside the cavity was not simulated. Atmospheric pressure prior to the event was the primary parameter that was varied. Debris velocities at the cavity exit were measured with breakwires placed at various elevations above the cavity exit.

The pressure rise in the TDS tests increased with increasing atmosphere pressure (prior to the DCH event); however, the efficiency (η = measured ΔP/maximum possible ΔP) of debris–gas interactions was independent of initial atmosphere pressure. Metal oxidation by blowdown steam was not complete, and debris gave up only about half its energy to the atmosphere in these open geometry tests. Debris velocities were insensitive to containment pressure.

3. Limited Flight Path (*SNL/LFP*) Tests

These six experiments were also performed in the Surtsey facility with an inert (argon) atmosphere. As in SNL/TDS, a 1/10-scale model of the Surry cavity and chromium-laced thermite melts were used. The LFP experiments were motivated by the observation that, in the majority (but not all) of U.S. PWR containments, the dominant exit path from the cavity does not communicate directly with the main volume of the upper containment. Instead, the dominant path is often a keyway or instrument tube tunnel which communicates to a compartmentalized lower containment, the structure of which present additional barriers to debris transport to the main volumes of the containment. This compartmentalized lower-containment region is commonly referred to as "the subcompartments."

The purpose of the SNL/LFP tests was to examine sensitivity to the length of unobstructed flight path. This was motivated by a belief that substantially lower pressures were expected in small-scale experiments because the compressed time scale for heat and mass transfer would truncate interactions before their completion. In the LFP series a concrete slab was positioned above the cavity exit chute to limit the unobstructed upward flight of debris dispersed from the cavity. The slab included a vertical steel plate extending downward from the edge of the slab in order to intercept debris splashed horizontally following its initial impact with the slab. The slab effectively blocked direct vertical transport of debris and inhibited horizontal transport, but there was ample space around the edges of the slab to permit unrestricted flow of gases to the volume above the slab. The slab effectively divided the Surtsey volume into a lower compartment and an upper compartment, but in no way were the details of any actual containment geometry simulated.

Two of the LFP tests were performed with the slab 0.91 meter (m) above the cavity exit, three tests with the slab at 1.85 m, and one test (LFP-8A) with the slab at 7.7 m; since the height of the Surtsey vessel is about 10 m, most of the Surtsey vessel volume is below the slab in the latter test and this experiment should be classified as an "open geometry" experiment rather than a "compartmentalized geometry" experiment. In addition to flight path, melt generator hole size was varied. The LFP tests demonstrate that even simple structures are effective at preventing most debris from entering the dome region. The SNL/LFP tests suggest that melt–gas thermal and chemical interactions occur predominantly in the cavity for subcompartment geometries of interest in many NPPs. A factor of four variation in flow area from the RPV mockup in these compartmentalized geometry tests produced only a ~30% increase in loads.

4. Wet Cavity (SNL/WC) Tests

These three experiments were similar to SNL/LFP except that the 1/10-scale Surry cavity was replaced with a 1/10-scale Zion cavity and the concrete slab was at the 7.7 m level; hence, these are "open geometry" experiments. WC-1 and WC-2 were very similar except that WC-2 had water in the cavity. WC-3 was similar to WC-1 except that it had a considerably larger melt generator hole size, resulting in correspondingly more rapid melt ejection, vessel blowdown, and melt dispersal from the cavity. The velocity of debris exiting the cavity was measured with break-wires placed at various elevations above the cavity exit. Pressure rise in the vessel was insensitive to water in the cavity, but ∼30% more hydrogen was produced. In these open-geometry tests, a factor of ∼6 increase in flow area from the melt generator produced negligible differences in pressure rise while increasing the velocity of debris at the cavity exit by a factor of three.

5. SNL Integral Effects Tests, Zion Geometry (SNL/IET Zion)

In these experiments, the thermite melts were ejected into a 1/10-scale model of the Zion cavity that was connected via a chute to the Surtsey vessel. Scale models (1/10-scale) of the Zion lower containment subcompartments and structures were included. The modeling of the Zion lower containment structures was quite detailed, in contrast to previous experiments in which the containment geometries were simple and nonprototypic. The length of the chute connecting the cavity to Surtsey was overscaled by a factor of about 2.7. Some of these experiments were intended to be counterparts to the smaller 1:40 scale ANL/IET experiments, which dictated distortions in the chute length because of geometrical constraints. Comparison of the SNL/ANL counterpart experiments, as well as other experiments at different physical scale, have not revealed any strong effect of physical scale on DCH loads.

The thermite mass of 43 kilogram (kg) was scaled to the "most probable" estimate of melt masses and compositions developed in support of the SASM effort [20] and thus did not represent an attempt to simulate highly conservative or bounding DCH scenarios. In more recent assessments of initial conditions [18], it was concluded that reactor melts would be predominantly oxidic for scenarios of interest to DCH. Thermite melts are half metallic and are thus conservative relative to the more recent NPP assessments.

The Surtsey atmosphere was inert (nitrogen) in the first two SNL/IET experiments. In all others except SNL/IET-5, a nitrogen–air mixture was included with an oxygen content of about 9–10% by volume (consistent with representative reactor accident sequences). The experiments with the nitrogen–air mixture were the first experiments in which DCH-produced hydrogen could burn, as all previous experiments either employed an inert atmosphere or did not include steam as the driving gas. In these experiments, the combustion of DCH-produced hydrogen contributed to the total pressure rise by an amount comparable to or even greater than direct heating of gas by hot debris.

In all these experiments, there was some water in the cavity: 3.48 kg (corresponding to estimated condensate levels) in all cases except SNL/IET-8A and SNL/IET-8B, in which the water amounts were much larger (63 kg). Other experimental parameters studied were the presence or absence of water on the basement floor, presence or absence of preexisting hydrogen in the Surtsey atmosphere, and inerting of the containment atmosphere (in SNL/IET-5). The results indicated that water on the basement floor did not contribute to containment loads, that preexisting hydrogen ($\sim 3\%$) did not contribute to loads, and that most hydrogen combustion was suppressed when the oxygen concentration was less than $\sim 5\%$ (i.e., SNL/IET-5). Melt ejection into a half-flooded cavity produced pressures in the cavity of 2.5 MPa, which may threaten the integrity of certain free-standing cavities.

6. ANL Integral Effects Tests (ANL/IET)

These experiments were designed to be scaled counterparts of the SNL/IET Zion geometry experiments. The linear scale factor was 0.0255 (approximately 1/40) relative to NPP scale. A major purpose of these experiments was the study of scale effects by comparing the results with the results of the larger-scale (1 : 10) SNL/IET experiments. Three of the experiments (ANL/IET-1RR, ANL/IET-3, and ANL/IET-6) were designed to be close counterparts of the corresponding SNL/IET tests. Test observations are similar to what was observed in the SNL/IET/Zion tests. These tests did not reveal any strong effect of physical scale when compared to the counterpart SNL/IET/Zion tests.

7. ANL/U Experiments

This series consisted of three experiments performed in the Zion geometry at 1/40-scale. In the ANL/IET experiments $Fe-Al_2O_3$ melts were used, whereas melts with prototypic core debris compositions (including UO_2 and metallic Zr) were used in the ANL/U experiments. The efficiency of atmosphere heating was somewhat reduced relative to the ANL/IET with $Fe-Al_2O_3$ melts and hydrogen production was somewhat high because the uranium based melts had a higher content of reactive melts. Overall, however, no significant differences with respect to experiments using $Fe-Al_2O_3$ melts were observed.

8. SNL Integral Effects Tests in Surry Geometry (SNL/IET/Surry)

In these experiments scaled models of the Surry NPP cavity and containment structures were used. Three experiments (SNL/IET-9, 10, and 11) were conducted in the Containment Technology Test Facility (CTTF) with a linear scale factor of 1/5.75 relative to the Surry NPP. The fourth experiment, IET-12, was performed at 1/10-scale in the Surtsey facility; although the structures in the latter experiment were replicas of the larger-scale CTTF experiments, the initial conditions were not designed to provide a scaled counterpart of any of the CTTF tests.

The three CTTF experiments are probably the most nearly prototypic of all the DCH experiments that have been performed. In addition to the large scale of these experiments, the atmosphere contained steam rather than the nitrogen diluents (CO_2 in SNL/IET-5) used in the SNL/IET and ANL/IET Zion experiments, and concentrations of preexisting hydrogen ranging from 2.0 to 2.4% were also present. RCS driving pressures of ~ 12 MPa were simulated. Furthermore, the melt generator was located inside the containment facility which permitted the study (in IET-11) of the effect of the annular gap between the reactor pressure vessel (RPV) and the biological shield wall. The IET-11 experiment also simulated the insulation, which was overscaled by a factor of two in mass, that surrounds the RPV and partially fills the annular gap between the biological shield wall and the RPV. These tests demonstrated that the RPV insulation was not an impediment to debris dispersal through the gap directly to the dome and that it may have contributed to hydrogen production. Lastly, Blanchat *et al.* [41] noted that impact of debris against the containment liner showed no signs of compromising its integrity.

9. FAI/DCH-Spain

Two experiments were conducted in cavity designs representing the Vandellos and ASCo nuclear power plants in Spain. The ASCo cavity was half full of water while the Vandellos cavity was dry. Melt ejection produced cavity pressures of ~ 3.0 MPa in the ASCo design. Melt was observed to be dispersed against the containment liner in both cavity designs with no indications of ablation or other threat to integrity. Measured containment pressures were well below containment-threatening levels; however, an overscaled containment volume, inert atmosphere, and no annular gap around the RPV to the dome all contributed to the low pressures in a nonprototypic fashion.

10. SNL/CE Tests

The SNL/CE tests were conducted at 1 : 10 physical scale in the Surtsey facility. The simulations were performed in Calvert Cliffs geometry with detailed representations of the reactor cavity, subcompartment structures, missile shield, and the biological shield wall. The tests had two main objectives: (1) to examine DCH in plants with annular cavities, and (2) to examine the impact of coejected water (i.e., coejected from the RCS with debris) on DCH loads.

Some Combustion Engineering plants with Bechtel annular cavities are quite small and any debris dispersal will occur predominantly through the annulus around the reactor pressure vessel. This annulus leads directly to the refueling pool and the containment dome. Low-temperature tests [58] had previously shown that dispersal would be nearly complete and that the annular gap was the dominant dispersal pathway in these geometries. The SNL/CE tests confirmed these observations using 33 kg of high-temperature iron–alumina thermite as the melt simulant. For tests with nonreactive atmospheres the observed DCH pressure rise (~ 0.3 MPa) was substantially higher than similar experiments (~ 0.1 MPa) conducted in Zion geometry (SNL/IET tests), where the primary dispersal pathway is into the containment basement containing the steam generators.

Many DCH-relevant scenarios would have substantial quantities of water in the RCS at the time of lower head failure. In such situations, this water would be coejected with melt from the RPV. As a practical matter, the coejection of melt with water could not be achieved experimentally, so the melt was prepared in the cavity and then subjected to the blowdown of flashing water or steam. The SNL/CES-1,3 experiments employed 100 kg of water (~ 300 K in CES-1 and saturated in CES-3) in the blowdown. CES-2 used steam as the driving fluid with no coejected water. The observed DCH loads and hydrogen combustion were comparable in the steam-driven and saturated-water-driven tests. The test with cold water resulted in $\sim 30\%$ less pressure rise and hydrogen production. Post-DCH temperatures dropped rapidly in the saturated-water test, less rapidly in the subcooled-water test, and only very slowly in the steam-driven test. DCH loads in the steam-driven and saturated-water-driven tests were essentially identical, suggesting that any potential pressurization from flashing water was not additive to the DCH load alone. Ongoing testing will employ reactive atmospheres.

III. Overview of Modeling Tools

This section summarizes NRC-sponsored and industry-sponsored analytical tools and calculation methodologies for quantifying containment loads in DCH events. The phenomenological complexities of containment pressurization during a DCH event have resulted in the development of a variety of containment analysis tools as listed in Table II. These methodologies quantify containment loads by representing DCH at various levels of detail, ranging from simple adiabatic equilibrium methods to modeling the physics associated with individual droplets. This table indicates the intended application, references for the basic model, and references for plant applications if available. A brief description of each of the models or codes that currently can be used to calculate DCH loads is given below. Implicit in this selection is an attempt to provide integral validation of the model or code against a significant portion of the DCH database discussed in Sect. II. MAAP is an exception in that it has been validated against only a limited portion of the database, but it has been used extensively by industry throughout the world.

Simple physics models attempt to represent or bound only the dominant processes contributing to DCH loads. There is no claim that they capture every detail of DCH phenomenology. The appeal of these models lies in their simplicity. The success of these models stems from the relatively simple systematics exhibited by the DCH database (Sect. II.B). When coded, these models can be conveniently and economically evaluated thousands of times to create a distribution of containment loads when coupled with a Monte-Carlo sampling of uncertainties in initial conditions and certain phenomenological parameters represented in the model. Pilch *et al.* [18, 23] have used the TCE and CLCH models in this fashion to resolve the DCH issue for the Zion and Surry NPPs. System-level codes, such as CONTAIN or CONTAIN/CORDE, attempt to represent a broader spectrum of DCH relevant phenomena with kinetics models, many of which are reviewed in Sect. IV through X. The treatment of DCH phenomena in the system-level codes MELCOR and MAAP are intermediate between simple physics models and CONTAIN.

A. SINGLE-CELL EQUILIBRIUM (SCE) MODEL

Single-cell equilibrium models were the first DCH models. Their utility in reactor applications is limited because they often predict pressures that are unrealistically high. The single-cell equilibrium (SCE) model of Pilch [23] is described here as being representative of all such models. The SCE model treats the entire containment volume as a single control volume in

TABLE II

SUMMARY OF DCH ANALYSIS METHODOLOGIES AND CODES

DCH analysis methodologies/codes	Basic characteristics	Model/code references	Plant analysis references
DHEAT2 model	A single-cell equilibrium model, debris–gas thermal interactions, chemical interactions, and water interactions	Williams [59]	
Single-cell equilibrium (SCE) model	Debris–gas thermal and chemical equilibrium	Pilch [60]	Pilch [61]
EBAL model	Debris–gas thermal and chemical equilibrium	Ginsberg and Tutu [62]	
Thermodynamic equilibrium model	Debris–gas thermal equilibrium and parametric chemical reactions with water interactions	Nourbakash et al. [63]	
Two-cell equilibrium (TCE)	Debris–gas thermal and chemical equilibrium treated separately in cavity (and sub-compartments) and upper dome	Pilch et al. [18]	Pilch et al. [18, 23] Pilch et al. [19] Schneider and Sherry [64]
Two-cell kinetic model	Extension of TCE allowing for incomplete thermal and chemical interactions through efficiencies related to interaction times	Pilch [61]	
PARSEC model	Transient stand-alone model with cavity interactions acting as source for interactions in dome; debris–gas interactions in dome tracked for groups of particles	Sienicki and Spencer [65]	
KIVA-DCH code	Sophisticated hydrodynamic code with particle tracking capability; debris–gas interactions in dome tracked for groups of particles	Tarbell et al. [26] Marx [66] Tarbell et al. [67] Sweet et al. [68]	
DIRHET model	Single-control-volume framework for debris–gas interactions	Corradini et al. [69]	
DCHCVIM model	Transient debris–gas interactions tracked for groups of particles in a single volume	Ginsberg and Tutu [70]	

continues

233

TABLE II

Summary of DCH Analysis Methodologies and Codes (*Cont.*)

DCH analysis methodologies/codes	Basic characteristics	Model/code references	Plant analysis references
Darwish code	Multivolume-control-volume code specific to DCH analyses	Darwish [71]	
Convection Limited Containment Heating (CLCH) model	Debris–gas thermal and chemical equilibrium treated as a flow process in the cavity	Yan and Theofanous [72]	Pilch et al. [23]
CONTAIN code -1.2 w/CORDE -1.12 Multifield -1.2 Standard prescription	System-level, multicell control volume code with debris–gas interactions tracked for groups of particles	Bergeron et al. [73] Gido et al. [74] Washington and Williams [75] Williams et al. [22]	Bergeron et al. [73] Williams et al. [76] Williams and Louie [77] Williams and Gregory [78] Gido et al. [79] Williams [80] Tutu et al. [81] Murata and Louie [82]
MAAP3B code MAAP 4 code	System-level code with debris–gas equilibrium assumed in containment cells where debris is dispersed	EPRI [83] EPRI [84]	Carter et al. [85] Fontana et al. [86]
CONTAIN/CORDE code	UK enhancement of CONTAIN code; CORDE provides phenomenological models for RPV discharge and cavity dispersal phenomena	Morris and MacBeth [87] Morris and Roberts [88] Morris [89]	Lowenhielm et al. [90] Gustavsson [91] Sweet and Roberts [92]
FUMO code	System-level, multicell control volume code	Fruttuoso et al. [93]	
MELCOR code	System-level, multicell control volume code; transient DCH interactions in cell with user-specified time constants	Summers [94]	Kmetyk [21] Kmetyk [95]

which heat transfer to structures is excluded. Water is the only potential energy sink that is sometimes considered in single-cell equilibrium models. The dispersed debris is assumed to mix completely with the entire containment atmosphere and to remain airborne long enough to enable all thermal and chemical interactions to come to equilibrium.

The maximum pressure rise in the containment resulting from DCH is obtained by combining the caloric equation of state for the atmosphere internal energy with the ideal gas law and the definition of γ as the ratio of gas specific heats. Doing so yields

$$\frac{\Delta P}{P^0} = \frac{\Delta U}{U^0} = \frac{\Sigma \, \Delta E_i}{U^0 (1 + \psi)} \quad \text{or} \quad \Delta P = \frac{\gamma - 1}{V} \frac{\Sigma \, \Delta E_i}{1 + \psi}. \tag{1}$$

This is the working equation for the single-cell equilibrium model.

The heat capacity ratio ψ, as defined in Table III, is important because $1/(1 + \psi)$ represents the fraction of debris thermal energy that can be transferred to the gas before debris–gas thermal equilibrium is achieved. On a containment-wide basis, ψ is usually small and most debris energy can be transferred to the atmosphere. The fraction $1/(1 + \psi)$ also appears when characterizing local debris–gas heat transfer (e.g., in the cavity). Locally ψ can be quite large (~ 10), indicating that melt can liberate only a small fraction of its energy in order to achieve local thermal equilibrium.

Some representative energy terms for the SCE model, which are easily adapted to any specific problem, are listed in Table III. A brief description of these terms follows. Blowdown of the RCS adds both mass and energy to the atmosphere. Steam blowdown occurs when the RCS coolant inventory is boiled away during core melt progression. Flashing blowdown can occur if the operator refloods the RCS with water (as occurred at TMI-II) or if residual water remains in the RCS at the time of vessel failure. Isentropic blowdown of the RCS inventory to containment conditions determines the amount of water that flashes to steam. Debris carries both latent and sensible heats that can be transferred to the atmosphere. Oxidation of the metallic components of debris increases the sensible heat of the debris. The analyst must specify whether oxidation reactions occur with steam or oxygen, and steam or oxygen limitations to metal oxidation can be easily included. Oxidation of the metallic components of debris can produce hydrogen that can subsequently burn. Core melt progression is accompanied by oxidation of clad material, producing hydrogen that is then released from the RCS into the containment prior to failure of the lower head. This preexisting hydrogen can also burn under some conditions. Oxygen limitations to hydrogen combustion are usually taken into account. DCH processes can also vaporize water, which adds both mass

TABLE III

ENERGY TERMS FOR SCE MODEL

Term	Expression	Representative values for reactor application
Steam blowdown RCS full of steam	$\Delta E_b = \dfrac{(PV)^0_{RCS}}{\gamma - 1}$	
Flashing blowdown RCS full of water	$\Delta E_b = M_{w,RCS} X_g h_g$	$M_{w,RCS} \sim 290$ mt-w $h_g = 0.744$ MJ/kg-g $X_g \sim 0.2 - 0.3$
Latent and sensible heat of debris	$\Delta E_t = M_d \Delta e_d$ $= \Sigma N_{d,i}\{C_p(T_d - T_a^0) + h_{js}\}_i$	$\Delta e_d \sim 1.3$ MJ/kg-d
Metal oxidation	$\Delta E_r = M_d \Delta e_r$ $= \Sigma M_{d,i} \Delta h_{r,i}$	$\Delta e_r \sim 0.20$ MJ/kg-d
Combustion of DCH produced hydrogen	$\Delta E_{H_2} = v_d M_d \Delta e_{H_2}$ $= \Delta e_{H_2} \Sigma (v_d M_d)_i$	$v_d \sim 1.26$ mole-H_2/kg-d $\Delta e_{H_2} = 0.24$ MJ/mole-H_2
Combustion of preexisting hydrogen	$\Delta E_{H_2} = X_{H_2} N_a^0 \Delta e_{H_2}$	$X_{H_2} \sim 0.04$
		$N_a^0 \sim 4.4 \times 10^6$ mole
Water vaporization water–energy limits	$\Delta E_w = -M_w(h_{fg} - h_g)$ $M_w = \min \left\{ \begin{array}{c} M_{w,cav} + M_{w,spray} \\ + (1 - X_g) M_{w,RCS} \cdot \dfrac{\Delta E_{DCH}}{h_{fg}} \end{array} \right\}$	$h_{fg} = 2.26$ MJ/kg-w $h_{fg} = 2.26$ MJ/kg-w $h_g \sim 0.744$ MJ/kg-w
Heat capacity ratio	$\psi = \dfrac{M_d C_d}{(N_a^0 + N_b + N_w)C_v}$	$C_d \sim 524$ J/kg/K $C_v \sim 28$ J/mole/K

and energy to the atmosphere; however, the penalty for vaporization is energy that otherwise would heat the atmosphere. The amount of vaporization can be limited either by the availability of water or by the amount of debris energy. The most important sources of water are water codispersed from the cavity with the debris, water suspended in the atmosphere resulting from the operation of containment sprays, and water coejected from the RPV (liquid component of flashing blowdown) with debris.

The bounding nature of the SCE model is seen in Fig. 4, which provides an assessment of the single-cell equilibrium model against the entire DCH database. Two statistical measures are shown on the plot: the relative bias and the relative RMS error (standard deviation) referenced to the bias

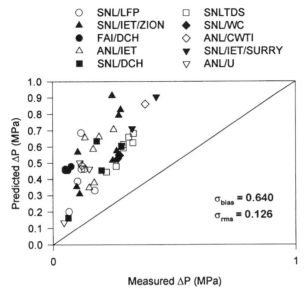

FIG. 4. Comparison of the single-cell equilibrium model with the existing DCH database.

line:

$$e_{\text{bias}} = \frac{\sum \left(\dfrac{\Delta P_{\text{pred},i} - \Delta P_{\text{meas},i}}{\Delta P_{\text{pred},i}} \right)}{N},$$

$$\sigma_{\text{rms}}^2 = \frac{\sum \left(\dfrac{\Delta P_{\text{pred},i} - \Delta P_{\text{meas},i}}{\Delta P_{\text{pred},i}} - e_{\text{bias}} \right)^2}{N-1}.$$

(2)

The SCE model predicts a containment-threatening pressure rise of ~ 0.70 MPa, which adds to an initial containment pressure of ~ 0.2 MPa, for a representative accident sequence in a PWR. Blowdown of RCS gases contributes $\sim 5\%$, latent and sensible heat in the debris contribute $\sim 51\%$, debris oxidation contributes $\sim 8\%$, combustion of DCH-produced hydrogen contributes $\sim 12\%$, and combustion of preexisting hydrogen contributes $\sim 24\%$ to the pressure rise. Recognizing the margin (exhibited by the DCH database) in SCE predictions, Sects. IV to X of this chapter address processes that have the potential to mitigate DCH loads. These processes include : (a) hydrodynamic processes that can limit the amount of core debris participating in DCH or hydrodynamic processes that can limit efficient mixing of the core debris with the entire containment atmosphere, (b) kinetic processes that can limit the amount of thermal and chemical energy that can be transferred from the debris to the atmosphere on the time scale of peak DCH loads, (c) kinetic or thermodynamic processes that can limit the amount of DCH-produced hydrogen that can be produced, (d) processes that can limit the combustion of DCH-produced hydrogen or preexisting hydrogen on DCH time scales, and (e) energy sinks (e.g., heat transfer to structures and vaporization of water) that drain energy from the atmosphere on DCH time scales. These processes are reflected to varying degrees in the models and codes summarized in the remainder of Sect. III.

B. Two-Cell Equilibrium Model (TCE)

The TCE model is a simple physics model that attempts to represent the dominant processes contributing to DCH loads. The TCE model extends the single-cell equilibrium model by accounting for the major effects of containment compartmentalization, which prevents efficient mixing of airborne debris with the entire atmosphere by confining the bulk of the debris to the subcompartment of the containment.

The TCE model divides the containment into two compartments: one for the cavity and subcompartments, and one for the containment upper dome. DCH governing phenomena are assumed to proceed independently in the two volumes. Therefore, the calculated loads are proportional to the sum of the energy given up in the two compartments. The primary heat sink for debris in the subcompartment is that portion of the blowdown gas that is coherent with the dispersal processes. The debris ejected from the RPV and dispersed from the cavity is partitioned between the subcompartment and dome to account for flow paths where the debris is carried to the containment upper dome without interacting significantly with the subcompartment (e.g., through the RPV annular gap). Debris dispersed into the subcompartment thermally equilibrates with blowdown steam, while debris dispersed to the dome thermally equilibrates with the dome atmosphere. The containment is treated as adiabatic for these processes. Metal–steam reactions that produce hydrogen during DCH are modeled in a hierarchical fashion (zirconium first, then steel) with thermodynamic limits applied to iron oxidation. Metal oxidation may be further limited by the amount of blowdown steam that is coherent with the debris dispersal process. All DCH-produced hydrogen is assumed to burn adiabatically as a hot hydrogen jet as it vents to the upper dome. Preexisting hydrogen in the containment atmosphere can burn as a deflagration (if the appropriate conditions exist) or as an autoignition event if DCH heats the atmosphere sufficiently. The energy released from a possible deflagration contributes to enhanced DCH loads only if the combustion rate exceeds the heat transfer rate to structures.

TCE avoids overly conservative loads predictions by taking credit for mitigating processes including: (1) accounting for limited combustion of preexisting hydrogen that occurs on DCH time scales, (2) confining most of the debris dispersed from the cavity to the subcompartment, and (3) limiting the amount of blowdown steam participating in DCH (characterized by the coherence ratio). The TCE model is used in conjunction with a phenomenologically based correlation for the coherence ratio, a model for the fraction of debris transported through the annular gap around the RPV to the dome, and a model for the extent of hydrogen combustion on DCH time scales.

The TCE model can be written as the product of an efficiency ($\eta = \eta_1 + \eta_2$) and the loads predicted by the SCE model,

$$\frac{\Delta P}{P^0} = (\eta_1 + \eta_2)\left(\frac{\Delta P}{P^0}\right)_{SCE} \quad \text{or} \quad \Delta P = (\eta_1 + \eta_2)\frac{\gamma - 1}{V}\frac{\Sigma \Delta E_i}{1 + \psi}. \quad (3)$$

The TCE model is analogous to a batch process where DCH interactions are assumed to occur independently in the subcompartment and the dome. Table IV lists the working formulas for the efficiencies of DCH processes in the subcompartment (η_1) and the dome (η_2). The fraction of cavity debris dispersed into the upper dome is given by Eq. (21) in Sect. VI. The efficiencies account for the dominant mitigating effects of compartmentalization (small value of f_{a2}) and the noncoherence of dispersal and RCS blowdown (large value of ψ_1). Physically, this corresponds to a small amount of debris energy being transferred to blowdown or subcompartment gases in order to achieve thermal equilibrium.

The TCE model is validated against the extensive DCH database in Fig. 5 for tests employing compartmentalized geometry. In open geometries, TCE tends to yield overly conservative results primarily because DCH interactions are limited by scale in open-geometry experiments. Integral validation has been performed at the 1:5.75, 1:10, 1:20, 1:30, and the 1:40 physical scales. Pilch et al. [23, App. E] argued that only experiments with inert atmospheres and no cavity or basement water should be used for validating model predictions of hydrogen produced on DCH time scales; for these experiments, (SNL/TDS, SNL/LFP, and two of the SNL/WC tests) the TCE model is adequately validated ($e_{bias} = -0.04$, $\sigma_{rms} = 0.11$). The TCE model underpredicts hydrogen data by a factor of ~ 2 for those tests that had reactive atmospheres or water in the cavity. Pilch et al. [23, App. E] have offered explanations for these discrepancies.

TABLE IV

EFFICIENCIES OF DCH PROCESSES IN THE SUBCOMPARTMENTS AND UPPER DOME
AS REPRESENTED BY THE TCE MODEL

Subcompartment

$$\eta_1 = f_{a1} \frac{1 + \psi}{1 + \psi_1} \left[1 - \left(\frac{\Delta E_{H_2}}{\Sigma \Delta E_i} \right)_{SCE} - \left(\frac{\Delta E_t}{\Sigma \Delta E_i} \right) \frac{T_1 - T_r}{T_d^0 - T_r} \right]$$

$$\psi_1 = \frac{f_{a1} N_d C_d}{\max \left[f_{a1} f_{coh} N_{RCS}^0 ; f_{v1} N^0 \right] C_v} \qquad T_1 = \frac{f_{a1} f_{coh} N_{RCS}^0 T_{RCS}^0 + f_{v1} N^0 T^0}{f_{a1} f_{coh} N_{RCS}^0 + f_{v1} N^0}$$

Dome

$$\eta_2 = f_{a2} \frac{1 + \psi}{1 + \psi_2} \left[1 - \left(1 - \frac{f_{burn}}{f_{a2}} \right) \left(\frac{\Delta E_{H_2}}{\Sigma \Delta E_i} \right)_{SCE} - \left(\frac{\Delta E_t}{\Sigma \Delta E_i} \right)_{SCE} \frac{T_2 - T_r}{T_d^0 - T_r} \right]$$

$$\psi_2 = \frac{f_{a2} N_d C_d}{\max \left[f_{a2} f_{coh} N_{RCS}^0 ; f_{v2} N^0 \right] C_v}$$

$$T_2 = \frac{f_{a2} f_{coh} N_{RSC}^0 T_{RCS}^0 + f_{v2} N^0 T^0}{f_{a2} f_{coh} N_{RCS}^0 + f_{v2} N^0} \qquad f_{burn} = \frac{N_{H_2, burn}}{\left(N_{H_2, burn} \right)_{SCE}}$$

○ SNL/LFP
▲ SNL/IET/ZION   ◇ ANL/CWTI
● FAI/DCH   ▼ SNL/IET/SURRY
△ ANL/IET   ▽ ANL/U

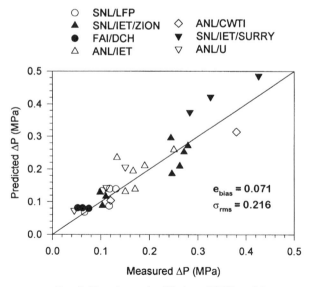

$e_{bias} = 0.071$

$\sigma_{rms} = 0.216$

FIG. 5. Experimental validation of TCE model.

C. CONVECTION-LIMITED CONTAINMENT HEATING (CLCH) MODEL

CLCH is a simple physics model that captures the basic physics as a flow process where debris–gas thermal and chemical equilibrium are assumed in the cavity. The CLCH model divides the containment into two compartments and treats the governing phenomena of DCH loading in these compartments independent of each other. The major feature mitigating DCH loads is the implicit assumption that debris is quickly trapped in the subcompartment. A correlation model for the coherence of debris dispersal and blowdown is provided. Metal–steam reactions that produce hydrogen are modeled in a hierarchical fashion with thermodynamic limits to iron oxidation. DCH-produced hydrogen is assumed to transport to the dome, where it burns as a diffusion flame. Combustion of preexisting hydrogen can occur as an autoignition event if DCH first heats the dome to some specified temperature.

The basic working equations for the CLCH model are given in Table V. Yan and Theofanous [72] give universal solutions in dimensionless form. They also present maps of DCH loads as a function of the coherence ratio, which illustrate the potential importance of autoignition events. CLCH and TCE predicted similar containment loads in an application to the Zion plant [23].

Integral validation of the CLCH model against the DCH database is shown in Fig. 6. In the CLCH model, hydrogen production and ΔP were both brought into approximate agreement by postulating that hydrogen produced by airborne debris reacting with coherent steam was augmented by two additional sources: debris remaining in the cavity was assumed to react with noncoherent blowdown steam, and debris dispersed beyond the subcompartments was assumed to react in the dome. With this interpretation, the CLCH model compared favorably with hydrogen production data ($e_{bias} = 0.07$, $\sigma_{rms} = 0.23$). In order to maintain agreement with ΔP, it was necessary to assume that neither source of additional hydrogen could burn in the DCH timeframe unless autoignition temperatures in the dome were exceeded, which did not happen in the experimental analyses. The gas analysis results implied that most of the hydrogen produced did burn; hence this interpretation also requires the assumption that substantial reaction occurred too late to affect DCH loads.

D. CONTAIN CODE

CONTAIN is a system-level computer code that has been developed for the U.S. Nuclear Regulatory Commission as a detailed analysis tool for evaluating containment response to severe reactor accidents. Modeling of

TABLE V

THE CLCH MODEL

(Vessel)	

$$\frac{P_{s,v}}{P_{s0,v}} = \left\{ 1 + \frac{\gamma - 1}{2} \Gamma \frac{t}{\tau_s} \right\}^{2\gamma/(1-\gamma)}$$

$$\dot{m}_{s,b} = \eta A_b \rho_{s0,v} a_{s0} \Gamma \left\{ 1 + \frac{\gamma - 1}{2} \Gamma \frac{t}{\tau_s} \right\}^{(\gamma+1)/(1-\gamma)}$$

$$T_{s,v} = T_{s0,v} \left(\frac{P_{s,v}}{P_{s0,v}} \right)^{(\gamma-1)/\gamma}$$

where

$$a_{s0} = \sqrt{\gamma R_s T_{s0,v}}, \Gamma = \left[\frac{2}{\gamma + 1} \right]^{(\gamma+1)/2(\gamma-1)} \text{ and } \tau_s = \frac{V_v}{\eta A_b a_{s0}}$$

(Cavity)

$$\dot{m}_m = \begin{cases} \dfrac{m_{m0}}{\tau_M} & t \le \tau_M; \\ 0 & t > \tau_M. \end{cases}$$

$$T_{s,c} = \frac{\dot{m}_m C_{p,m} T_m + \dot{m}_{s,b} C_{p,s} T_{s,v} + \Delta H_r \dot{m}_m \omega_{Zr}}{\dot{m}_m C_{p,m} + \dot{m}_{s,b} C_{p,s}} \quad t \le \tau_M.$$

(Containment)

$$(m_{s,a} C_{v,s} + m_a C_{v,a}) \frac{dT_a}{dt} = \dot{m}_s C_{p,s} T_{s,c} - \dot{m}_s C_{v,s} T_a t \le \tau_M,$$

$$T_{a,f} = \frac{m_{s,c}(\tau_M) C_{v,s} T_{s,v}(\tau_M) + m_{s,a}(\tau_M) C_{v,s} T_a(\tau_M) + m_a C_{v,a} T_a(\tau_M) + \Delta H_{H_2} m_{H_2}}{m_{s0} C_{v,s} + m_a C_{c,a}} t > \tau_M,$$

$$P_{a,f} = \left(\frac{m_{s0,v}}{V_v + V_a} R_s + \frac{m_a}{V_v + V_a} R_a \right) T_{a,f}.$$

Note. Nomenclature of Yan and Theofanous [72] used here.

DCH has been an important focus of the CONTAIN development program. The basic design strategy for CONTAIN has been to include detailed mechanistic models for phenomena that are sufficiently well understood to justify definition of a mechanistic model. Where phenomenological understanding is adequate for mechanistic model development, the approach used is to provide simple models and input flexibility to permit studying the effect of modeling uncertainties upon the results of interest.

CONTAIN is a lumped-parameter control-volume code. In typical analyses the containment is divided into a number of control volumes, commonly called "cells," with gas flows between the cells being calculated

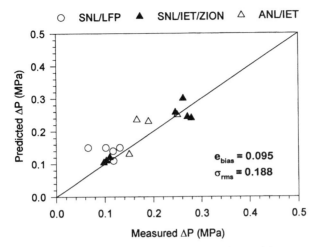

FIG. 6. Experimental validation of CLCH model.

using simple orifice flow models. Momentum of flow entering a cell is assumed to be dissipated within the receiver cell, that is momentum advection through interconnected cells is not modeled. Hence the code cannot model momentum-dominated jet effects. It is also unable to model gas stratification reliably in open volumes.

Models for DCH processes are defined in terms of time-dependent models for the rates of the controlling processes rather than in terms of the integral energy transfers estimated from equilibrium considerations in the simpler models described previously. The most recently documented version of the DCH model [75] includes relatively detailed models for the debris–gas heat transfer and chemical reaction, hydrogen combustion, debris transport and deentrainment due to debris–structure interactions, and atmosphere-structure heat transfer. More recent NRC sponsored versions of CONTAIN include models for melt ejection from the RPV and entrainment and fragmentation of melt in the cavity. However, the phenomena controlling these processes are considered especially uncertain and have not undergone validation against the more recent separate effects database.

Great Britain has independently developed DCH-specific RPV and cavity models that they call CONTAIN/CORDE. The CORDE models have undergone some limited separate effects validation. The CONTAIN/CORDE codes have been used extensively in Europe.

Modeling uncertainties are substantial for many of the phenomena related to DCH that CONTAIN attempts to model. CONTAIN DCH calculations are often presented as a range of containment loads where various assumptions are made regarding the uncertain phenomena. Predicted DCH loads, however, are usually sensitive to only a few of the phenomenological uncertainties in any given application.

Because CONTAIN models rates of debris–gas interactions rather than making equilibrium assumptions and because CONTAIN models mitigating effects such as atmosphere–structure heat transfer that are neglected by the simpler models, it is commonly expected that CONTAIN will predict lower loads than the simpler models. However, CONTAIN also includes models for possible contributors to DCH that are neglected by the simpler models, including interactions of nonairborne debris and some of the effects of debris–water interactions. Hence it does not always follow that CONTAIN will calculate lower loads than the simpler models, especially in sensitivity studies designed to provide a reasonably conservative allowance for modeling uncertainties. Limited comparisons of CONTAIN and TCE are given by Pilch *et al.* [18].

The CONTAIN code provides an integral analysis tool that includes phenomenological models for many of the individual processes that govern DCH, and it can therefore provide insights as to the importance of these processes in controlling the overall containment response. These insights are valuable because the harsh DCH environment generally precludes obtaining experimental information at this level of detail. It is obvious, however, that any such insights from the CONTAIN analyses are only as good as the models. The CONTAIN code has recently been subjected to an extensive independent peer review [96] that provides considerable insight as to how the code results should be interpreted. In addition to evaluating the CONTAIN models themselves, the peer review also evaluated the state of current physical understanding of the various phenomena involved, and these results are of interest in a broader context than just the CONTAIN code.

Integral validation of CONTAIN against the DCH database [22] is shown in Fig. 7. The figure shows that the CONTAIN model does a good job of accounting for the various differences between the Surry-geometry, Zion-geometry, and open-geometry experiments. It also reproduces the effects of hydrogen combustion well in the SNL/IET experiments, and comparison with the ANL/IET results reveals no major scale distortions in the model. The poorest results were obtained for some of the LFP experiments, in which the code overpredicted debris transport to the dome and hence overpredicted ΔP.

CONTAIN predictions for hydrogen production ($e_{bias} = -0.029$, $\sigma_{rms} = 0.174$) are in good agreement with all the tests considered in its assessment. Williams *et al* [22] considered the ability to reproduce both the ΔP

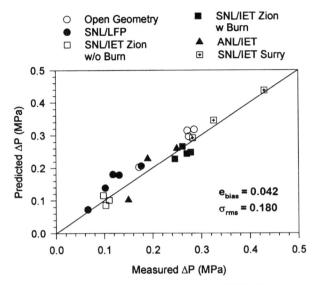

FIG. 7. Experimental validation of CONTAIN.

data and the hydrogen production data to be important to CONTAIN validation because rates of chemical reactions producing hydrogen are calculated using a heat–mass transfer analogy similar to that described in Sect. VII.C. Debris–gas heat transfer and hydrogen production are therefore closely coupled in this model, and this relationship was not found to be sensitive to most of the acknowledged modeling uncertainties.

E. MELCOR

MELCOR is a system-level computer code whose phenomenological level of DCH modeling is intermediate between CONTAIN and TCE. In the MELCOR code, dispersed debris is initially partitioned among suspension in one or more compartment atmospheres and direct deposition (by impaction) to one or more surfaces (including that of a debris pool, if present) under user control. If there is a water pool in the compartment into which the debris is injected, water from that pool will be converted to water droplets in the atmosphere at a rate proportional to the debris injection rate; the constant of proportionality is user-controllable. The atmosphere of the compartments can communicate with each other through flow paths, and suspended water will be transported with the gases.

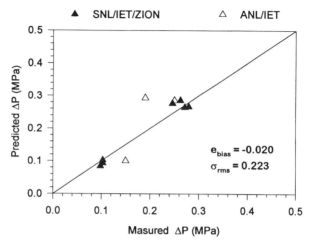

FIG. 8. Experimental validation of MELCOR.

Hydrogen combustion can be simulated by selecting appropriate model parameters.

Debris initially allocated to a compartment atmosphere does not further transport among compartments, but interactions within the compartment are considered. Suspended debris can exchange thermal energy with the atmosphere, any metals it contains can oxidize, and it can settle onto a specified surface (or into a specified debris pool) in the volume; each process proceeds at a rate governed by a user-supplied time constant. Both steam and oxygen are included in the oxidation calculation.

When debris is deposited onto a surface, whether by initial deposition or by settling, its sensible heat is rapidly transferred to that surface. Any remaining metal content can continue to oxidize, governed by a user-supplied time constant. The resulting chemical heat and any decay heat associated with the debris is communicated as a source to the surface, and the thermal response of the structure is calculated by solving an appropriate conduction equation.

Assessment of the MELCOR code against the DCH database [95] is given in Fig. 8. Loads predictions are consistent with the DCH database. MELCOR generally overpredicts the amount of hydrogen produced (e_{bias} = 0.17, σ_{rms} = 0.12). For the SNL/IET tests with reactive atmospheres, MELCOR predicts that ~ 31% of the apparent hydrogen production is actually direct oxygen uptake by the dispersed debris. Assessments indicate that the various parameters and time constants used in the MELCOR code are difficult to select for geometries where experimental data are not available.

F. MAAP Code

The Modular Accident Analysis Program (MAAP) is an industry sponsored integral systems code to assess the response of nuclear power plants to potential accident situations. MAAP had its origin in the analyses performed for the Zion Probabilistic Safety Study [2] and the IDCOR program [97] and early versions (MAAP 1.0, 2.0, and 2B) considered conditions resulting from a high-pressure melt ejection in a manner similar to which they were addressed in the ZPSS.

Many of the U.S. utilities, in addition to several utilities in Europe and the Far East, used the MAAP 3B program to support their individual plant examinations (IPEs). The MAAP 3B model evaluates the potential for melt dispersal from the reactor cavity through an entrainment model based on a Kutatuladze criterion, with the constant being determined by ANL experiments. Once debris is dispersed from the reactor cavity, it is assumed to heat the atmosphere within the first containment subcompartment, which is generally the steam generator compartment. As this containment region comes to equilibrium with the core debris, natural circulation will transport some but not all of the energy in the debris to other parts of the containment. This modeling approach is documented in the MAAP 3.0B User's Manual [83]. Predicted containment pressurization due to HPME/DCH does not approach the design basis pressure in typical plant evaluations.

The MAAP 4 approach [84] to assessing debris dispersal from the reactor cavity–instrument tunnel uses the integral model proposed by Henry [98] to determine the extent of debris that was entrained and which was removed from the reactor cavity more as a liquid film than entrained droplets. This differentiation is important in representing the extent of material that would be available for distribution throughout the containment atmosphere for direct heating as well as that fraction of the melt that would have a small characteristic dimension that could influence the release of fission products while this finely particulated melt is at elevated temperatures. Calculated results were compared with the early SNL/DCH experiments, including those that had simulations of the fission products in the iron thermite melt. This model gave reasonable representations of the vessel pressurization found in these initial Surtsey tests as well as reasonable characterization of the fission products released during the time that some of the debris could be finely particulated. Systematic validation against the much broader DCH database has not been attempted.

The MAAP 4 model predicts containment pressurization comparable to MAAP 3B.

IV. Discharge Phenomena

Discharge of the molten material from the RPV is initiated when the lower head fails by a thermally induced rupture (~ 0.4 m) [23] or by expulsion of an incore instrument guide tube from the lower head (~ 0.03 m) [1]. Molten core material is considerably hotter than the melting point of RPV steel; consequently, corium discharge could be accompanied by rapid ablation of the hole that was initially formed in the lower head. Hole ablation not only accelerates corium discharge, but the resulting hole size has a direct influence on RCS blowdown.

For a failure on the bottom of the RPV, single-phase corium discharge progresses until, at some critical liquid level, a funnel-like depression forms over the outlet allowing gas to vent simultaneously with the remaining melt (Fig. 9). This is termed gas blowthrough, which has in the past been identified as a mechanism that potentially could allow complete blowdown of the RCS before all the melt is ejected into the cavity. However, experimental data and analysis now suggest that nearly complete ejection of melt (at high pressure) can be expected if the failure site is at the bottom of the RPV. Side failure of the RPV is also possible (Fig. 10). Under these conditions, melt could be entrained out the hole until the level drops to some critical distance below the failure site, leading to permanent retention of significant melt quantities in the RPV. The melt mass remaining in the RPV at the end of RCS blowdown is important

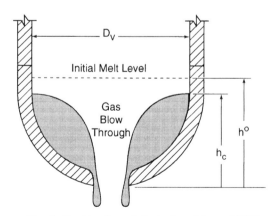

FIG. 9. Gas blowthrough for bottom failure of RPV.

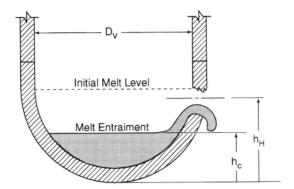

FIG. 10. Melt entrainment through a side failure.

because this material (although it may ultimately drain into the cavity) will never be subjected to the dispersive influence of the blowdown gases in the cavity. Consequently, this material may not contribute significantly to DCH.

Blowdown of RCS gas through failure in the lower head governs both debris dispersal from the cavity and the attendant debris fragmentation. The quantity and size of dispersed debris control the magnitude and rate of important DCH processes. The thermodynamic extremes of isentropic blowdown and isothermal blowdown yield similar results, so both perspectives have been employed in DCH analyses.

A. HOLE ABLATION

The earliest hole ablation model was developed as part of the Zion Probabilistic Safety Study [2]. A similar model was used by Sienicki and Spencer [99] to explore the effect of melt superheat on hole ablation. Moody (see [20, App. 3]) presented an ablation model and then proceeded to examine the scaling implications of potential experiments. Morris and MacBeth [87] implemented an ablation model into the CORDE code. Pilch [23, App. J] extended an earlier model [100] and validated the model against the available database. Although similar in their basic concept, these models are distinguished by their treatment of the melt–substrate heat transfer coefficient and by a modeling assumption concerning the existence of an oxide crust floating above the ablating steep interface. More recently, Sehgal [101] and Dinh *et al.* [102] discussed an ablation model that replaces the correlation approach used in previous models with a numerical simulation of the temperature-dependent (viscosity and thermal conductivity) boundary layers near the ablating interface. Results of this more detailed model are similar to the results obtained with the simplified model of Pilch. Some details of the simplified models will now be summarized.

Albation of the orifice will occur when the interface temperature between wall material and the flowing core debris reaches the melting point of the RPV steel. For any conditions of interest there is a short delay before wall ablation is initiated. The instantaneous contact temperature between the walls of the orifice and the melt is typically ~ 1550 K, which is below the melting point of the RPV wall (steel) and the freezing point of the oxide components in the core melt. Conditions at instantaneous contact favor crust formation (i.e., frozen core material) without melting the RPV substrate. The convective heat flux from the flowing core debris to the crust will cause the RPV substrate to heat up and begin melting [103] in a time given approximately by

$$t_1 = \frac{\pi}{4\alpha_w} \left[\frac{k_w(T_{mp,w} - T_w)}{h_{d,w}(T_d - T_{mp,d})} \right]. \tag{4}$$

This delay in wall melting is small compared to the melt ejection time (assuming no ablation); consequently, ablation can be assumed to start with initiation of flow through the hole. If the lower head fails by a sufficiently large rupture, then the melt ejection time could be comparable to the delay in wall melting. Assuming an immediate onset of ablation is conservative, while having a negligible impact on the final hole size.

Once wall melting is initiated, the ablation rate quickly reaches a steady state that can be obtained by balancing the convective transfer of energy to the substrate with the energy required to heat and melt the wall:

$$\frac{dD_h}{dt} = \frac{2h_{d,w}\,\Delta T_r}{\rho_w \left[C_{p,w}(T_{mp,w} - T_w) + h_{f,w} \right]} \approx \dot{D}_{\text{ref}}. \tag{5}$$

The melt discharge (with a changing hole size) determines the time over which ablation occurs; thus, hole ablation and melt discharge are coupled phenomena that can be treated numerically or analytically.

The heat transfer coefficient, $h_{d,w}$, and the reference temperature difference, ΔT_r, are important relations that distinguish the various models (Table VI). The ZPSS study assumed that a thin oxide crust floating above the ablating interface was generated and replenished; consequently, the melting point of the crust then controls convective heat transfer to the wall. Sienicki and Spencer made similar assumptions. Pilch, Moody, and Morris and Macbeth all assume that any insulating crust and all the

TABLE VI
CONSTITUTIVE RELATIONS FOR HOLE ABLATION MODELS

Reference	ΔT_r	$h_{d,w}$
Pilch et al. [23]	$T_d - T_{mp,w}$	$h_{d,w} = h'_{d,w} \dfrac{\beta}{e^\beta - 1}$
Pilch and Tarbell [100]		$\beta = \dfrac{\rho_w C_{p,w} \dot{D}'_{ref}}{2 h'_{d,w}}$
		$h'_{dw} = 0.0292 \dfrac{K_d}{L} \mathrm{Re}_L^{0.8} \mathrm{Pr}^{0.33}$
ZPSS [2] Sienicki and Spencer [104]	$T_d - T_{mp,c}$	
Moody [105]	$T_d - T_{mp,w}$	$h_{d,w} \sim 20 \times 10^3 \dfrac{w}{m^2 K}$
Morris and MacBeth [87]	$T_d - T_{mp,w}$	$0.036 \dfrac{K_d}{D} \left(\dfrac{D}{L}\right)^{1/18} \mathrm{Re}_D^{0.8} \mathrm{Pr}^{0.33}$

ablating steel are stripped from the interface by the flowing corium. Thus, the steel melting point (not the oxides) determines the potential for convective heat transfer to the ablating interface.

The L/D_h ratio for the hole is typically ≤ 1, so entrance region effects are important in evaluating the heat transfer coefficient, $h_{d,w}$. This is recognized in both the Pilch and CORDE models. Note that the heat transfer coefficient is constant or nearly constant during melt discharge. Analogous to heat transfer and transpiration [108], Pilch applies a correction that is important when the ablation rate is large.

Assuming that $h_{d,w}$ and ΔT_r are constant (or nearly constant), the final hole size can be computed from (see [23, App. J])

$$\frac{\Delta D_h}{D_h^0} = \frac{\dfrac{\tau_M}{\tau_D}}{1 + 0.6934 \left(\dfrac{\tau_M}{\tau_D}\right)^{2/3}}, \tag{6}$$

where

$$\tau_M = \frac{M_d^0}{\dot{M}_d} = \frac{M_d^0}{\rho_d C_d \dfrac{\pi \left(D_h^0\right)^2}{4} \left(\dfrac{2\left(P_{RCS}^0 - P_c^0\right)}{\rho_d}\right)^{1/2}} \tag{7}$$

is the characteristic time to eject all the melt from the RPV in the absence of ablation and where

$$\tau_D = \frac{D_h^0}{\dot{D}_h} = \frac{D_h^0}{\left(\dfrac{2h_{d,w}\left(T_d^0 - T_{mp,w}\right)}{\rho_w\left[C_{p,w}\left(T_{mp,w} - T_w\right) + h_{f,w}\right]}\right)} \tag{8}$$

is the characteristic time to double the initial hole size by ablation. Figure 11 validates the model against the existing database. The bias and standard deviation are defined in a manner analogous to Eq. (2). This figure also illustrates that ablation increases the hole size only slightly for initial hole sizes characteristic of lower head rupture.

Theofanous has raised the issue of whether ablation following multiple simultaneous penetration failures could cause the growing holes to interact and combine in such a way as to cause a yet larger portion of the head to fail. Table VII addresses the issue of multiple penetration failures using the Pilch model. The Sienicki and Spencer model predicts less ablation. As an extreme, 90 metric ton (mt) of melt is considered. The melt mass (M_d^0) available to flow through each hole decreases with each additional failure; consequently, the final hole size $(D_{h,f})$ of any given penetration also decreases with each additional failure. A minimum of three simultaneous failures in a triangular pattern is required if the holes are to interact in

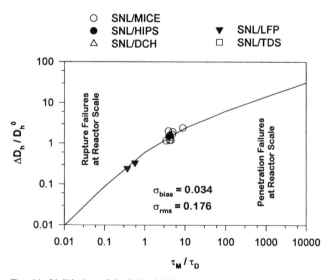

Fig. 11. Validation of the hole ablation model against experiment data.

TABLE VII

SENSITIVITY TO THE NUMBER OF SIMULTANEOUS PENETRATION FAILURES

Number of failures	M_d^0/failure (mt)	$D_{h,f}$ (m) each failure	$D_{h,f}$ (m) equivalent	t_d (s)
1	90	0.502	0.502	5.1
2	45	0.402	0.569	4.1
3	30	0.353	0.611	3.5
4	22.5	0.322	0.644	3.2

such a way to cause a larger portion of the head to fail. For three simultaneous failures, the maximum size of one failure is only 0.353 m while the spacing between penetrations is about 0.50 m; consequently, multiple penetration failures cannot interact directly without structural failure of the remaining metal connecting the holes.

Table VII does show that the final equivalent diameter of all multiple failures increases with the number of failures, although the sensitivity is not excessive. The last column in Table VII lists the time required to eject the extreme of 90 mt of melt through all available holes (t_d). For only one penetration failure, the discharge time is only 5.1 s. The coherence of multiple failures only occurs if each failure occurs within an interval small compared to 5.1 s. We believe that this is unlikely.

B. MELT DISCHARGE AND GAS BLOWDOWN

Gas blowthrough during melt expulsion from a hole in the bottom of the RPV (Fig. 9) could allow the RCS to depressurize before all the melt is expelled from the RPV. Blowthrough becomes more important as the melt flow rate (larger driving pressure or bigger hole size) becomes larger. For typical reactor conditions Pilch and Griffith [109] concluded that the onset of blowthrough is predominantly a function of the Froude number and only a weak function of the size of the vessel. Under these conditions, the correlation of Reimann and Khan [106] (Table VIII) is representative of many similar correlations. Using this correlation, the fraction of melt remaining in the RPV at the onset of gas blowthrough is given by

$$f = 4 \left(\frac{h_c}{D_v} \right)^2 \left(\frac{3}{2} - \frac{h_c}{D_v} \right), \qquad (9)$$

which predicts that $\sim 25\%$ of 75 mt would remain in the vessel at the onset of blowthrough for representative reactor conditions ($P = 16$ MPa,

TABLE VIII

BLOWTHROUGH AND FLOW QUALITY CORRELATIONS FOR SIDE AND
BOTTOM FAILURES OF THE RPV

Phenomenon	Bottom failure (Reimann and Khan [106] gas blowthrough)	Side failure (Ardon and Bryce [107], liquid entrainment)		
Critical depth	$\dfrac{h_c}{D} = 0.62 \mathrm{Fr}_D^{0.4}$ $$\mathrm{Fr}_D = \dfrac{v}{(gD)^{1/2}}$$	$\dfrac{h_c}{D} = 0.63 \mathrm{Fr}_D^{0.4}\left(\dfrac{\tau_g}{\tau_d}\right)^{0.2}$		
Void or flow quality	$a_g = 1 - \left(\dfrac{h}{h_c}\right)^{1.62}$	$X_g = X_0\left(1 + \dfrac{h}{h_c}\right)$ $$\times\left(1 - 0.5\dfrac{h}{h_c}\left(1 + \dfrac{h}{h_c}\right)X_0^{(1-h/h_c)^{0.5}}\right)$$ $$X_0 = \dfrac{1.15}{1 + \left(\dfrac{\rho_d}{\rho_g}\right)^{0.5}}$$		
Range of validity	$2 < \mathrm{Fr}_D < 60$	$25 < \mathrm{Fr}_D < 115$	$0.001 < \dfrac{\rho_g}{\rho_d} < 0.035$	
Expected NPP range	$7 < \mathrm{Fr}_D < 13$	$7 < \mathrm{Fr}_D < 13$	$0.001 < \dfrac{\rho_g}{\rho_d} < 0.013$	

$D_h = 0.5$ m, $D_v = 4.2$ m). The fraction retained decreases slightly with decreasing system pressure because of the velocity (hence pressure) dependence of the Froude number.

Simultaneous discharge of both gas and liquid follows blowthrough assuming isothermal blowdown. A mass balance,

$$\frac{dM_d^*}{dt^*} = \left(1 - \alpha_g\right)\left(M_g^*\right)^{1/2}\frac{\tau_g}{\tau_M} \quad \text{and} \quad \frac{dM_g^*}{dt^*} = -\alpha_g M_g^*, \quad (10)$$

can be written for both the melt and gas inventory in the RCS, where α_g is the transient gas void fraction in the orifice (Table VIII). The gas equation is normalized so that every term is order unity. The magnitude of the liquid equation is determined by τ_g/τ_M, where τ_g and τ_M are the single-phase gas and liquid discharge times as if each phase flowed independently and completely filled the entire breach area. When τ_g/τ_M is large compared to unity (~ 27 for the blowthrough calculations evaluated above), then liquid discharge will be complete long before gas blowdown is complete. In DCH experiments [40, 52], 90–95% of the melt is ejected from the RPV mockup; consequently, it is common to assume complete melt ejection from the RPV for a bottom failure.

Enhanced retention of melt in the RPV is possible if the failure occurs on the side of the RPV. Volumetric heating (decay heat) of molten core material on the lower head could create a side-peaked heat flux that might favor a side failure of the RPV near the upper surface of the melt. A side failure of the RPV resulting from localized stress concentrations might also occur in those RPVs that are supported by a skirt.

For a side failure of the RPV located at the melt surface (Fig. 10), simultaneous gas–liquid discharge will be initiated and continue until the gas inventory is exhausted or until the melt has drained to a critical level below the level of the breach. Equation (10) is still applicable here, so we expect that the melt will drain down to the critical level for entrainment without significant depressurization of the RCS. The critical condition for liquid entrainment through a side failure has been addressed by Ardon and Bryce [107] (Table VIII). The fraction of the initial melt invectory that cannot be swept out a side failure is obtained from geometric considerations as

$$
f = \left(1 - \frac{h_c}{h_H}\right)^2 \left(\frac{1}{1 - \frac{2}{3}\frac{h_H}{D_v}} - \frac{\frac{2}{3}\frac{h_H}{D_v}}{1 - \frac{2h_H}{37D_v}}\left(1 - \frac{h_c}{h_H}\right)\right) \approx \left(1 - \frac{h_c}{h_H}\right)^2,
$$

(11)

which predicts that $\sim 60\%$ of the initial melt inventory would remain permanently in the vessel for representative conditions.

The assessments of melt retention in bottom and side failures of the RPV are based solely on hydrodynamic considerations. The free-surface flows associated with gas blowthrough in a bottom failure and liquid entrainment in a side failure likely would be affected by the oxidic crust that forms on the top surface of the melt. The presence of such an oxide crust is likely to enhance melt retention on a side failure. Consequently, DCH loads might be significantly mitigated should the RPV fail on the side. This mitigation mechanism has not been tested as part of the DCH database.

The above discussion shows that the period of simultaneous melt–gas discharge is small compared to the gas blowdown time itself. To first order then, the two-phase period can be neglected when examining gas blow-down of the RCS. The integral DCH experiments (Table I) all show that peak DCH loads are achieved on a time scale comparable to or less than the blowdown time of the RCS. Since RCS blowdown is the driving force for all DCH processes, it is useful to briefly examine the process.

Wulff (see [20, App. K]) examined the scaling implications of RCS blowdown from two bounding perspectives. The isentropic path constitutes the limiting case of rapid expansions without heat transfer, while the isothermal path constitutes the opposite limit of extremely intensive heating of the vapor by internal RCS structures and by core material stacked up near the break hole.

Table IX summarizes some key results for these two limiting cases. The characteristic blowdown time recommended here corresponds to one *e*-folding time (approximately for the isentropic path also) for the depressurization transient. This definition of the blowdown time constant differs by $\sim 4\%$ in the isentropic and isothermal definitions. In general, Wulff showed that the difference is small between the extremes of fully isentropic and fully isothermal thermodynamic paths. Thus, the selection of one over the other is a matter of personal preference and convenience, and elements of both these perspectives have been cited or used in various correlations or models.

Lastly, we note that alternate definitions of the characteristic blowdown time sometimes appear (e.g., $\tau_b = V/Aa_0$, where $a_0 =$ sonic velocity). Such definitions differ from our recommendation by a constant factor and are acceptable provided care is taken to be consistent.

V. Cavity Dispersal Phenomena

Debris dispersal from the reactor cavity has been the focus of many programs since the Zion Probabilistic Safety Study [2], where dispersal was assumed to eliminate the possibility of core concrete interactions in the reactor cavity. This acknowledgement led to the recognition that HPME followed by dispersal from the cavity could lead to DCH and short-term pressurization of the containment.

Three questions must be answered when investigating cavity dispersal:

1. How much melt is dispersed from the cavity? To first order, this has been traditionally taken as a measure of how much material can participate in DCH. More recent CONTAIN modeling parametrically allows for possible DCH contributions from nonairborne debris.

TABLE IX
BLOWDOWN OF THE REACTOR COOLANT SYSTEM

Process	Thermodynamic path	
	Isentropic	Isothermal
Vapor expansion in RCS	Isentropic	Isothermal
Vapor acceleration through break	Isentropic	Isothermal
Mass flow rate	$\dot{M}^* = \dfrac{M^0}{\dot{M}_{\text{ref}}} = \left(1 + \dfrac{\gamma-1}{2}\dfrac{t}{\tau_b}\right)^{(\gamma+1)/(\gamma-1)}$	$\dot{M}^* = \dfrac{M^0}{\dot{M}_{\text{ref}}} = e^{-t/\tau_b}$
Depressurization history	$P^* = \dfrac{P}{P^0} = \left(1 + \dfrac{\gamma-1}{2}\dfrac{t}{\tau_b}\right)^{-2\gamma/(\gamma-1)}$	$P^* = \dfrac{P}{P^0} = e^{-t/\tau_b}$
Characteristic mass flow rate	$\dot{M}_{\text{ref}} = C_d P^0 A_h \left(\dfrac{MW_g}{R_u T^0}\left(\dfrac{2}{\gamma+1}\right)^{(\gamma+1)/(\gamma-1)}\right)^{1/2}$	$\dot{M}_{\text{ref}} = C_d P^0 A_h \left(\dfrac{1}{e}\dfrac{MW_g}{R_u T^0}\right)^{1/2}$
Characteristic blowdown time	$\tau_b = \dfrac{M^0}{\dot{M}_{\text{ref}}} = \dfrac{1}{C_d}\left(\dfrac{\gamma+1}{2}\right)^{(\gamma+1)/2(\gamma-1)}\dfrac{V}{A_h}\left(\dfrac{MW_g}{\gamma R_u T^0}\right)^{1/2}$	$\tau_b = \dfrac{M^0}{\dot{M}_{\text{ref}}} = \dfrac{1}{C_d}\dfrac{V}{A_h}\left(\dfrac{eMW_g}{R_u T^0}\right)^{1/2}$

2. Is the dispersed material highly fragmented? This gives insight into whether kinetic limitations to debris–gas interactions are possible.
3. Is the dispersal process coherent with blowdown of RCS gases into the cavity? The noncoherence of dispersal and blowdown means that debris can interact with only a limited portion of the blowdown steam. This limits the heat sink for debris–gas heat transfer in plants with compartmentalized geometry and provides a potential steam limitation of oxidation of the metallic components in the melt. Again, CONTAIN's more recent modeling of nonairborne debris allows parametrically for gas interactions with debris that remains in the cavity.

As a practical matter, no existing cavity experimentally studied to date has been shown to be retentive, so nearly complete dispersal can be expected under conditions typical of many DCH events. Given extensive dispersal, the issue then reduced to characterizing where the debris ultimately comes to rest. Specifically, we must examine if dispersed debris can accumulate against the containment liner, plug the containment sumps, or otherwise compromise safety equipment. Such assessments are plant-specific and have been performed to a limited extent as part of the IPEs for U.S. plants.

Integral experiments and more recent separate effects experiments indicate that fragmentation is extensive enough that efficient debris–gas interactions cannot be precluded and are often expected. This has led researchers in the past to seek a low-pressure cutoff below which extensive dispersal and fragmentation would not be expected. As a practical matter, the low-pressure cutoff is ~2MPa for many cavity designs. However, for plants with compartmentalized geometries depressurization to such low pressures is not required before nonthreatening loads are realized because the steam supply itself becomes limiting (Sect. II.B). The low-pressure cutoff for plants where most dispersed debris enters the dome may be of more interest because debris will then have most of the containment atmosphere with which to interact. The bounding nature of dispersal in the regimes of greatest interest is fortunate because our current understanding of dispersal phenomena precludes quantitative predictions for the great variety of cavity geometries that exist in nuclear power plants. It is useful to review the work that culminates in these perspectives.

A. DISPERSAL THRESHOLD AND DISPERSAL FRACTION

Spencer *et al.* [3, 4] experimentally confirmed that the threshold for debris dispersal from the Zion cavity is reasonably predicted by the Kutatuladze criterion proposed in the Zion Probabilistic Safety Study, as well as the entrainment inception criterion by Ishii and Grolmes [110]. Since that time, debris dispersal correlations have become available for a number of cavity designs, and these correlations can be used to define a low-pressure cutoff such that the RCS pressure is insufficient to cause significant dispersal from the cavity.

Table X summarizes simulant fluid experiments that have been performed to address debris dispersal issues. Table XI provides a complete accounting of correlations for the low-pressure cutoff that have been developed using this database. We note that all the correlations for the low-pressure cutoff are independent of physical scale except the Chun *et al.* [122] correlation, which predicts that the low-pressure cutoff decreases with increasing physical scale. Nearly complete RCS depressurization (to < 2 MPa) is often required to ensure significant melt retention in the cavity. At these low RCS pressures the limited steam supply plays a more dominant role than dispersal in limiting DCH loads; consequently, the low-pressure cutoff is not always a useful indicator of conditions where DCH might become insignificant. The low-pressure cutoff for dispersal is a more relevant criterion for plants where most debris is dispersed to the dome and can thus interact with a large fraction of the containment atmosphere.

Table XII presents a trends chart showing *experimentally* observed dependencies of debris dispersal on key variables. We note that no model correlation embodies all these dependencies, even for a single well-studied cavity geometry. Consequently, we cannot claim a complete predictive or correlational capability for any cavity in the intermediate pressure range where partial dispersal is expected.

Debris dispersal in the intermediate range of partial dispersal has been approached in three different ways: pure correlation, phenomenologically based correlation, and phenomenologically based modeling. The pure correlation approach forms dimensionless groups from a list of parameters thought to be important. This approach has been adopted by MacBeth *et al.* [113], Chun *et al.* [122], Kim *et al.* [123], and Kim *et al.* [127], and their correlations are listed in Table XIII. Tutu *et al.* [118] employed a phenomenological correlation approach by extracting the necessary dimensionless groups from a simple one-dimensional formulation of the transport equations. Tutu's correlations have been developed for four different cavity designs as shown in Table XIV. Neither of the correlation approaches give insight into the functional form of the correlation; consequently, the working equations reflect the experience and insight of the individual investigator.

Levy [128] pursued a phenomenological modeling approach to debris dispersal by assuming that surface entrainment from a pool of melt was the controlling process. The Kataoka and Ishii [129] entrainment rate

TABLE X

SURVEY OF DISPERSAL EXPERIMENTS USING LOW-TEMPERATURE SIMULANT FLUIDS

Reference	Cavity type (country)	Scale	Cavity structures	Melt simulant	Melt delivery	Driving gas	RCS press (MPa-abs)	Hole size full scale equiv. (m)
Spencer et al. [3, 4]	Zion (US)	1:40	No inst. struct. No RPV gap	Water, Cerrelow, SS shot	SS	N_2	—	1.0
MacBeth and Trenberth [111]	Sizewell (UK)	1:25	No inst. struct. No RPV gap	Water Flutec ethylene–water, silicon oil	SS	Air, He	0.57–0.75	0.20–0.31
Rose [112]	Sizewell (UK)	1:25	Inst. struct. No RPV gap	Water	SS	Air	0.17–0.51	0.31
MacBeth et al. [113]	Sizewell-like (UK)	1:132–1:21	No inst. struct. No RPV gap	Water	SS	Air	0.35–0.95	0.32
Hall [114]	Sizewell (UK)	1:25	W & w/o inst. struct. W & w/o RPV gap W & w/o manway	Water	Trans.	Air	1.0–13.0	0.24
MacBeth and Rose [115]	Ringhals (Sweden)	1:25	No inst. struc. No RPV gap	Water, gallium–indium–tin	SS	Air	0–0.4	0.31
Nichols and Tarbell [116]	Zion (US)	1:10	W & w/o inst. struc. No RPV gap	Water	Trans.	Air, He	0.0–4.0	0.38

continues

TABLE X

SURVEY OF DISPERSAL EXPERIMENTS USING LOW-TEMPERATURE SIMULANT FLUIDS (*Cont.*)

Reference	Cavity type (country)	Scale	Cavity structures	Melt simulant	Melt delivery	Driving gas	RCS press (MPa-abs)	Hole size full scale equiv. (m)
Pilch *et al.* [117]	Zion, Surry (US)	1:10	W & w/o inst. struc. No RPV gap	Water	Trans.	Air–He	0.0–4.0	0.25–0.38
Tutu *et al.* [118]	Zion, Surry Watts Bar (US)	1:42	No inst. struc. No RPV gap	Water, Wood's metal	Trans.	N_2, He	2.7–5.3	0.29–0.48
Tutu *et al.* [119]	Surry (US)	1:42	No inst. struc. No RPV gap	Water, Wood's metal	Trans.	N_2, He	—	0.20–0.40
Tutu and Ginsberg [120]	Surry, Zion (US)	1:42	W (Surry) & w/o (Zion) Inst. struct. No RPV gap	Water	Trans.	N_2, He		
Tutu *et al.* [121]	Watts bar (US)	1:42	No inst. struc. No RPV gap	Water, Wood's metal	Trans.	N_2, He	0.38–5.27	0.20–0.48
Chun *et al.* [122]	Young-Kwang (Korea)	1:25–1:41	No inst. struc. No RPV gap	Water, Wood's metal	Trans.	N_2, CO_2	0.3–3.0	0.23–0.40
Kim *et al.* [123]	Kori, Young-Kwang (Korea)	1:20–1:30	No inst. struc. No RPV gap	Water	Trans.	—	0.3–2.6	0.3–0.6 0.2–0.4

TABLE XI

Low-Pressure Cutoff for Debris Dispersal from Various Reactor Cavities

Reference	Equation	Comments
Spencer *et al.* [3, 4]	$Ku = \dfrac{\rho_g V_g^2}{(\rho_d g \sigma_d)^{0.5}} \geq 6 - 13$	Zion cavity (US)
Spencer *et al.* [3, 4]	$\dfrac{\rho_g V_g^2}{\sigma^2 \rho_d \mu_d} \geq N_\mu^{0.16}$ $N_\mu = \dfrac{\mu_d}{\left(\rho_d \sigma_d \left(\dfrac{\sigma_d}{g \Delta \rho} \right)^{0.5} \right)^{0.5}}$	Zion cavity (US)
MacBeth *et al.* [113]	$Eu_c = \dfrac{\rho_g V_g^2}{2 P_{cav}} \geq 4 \times 10^{-5}$	Sizewell cavity (UK)
Tutu *et al.* [119]	$\dfrac{\rho_g V_g^2}{(\rho_d g \sigma_d)^{0.5}} \left(\dfrac{L}{D_h^0} \right)^2 \left(\dfrac{\rho_g}{\rho_d} \right)^{0.5} \geq 7$	Surry 1:42 scale cavity (US) $L = 4.14$ m $A_{ref} = 12.6$ m^2
Chun *et al.* [122]	$\left(\dfrac{\rho_g V_g^2 L}{\sigma} \right)^{0.927} \left(\dfrac{\rho_g V_g^2}{\rho_d g L} \right)^{0.125}$ $\times \left(\dfrac{\rho_g}{\rho_d} \right)^{0.377} \dfrac{L}{D_h^0} \geq 200$	Young-Kwang (Korea) 1:25 and 1:41 scale ~95% retention $L = 4.72$ m $A_{ref} = 20.15$ m^2
Kim *et al.* [123]	$\dfrac{\tau_b V_g}{L_p} \geq 1.47$ See Table IX for τ_b	Kori and Young-Kwang Cavities (Korea) 1 : 20 and 1 : 30 scale $L_p = 16.95$ m, $A_{ref} = 18.18$ m^2 Young-Kwang $L_p = 11.26$ m, $A_{ref} = 6.48$ m^2 Kori Water transient exps.

correlation was employed in the analysis. Levy was forced to introduce parameter dependencies to account for velocity stratification effects, which were not addressed explicitly in the model. Even then the final equations employ a material-dependent "constant" to account for differences in melt simulants employed in the experiments. Levy's model was developed for four different cavity designs as shown in Table XV.

TABLE XII

TREND CHART SHOWING EXPERIMENTALLY OBSERVED DEPENDENCIES OF DEBRIS DISPERSAL ON KEY VARIABLES

Increasing variable	Observed trend in dispersal	References	Explanation/comment
RCS parameters			
Vessel pressure	Increase	All Ishii et al. [124] Wu [125]	Increasing RCS pressure means higher gas velocities in the cavity, which leads to higher dispersal. However, Ishii observed that dispersal can occur simultaneously as a film sweepout process and as a surface entrainment process.
Vessel hole size	Increase	MacBeth and Trenberth [111] Tutu et al. [118, 119, 121] Chun et al. [122] Kim et al. [123]	Increasing hole size means higher gas velocities in the cavity. This is offset partially by a decrease in the RCS blowdown time. Increasing hole size also favors greater velocity stratification (i.e., higher than mean velocity) along the cavity floor.
Gas volume	Increase	Pilch et al. [117] Tutu and Ginsberg [120]	The gas blowdown time becomes comparable with the characteristic dispersal time. A dramatic decrease in gas volume resulted in no dispersal.
Blowdown time	Increase	MacBeth et al. [113] MacBeth and Rose [115]	Longer blowdown times means a longer entrainment interval. This is consistent with the bulk of the database; however dispersal from the Ringhals cavities was predominantly as a slug or wave, which was found to be independent of blowdown time.
Gas molecular weight	Increase	MacBeth and Trenberth [111] Nichols and Tarbell [116] Pilch et al. [117] Tutu et al. [118, 119, 121] Chun et al. [122]	The gas molecular weight has negligible impact on dispersal in steady-state tests because the lower gas density is compensated for by a higher velocity. Increasing the gas molecular weight increases the blowdown time in transient experiments and consequently increases dispersal.
Melt parameters			
Melt volume	No change	Nichols and Tarbell [116] Hall [114]	This implies that the dispersal force is proportional to the amount of melt, which is consistent with the Wallis formula for drag on a liquid film.

continues

263

TABLE XII

TREND CHART SHOWING EXPERIMENTALLY OBSERVED DEPENDENCIES OF DEBRIS DISPERSAL ON KEY VARIABLES (*Cont.*)

Increasing variable	Observed trend in dispersal	References	Explanation/comment
Melt density or surface tension	Decrease	Spencer *et al.* [3, 4] Tutu *et al.* [118, 119, 121] Chun *et al.* [122]	Experimentally increases in density were always correlated with increases in surface tension, so the observed trend cannot be uniquely ascribed to density or surface tension.
Melt viscosity	Decrease (weak)	MacBeth and Trenberth [111] MacBeth *et al.* [113]	Only a small decrease in dispersal has been observed in going from water ($\mu = 0.0011$ Pa · s) to silicone oil ($\mu = 0.012$ Pa ·s).
Cavity parameters Physical scale	Increase or no change	MacBeth *et al.* [113] Chun *et al.* [122] Kim *et al.* [123]	In steady-state experiments, MacBeth *et al.* found that the dispersal fraction remained constant for a 6-fold increase in physical scale when the entrainment time was held constant. In a transient situation the blowdown time increases with physical scale, so we expect dispersal to also increase. This was confirmed by Chun *et al.* in transient tests. The data of Kim *et al.*, however, show no strong dependence on physical scale.
Off-centered orifice	Increase	UK	The swirl induced by an off-centered orifice increases dispersal.
Orifice–floor height	Decrease	Rose [112]	Gas stratification along the floor decreases as the orifice–floor height increases. One implication of this is that dispersal has been observed to vary even though $P_{RCS}A_h = $ const.
Addition of guide tube structures	Decrease or no change	Rose [112] Nichols and Tarbell [116] Pilch *et al.* [117] Hall [114] Allen *et al.* [126] Tutu and Ginsberg [120]	The addition of cavity structures significantly alters the flow patterns in the cavity, which tends to decrease dispersal in tests using low temperature simulant fluids. However, Allen *et al.* have shown that the structures can be dispersed with the melt when high-temperature melts are employed.

continues

TABLE XII

TREND CHART SHOWING EXPERIMENTALLY OBSERVED DEPENDENCIES OF DEBRIS DISPERSAL ON KEY VARIABLES (*Cont.*)

Increasing variable	Observed trend in dispersal	References	Explanation/comment
Addition of an additional manway	Decrease (weak)	Hall [114]	Addition of another manway decreases the gas flow through the cavity exit, which was observed to decrease dispersal slightly.
Different cavity geometry Watts bar → Surry → Zion	Increase	Tutu et al. [118, 119, 121] Pilch et al. [117]	Debris dispersal can be cavity-specific with the absence of dead-end spaces and right angles increasing dispersal.
RPV annular gap	Increase	Hall [114] Blanchat et al. [41]	The addition of an RPV annular gap increased dispersal in simulant fluid experiments where dispersal was already poor because of cavity structures (Hall [114]). Blanchat et al. showed that RPV insulation that partially fills the gap will be ablated by high-temperature melts and swept from the gap.
Containment pressure	Unknown or no change	Allen et al. [40]	The containment pressure has not been systematically varied in any simulant fluid experiments, although this parameter appears in some correlations. Dispersal data taken with high-temperature melts at three different containment pressures reveals no obvious or strong trend; however, these data and some DCH experiments with high-temperature melts lead to choked flow from the cavity over some portion of the dispersal period. Under these conditions, we would expect flow through the cavity, and hence dispersal, to be independent of the containment pressure.

TABLE XIII

CORRELATIONAL-BASED APPROACHES TO DEBRIS DISPERSAL

Cavity/Reference	Correlation
Sizewell (UK) MacBeth et al. [113]	$f_{\text{disp}} = 10^{-4} \left(\dfrac{2 P_{\text{RCS}}^0 g t^2}{\sigma} \right)^{0.375} \left(\dfrac{X}{0.75 - BX} - 0.05 \right)$ $X = \dfrac{\rho_{g,c} V_{g,c}^2}{2 P_c} \times 10^3 = \text{Eu} \times 10^3$ $B = 16.4 \times 10^{-9} \dfrac{4 P_c^2}{\sigma_d P_d g} \left(\dfrac{\rho_g \sigma_d^2}{2 P_c \mu_d^2} \right)^{0.25}$
Young-Kwang (Korea) Chun et al. [122]	$f_{\text{disp}} = 0.45\{1 + \tanh(3.79(\log_{10} X - 2.87))\}$ $X = \text{We}^{0.927} \text{Fr}^{0.125} \left(\dfrac{\rho_{g,c}}{\rho_d} \right)^{0.377} \left(\dfrac{L_R}{D_h^0} \right)^{1.764}$ $\text{We} = \dfrac{\rho_{g,c} V_{g,c}^2 L_R}{\sigma_d} \quad \text{Fr} = \dfrac{\rho_{g,c} V_{g,c}^2}{\rho_d g L_R}$ $L_R = 4.72 \text{ m}$
Young-Kwang (Korea), Kori (Korea) Kim et al. [123]	$f_{\text{disp}} = 3.4 \dfrac{\tau_b V_{g,c}}{L_p} - 5.0$ $L_p = 16.95 \text{ m Young-Kwang}$ $L_p = 11.26 \text{ m Kori}$
YGN (Korea), Kori (Korea), Watts bar (US) Kim et al. [127]	$f_{\text{disp}} = 40\left\{ 1 + \tanh\left(3.79 \log \dfrac{t^*}{15} \right) \right\}$

Other dispersal models based on surface entrainment have appeared in the literature. Henry [98] employed the Ricou–Spalding [130] correlation, and the CORDE model [131] uses a similar correlation modified to address the low-pressure cutoff. Williams and Griffith [132] employed a modified form of the Whalley–Hewitt [133] entrainment rate correlation for annular two-phase flow. The CORDE models are unique in that they explicitly model stratification of the gas flow (with some limited validation) along the floor of the cavity. We note that any dispersal model based on surface entrainment alone is not consistent with experimental observations (Zion geometry [125]) that film sweepout competes with surface entrainment as a dispersal mechanism over some range of conditions.

We note from the above discussions that each correlation or model is usually tailored to the individual cavity design. Only the Kim et al. [123, 127] correlations apply simultaneously to several different cavities. We

TABLE XIV

Tutu–Ginsberg Phenomenologically Based Correlations for Debris Dispersal

$$f_{\text{disp}} = 1 - \frac{C_1}{1 + C_2 \left|\log_{10} X - C_3\right|^{C_4}} - \frac{C_5}{\exp\left(\dfrac{\left|\log_{10} X - C_3\right|}{C_6}\right)^{C_7}}$$

$$X = \frac{N_5 N_{11}^{C_8} N_2^{C_9}\left(1 + C_{10} N_4^{C_{11}}\right)}{1 + C_{12} N_1^{C_{13}}}$$

$$N_1 = \frac{\sigma \rho_d}{\rho_R^2 U_R^2 L} \qquad N_2 = \frac{\rho_R}{\rho_d} \qquad N_4 = \frac{P^0 V^0}{P_{\text{con}}}$$

$$N_5 = \frac{\rho_R U_R^2}{\left(\rho_d g \sigma_d\right)^{0.5}} \qquad N_{11} = \frac{L}{D_h^0}$$

$$\rho_R = \frac{P_{\text{con}}}{RT_{\text{RCS}}^0} \qquad U_R = \left(\gamma\left(\frac{2}{\gamma + 1}\right)^{(\gamma + 1)/(\gamma - 1)}\right)^{0.5} \frac{A_h}{A_R} \frac{P^0}{\rho_R\left(RT_{\text{RCS}}^0\right)^{0.5}}$$

$$P^0 = P_{\text{RCS}}^0\left(\frac{T_{g,c}}{T_{\text{RCS}}^0}\right)^{0.5} \qquad V^0 = V_{\text{RCS}}^0\left(\frac{T_{g,c}}{T_{\text{RCS}}^0}\right)^{0.5}$$

Parameter	Zion (U.S.)	Surry (U.S)	Watts Bar (U.S.)	Sizewell (UK)
C_1	0.296	0.472	0.497	0.7443
C_2	0.0813	0.688	0.858	0.875
C_3	0.839	0.481	0.263	1.021
C_4	3.6	2.75	5.0	3.5
C_5	0.704	0.528	0.503	0.2557
C_6	1.351	0.958	3.451	0.928
C_7	6.6	8.75	3.0	4.75
C_8	1.512	1.818	1.618	0.89
C_9	0.464	0.652	0.440	0.319
C_{10}	2.32	0.575	0.273	1.399
C_{11}	0.324	0.680	1.630	0.625
C_{12}	0.0	0.350	3.783	1.847
C_{13}	0.0	0.121	0.522	0.109
L (m)	4.536	4.14	9.6	15.7
A_R (m^2)	5.13	12.6	18.7	12.0

Note. The terms characterizing plant geometry (L, A_R) are listed here at their full-scale values.

TABLE XV

LEVY'S PHENOMENOLOGICALLY BASED MODEL FOR DEBRIS DISPERSAL

$$Y = K_c \left(\frac{D_{hg} s}{D_h} \right)^2 \left(\frac{R_S T_S}{RT^0_{RCS}} \right)^{0.5} \frac{0.36 V_{RCS}}{A_h \left(RT^0_{RCS} \right)^{0.5}} Eu^{2.3} \frac{2Pc}{\sigma_d} \left(\frac{2P_c}{\rho_d} \right)^{0.5} \left(\frac{\mu_{g,c}}{\mu_d} \right)^{0.26}$$

$$Y = \log_{10} \left(\frac{\left(1 + 300 \frac{\delta_i}{D_c} \right)^{0.5} - 1}{\left(1 + 300(1 - f_{disp}) \frac{\delta_i}{D_c} \right)^{0.5} - 1} \frac{\left(1 + 300(1 - f_{disp}) \frac{\delta_i}{D_c} \right)^{0.5} + 1}{\left(1 + 300 \frac{\delta_i}{D_c} \right)^{0.5} + 1} \right)$$

$$Eu = \left(\frac{A_h}{A_c} \right)^2 \left(\frac{P^0_{RCS}}{P_c} \right)^2 \frac{\gamma}{2} \left(\frac{2}{\gamma + 1} \right)^{(\gamma + 1)/(\gamma - 1)}$$

Parameter	Zion (U.S.)	Surry (U.S.)	Watts bar (U.S.)	Sizewell (UK)
K_c				
Water	—	0.8	1.0	—
Wood's metal	0.01	0.016	0.06	0.02
D_c (m)	2.72	3.51	4.24	3.5
D_s (m)	0.38	0.40	0.40	0.313
$R_s \left(\frac{J}{kg\ K} \right)$ air	297	297	297	297
T_s (K)	293	293	295	293

note also that the definition of the geometric reference lengths and areas are not standardized between different cavity designs, even within a given correlation, so care is required in their application. Also, the correlations and models are all tied to experiments employing low-temperature nonreactive melt simulants. This introduces two complications in application. The composition (i.e., molecular weight) of the flowing gas can change due to metal–steam reactions and debris–gas heat transfer will accelerate the gas flow through the cavity. However, lacking a fully predictive correlation or model is no handicap to plant calculations because all the correlations and models, coupled with the high-temperature DCH database (Table I), all suggest that debris dispersal from the cavity is complete or nearly complete for existing cavity designs for the RCS pressures of interest to DCH. This realization has motivated attempts to design cavities that are immune to dispersal.

The Brockmann [134] design employs a deep layer of large gravel to "absorb" the melt during HPME, thus quenching and separating the debris from the subsequent blowdown. The specific Brockmann design has never been tested. However, Tarbell *et al.* [5] successfully used gravel beds to absorb high-pressure melt streams in the SPIT-7,8 experiments. Allen *et al.* [40] also used gravel beds successfully to absorb melt in separate effects tests aimed at characterizing the melt temperature.

The Carter *et al.* [85] design was proposed for U.S. advanced pressurized water reactors. The design is characterized by a collection volume recessed in the cavity wall where the cavity tunnel turns vertical. The offset volume should be at least twice the volume of the total inventory of core material. Carter *et al.* did not test the design, but analyses were performed to show that ~90% of the melt would deentrain in the capture volume as the gas turns from horizontal to vertical before venting from the cavity. Kim *et al.* [127] have performed some experiments (using N_2 and water) of the capture volume concept for the Korean YGN cavity design. Some enhanced retention was observed in the regime of incomplete dispersal, but the enhanced retention, however, was far less than what would be expected from simple deentrainment calculations of the type that Carter *et al.* performed. It remains untested that the capture volume concept will actually lead to enhanced retention and minimal debris–gas interactions at the high RCS pressures of interest to DCH. It should be noted that Kim *et al.* also tested a variation of the capture volume concept that involves an overhang placed in the vertical portion of the cavity exit. Kim *et al.* found that the overhang actually increased dispersal relative to a cavity without the overhang.

The Tutu *et al.* [135] design is characterized by a very large cavity with a very large catch volume (much larger than the Carter *et al.*, or Kim designs) at one end of the cavity and below the cavity floor. Additional plates and vanes are used to separate melt from gas. The Tutu *et al.* design has been tested with low-temperature simulants (N_2, water) as part of its patent application. Debris dispersal to the containment is significantly reduced in the Tutu *et al.* design. The experiments showed rapid separation of melt from the flowing gas suggesting that debris–gas interactions in the cavity would also be reduced, but this remains to be tested with high-temperature materials.

Debris dispersal is never complete in DCH experiments using high-temperature simulants even though the purely hydrodynamic correlations would predict otherwise. One reason is that melt can freeze on cavity surfaces during the dispersal interval when high-temperature melts are employed. Melt retention in experiments is predominantly observed as a thin frozen crust on all cavity surfaces. Assuming that freezing is the only retention mechanism [24] suggests that the fraction of melt dispersed from

TABLE XVI

MELT RETENTION BY FREEZING DURING CAVITY DISPERSAL

Parameter	SNL/IET-1 to 8B (Allen *et al.* [40])	ANL/IET-1R to 8 (Binder *et al.* [52])	SNL/IET-9 to 11 (Blanchat *et al.* [41])
Cavity	Zion	Zion	Surry
Scale	1 : 10	1 : 40	1 : 5.75
Melt simulant	$Fe-Al_2O_3$	$Fe-Al_2O_3$	$Fe-Al_2O_3$
f_{disp} observed	0.62–0.89	0.69–0.80	0.73–0.89
f_{disp} Eq. (12)	0.91	0.85	0.88

the cavity can be approximated by

$$f_{disp} \sim 1 - \frac{2\lambda(\alpha_c R_\tau \tau_b)^{1/2} 6 V_{CAV}^{1/3} \rho_d}{M_d^0}, \tag{12}$$

where λ is a growth rate constant for freezing of a superheated melt on concrete. Exothermic reactions of steam with the film were not considered in this formulation. For thermite on concrete, $\lambda \sim 0.137$; for corium on concrete, $\lambda \sim 0.48$. Note that freezing occurs over the dispersal interval $\sim R_\tau \tau_b$, where τ_b is the characteristic blowdown time and R_τ is the coherence ratio (see Sect. V.C). Table XVI shows that this simple model qualitatively predicts melt retention in experiments for a wide range of physical scales. Similar dispersal fractions are predicted for representative DCH applications at plant scale [19].

B. MELT FRAGMENTATION

Two issues arise as to whether melt fragmentation is sufficient to ensure efficient heat and mass transfer interaction between dispersing debris and blowdown gases. First, how much of the material is dispersed as a liquid film? The premise is that material dispersed as a liquid film may break up on exiting the cavity but that it will not achieve the same level of fine fragmentation associated with surface entrainment processes. The second issue addresses particle sizes associated with surface entrainment. Before addressing these issues, we note that any approach is necessarily a simplified subset of a great many processes that can occur in the cavity. These include: hydrodynamic jet breakup [124] during single-phase melt ejection, pneumatic atomization of the melt during two-phase discharge from the RCS [100], entrainment and deentrainment in both horizontal and vertical sections of the cavity, jet fragmentation due to dissolution of gases in the melt during HPME [136], splashing and droplet–structure interactions

[137], and droplet–droplet interactions. The net effect of these processes is to promote heating of the blowdown gases, which accelerates the gas flow through the cavity. This has a feedback effect on dispersal with a potential to increase entrainment. This feedback effect is not reflected in separate effects tests using low-temperature simulant fluids.

Wu and Ishii [138] give

$$f_m = \frac{1}{3}\left(1 - \frac{\tau_{\text{tran}}}{\tau_m}\right) \tag{13}$$

for the fraction of melt that can leave the cavity under its own momentum during HPME before the onset of gas blowdown into the cavity. Here, τ_{tran} is the characteristic time for melt to reach the cavity exit and τ_m is the characteristic melt ejection time from the RCS. For representative conditions, most of the melt will be subject to the dispersal and possible entrainment action of the blowdown gases; that is, significant dispersal by momentum prior to gas discharge is not likely.

The dispersal action of blowdown gases manifests itself as a competition between film sweepout and surface entrainment processes. This competition has been addressed explicitly by Henry [98] and Ishii et al. [124] (Table XVII). Separate effects experiments (Ishii et al. [124], Zhang et al. [140], Wu [125]) using low-temperature simulant fluids qualitatively confirm predictions of the Ishii model. In these air–water experiments, it was found that ~ 20–40% of the water was dispersed from a scaled Zion cavity as entrained particles. This observation was sensitive to the gas pressure and hole size. The entrainment fraction could be higher, however, when hot melt heats the blowdown gases as expected. The Ishii and Mishima [139] correlation, which is tied closely to an annular two-phase flow database, is another way to predict the fraction of liquid flux that is entrained in the core region of the flow.

Surface entrainment results in fine fragmentation of the melt. Particle sizes have been measured with sieves in some high-temperature DCH experiments using either an Fe–Al_2O_3 thermite or a uranium thermite (Blomquist et al. [141, 142], Tarbell et al. [25], Allen et al. [28, 29, 30, 31]). Particle sizes are distributed lognormally with a mass mean diameter of ~ 0.5–1.5 mm and a geometric standard deviation, $\sigma_g' \sim 4$. The Sauter mean diameter (rather than the mass mean) best characterizes the initial debris–gas interaction rates. The Sauter mean and mass mean sizes are related by

$$\log_{10} d_{32} = \log_{10} d_{30} - 1.1513 \log_{10}^2 \sigma_g'. \tag{14}$$

Sienicki and Spencer [143] concluded that the mass mean sizes in the ANL/CWTI-13 experiment are qualitatively consistent with the correla-

TABLE XVII
Contribution of Surface Entrainment to Debris Dispersal From a Reactor Cavity

Parameter	Henry [98]	Ishii et al. [124]	Ishii and Mishima [139]
Film sweepout or carryover time t_f (s)	Pressure-driven film expulsion: $$\left(\frac{10.4 L_p L_c A_c^2}{P_{RCS}^0 A_s A_h}\right)^{0.5} (\rho_d \rho_{g,c})^{0.25}$$	Momentum-driven carryover: $$\frac{4}{\pi}\,\frac{M_d^0}{\rho_d D_j^2 V_j}$$	
Entrainment flux $\dot{m}_e''$ (kg/m^2/s)	Ricou and Spalding [130]: $$0.1\left(\frac{\rho_d}{\rho_{g,c}}\right)^{1/2}\frac{\dot{m}_g}{A_c}$$	Kataoka and Ishii [129]: $$6.6 \times 10^{-7}\frac{\mu_d}{D_c}(Re_f We)^{0.925}\left(\frac{\mu_{g,c}}{\mu_d}\right)^{0.25}$$ $$Re_f = \frac{\rho_d j_{d,c} D_c}{\mu_d}$$ $$We = \frac{\rho_{g,c} j_{g,c}^2 D_c}{\sigma_d}\left(\frac{\Delta\rho}{\rho_{g,c}}\right)^{1/3}$$	
Entrained fraction f_e	$$\frac{\dot{m}_e'' A_s}{m_d^0} t_f$$	$$\frac{\dot{m}_e'' A_s}{m_d^0} t_f$$	$$f_e = 1 - \exp\left(-10^{-5}\left(\frac{Z}{D_c}\right)^2 \frac{Re_f}{j_g^*}\right)$$ $$\tanh(7.25 \times 10^{-7} We^{1.25} Re_f^{0.25})$$ $$j_g^* = \frac{\rho_{g,c} j_{g,c}^2}{(\sigma_d g\,\Delta\rho)^{0.5}}\left(\frac{\Delta\rho}{\rho_g}\right)^{1/3}$$

tion of Kataoka and Ishii [129]:

$$d_{30} = 0.028 \frac{\sigma_d}{\rho_{g,c}(V_{g,c} - V_f)^2} \, \mathrm{Re}_f^{-1/16} \, \mathrm{Re}_g^{2/3} \left(\frac{\rho_d}{\rho_{g,c}}\right)^{1/3} \left(\frac{\mu_{g,c}}{\mu_d}\right)^{2/3}, \quad (15)$$

where

$$\mathrm{Re}_f = \frac{\rho_d j_d D_c}{\mu_d} \quad \text{and} \quad \mathrm{Re}_g = \frac{\rho_{g,c} V_{g,c} D_c}{\mu_{g,c}}. \quad (16)$$

Ishii *et al.* [124] and Zhang *et al.* [140] have performed separate effects tests with water and Wood's metal and found that the Kataoka and Ishii [129] correlation appeared to underpredict the particle size by as much as a factor of two. Zhang and Ishii [144] attribute this to entrance region effects. More recently, Wu [125] discovered that particle sizes in the Zhang and Ishii experiments were somewhat distorted by the delivery system and that more careful testing by Wu in the same facility produced entrainment sizes that were more consistent with the Kataoka *et al.* correlation.

The Kataoka and Ishii correlation sometimes predicts particle sizes from entrainment that are sufficiently large that secondary fragmentation (We $\geq$ 12) can occur in the gas core (Sienicki and Spencer [143], Ishii *et al.* [124]). Accounting for secondary fragmentation, Sienicki and Spencer were able to predict particle sizes for the SNL/DCH-1 experiment when initial entrainment sizes were given by the Kataoka and Ishii correlation. Secondary fragmentation was predicted by a Pilch *et al.* [145] correlation:

$$\frac{d_{32}}{d^0} = \left(\frac{\rho_g (V_g - V_d)^2}{\delta}\right)^{-1/2}. \quad (17)$$

Using mean gas velocities and ignoring heating of the blowdown gas, Ishii *et al.* [124] and Wu [125] predict a mass mean particle size of ~ 0.6 mm for a representative set of DCH conditions. Similar sizes are predicted by Sienicki and Spencer if differences in hole size are taken into account. Larger particles are expected if secondary fragmentation does not go to completion or if entrained debris slows the gas. On the other hand, smaller particles are expected if localized gas velocities near the entrainment surface significantly exceed the mean gas velocity or if heating of the blowdown gas is significant. Chemical oxidation can have an opposing effect, however, by converting steam to hydrogen.

These types of uncertainties, coupled with potential differences due to variations in existing cavity geometries, are common to all dispersal phenomena. Fortunately, for high RCS pressure conditions of most interest to DCH, we can expect and certainly cannot rule out that dispersal is nearly complete, that entrainment is a significant part of the dispersal so the melt is highly fragmented, and that the mass mean particle sizes are ~1 mm. Under such conditions, efficient debris–gas interactions are plausible in the cavity, and Sect. VII will cite arguments that near-equilibrium conditions can exist in the cavity for at least some DCH scenarios.

C. COHERENCE

Interactions of airborne debris with blowdown gas in the cavity region are largely limited to that portion of the blowdown gas that is coherent with the dispersal process. The ratio of the characteristic dispersal time to the characteristic blowdown time is termed the coherence ratio (R_τ). Smaller values of the coherence ratio mean that the primary heat sink for debris–gas thermal interactions is smaller and that metal–steam reactions are more likely to be steam-limited. For isentropic blowdown, the fraction of the blowdown gas that is coherent with debris dispersal is given by Pilch (see [23, App. E]) as

$$f_{\text{coh}} = 1 - \left(1 + \frac{\gamma - 1}{2} R_\tau\right)^{-2/(\gamma - 1)}. \tag{18}$$

For $R_\tau \leq 0.5$, $f_{\text{coh}} \sim R_\tau$, so that R_τ is directly proportional to the amount of blowdown gas that can react with airborne debris. Tutu and Ginsberg [146] were the first to recognize the potential importance of coherence in limiting debris oxidation reactions during the dispersal process.

Two DCH models have independently developed coherence ratio correlations as an integral part of their formulations; in application the correlation specific to the model should be used. The first correlation, developed by Pilch (see [23, App. E]) using measured estimates of R_τ, is given by

$$R_\tau = Cf_{\text{disp}} \left(\frac{T_{\text{RCS}}^0}{T_d^0}\right)^{1/4} \left(C_d \frac{M_d^0}{M_{g,\text{RCS}}^0} \frac{A_h V_{\text{cav}}^{1/3}}{V_{\text{RCS}}}\right)^{1/2}, \tag{19}$$

where $C = 9.611$ and $\sigma_{\text{rms}} = 0.24$ ($e_{\text{bias}} = 0.014$) for the Zion cavity, $C = 12.2$ and $\sigma_{\text{rms}} = 0.154$ ($e_{\text{bias}} = 0.014$) for the Surry cavity, and $C = 10.97$ and $\sigma_{\text{rms}} = 0.234$ ($e_{\text{bias}} = 0.037$) for the combined data set. Here the bias and standard deviation are defined in a manner analogous to Eq. (2). The relative bias (e_{bias}) and relative standard deviations (σ_{rms}) are obtained from a nearly all-inclusive consideration of the database, which includes experiments at scales 1:40, 1:30, 1:20, 1:10, and 1:5.75. The data thus confirm model predictions that there is no effect of physical scale. Experiments have been conducted at driving pressures ranging from 3 to 13 MPa, with hole sizes ranging from 0.4 to 1.0 m (full-scale equivalent), and melt densities ranging from 4000 to 8000 kg/m³. These dependencies are adequately accounted for by the model; consequently, application of the correlation interpolates on the database.

The second correlation for the coherence ratio is attributed to Yan and Theofanous [72] and is given by

$$(R_\tau)_{Y,T} = 0.2 C_d \frac{A_h V_c^{1/3}}{V_{RCS}} \frac{T_{RCS}^0}{P_{RCS}^0} (\gamma R_{STM})^{1/2}, \tag{20}$$

which is intended to apply to both Zion-like and Surry-like cavities. We note that the lead constant (0.2) has dimensions of $m^2 k^{1/2} s/kg$, as some dimensional parameters have been absorbed into the fitting constant. The relative standard deviation is $\sigma_{rms} = 0.125$, as indicated by a more restrictive examination of the database. Equation (20) should be restricted to primary system pressure conditions in the ~ 6–13.5 MPa range, primary system temperatures in the ~ 571–787 K range, and breach diameters (reactor scale) in the ~ 0.40–0.56 m range.

The CONTAIN codes includes models for debris dispersal that in principle could be used to calculate coherence; however, Williams *et al.* [22] considered them insufficiently validated to recommend them for general use. Instead, it was recommended that coherence be estimated using a simplified version of Eq. (19) in which the factor $f_d (T_{RCS}^0/T_d^0)^{1/4}$ was dropped and the constant C was taken to be ~ 4. This value was based upon an analysis of the experimental coherence data using a technique considered by Williams *et al.* to be more appropriate in defining coherence for use with the CONTAIN code. In contrast with the results of Pilch *et al.* [23], they concluded that R_τ in the $1/40$-scale ANL/IET experiments was higher (by about a factor of two) than in the $1/10$-scale SNL counterpart experiments. They therefore recommended allowing for substantial uncertainties in R_τ, in part related to scaling uncertainties. However, in CONTAIN calculations, Williams *et al.* also noted that calculated loads are rather insensitive to coherence, with factor-of-two variations in R_τ typically resulting in loads variations of $\leq 10\%$ for both NPP analysis (Williams and Louie [77]) and experimental analysis (Williams *et al.* [22]). Hence, they did not consider uncertainties in coherence to be a major contributor to uncertainties in predicting DCH loads.

VI. Transport Phenomena

Many reactor containments are highly compartmentalized so that dispersed debris will interact with structures, which normally prevents efficient mixing of the dispersed melt with the entire atmosphere. The dispersal and transport of debris through the containment is important for four reasons: (1) the containment sump can become plugged, (2) other safety equipment can be compromised, (3) debris may impinge and accumulate against the containment liner, and (4) debris–gas interactions can be more efficient in the dome. Debris transport must be evaluated on a plant by plant basis; however, the existing database for Zion-like and Surry-like geometries offers some insight into some of the key processes.

The transport of debris throughout the containment could lead to direct thermal attack (not overpressure failure as DCH is typically envisioned) on the containment liner. Pilch et al. [147] noted that failure of the seal table, which caps a deal-end region near the cavity exit, in ice condenser plants could allow debris to accumulate against the containment linear. The ANO-1 NPP cavity ducts material directly to the containment boundary. Hammersley et al. [56] note the potential for dispersed debris to be directed against the liner of the ASCo and Vandellos (both in Spain) plants. In the Surry plant, debris exiting the cavity can be deflected from the roof of the RHR platform towards the containment liner. Other plant geometries also may be conducive to impingement and accumulation of debris against the liner. Focused modeling efforts on impingement and long-term thermal attach do not exist; however, limited experimental evidence (Surry, Blanchat et al. [41]; ASCo and Vandellos, Hammersley et al. [56]) suggest that impingement of debris on the liner during dispersal will not lead to its immediate melt through. The debris seems to freeze on the steel and flake off easily.

There are two primary debris transport pathways from the reactor to the containment dome in most nuclear power plants: (1) through the annular gap between the RPV and the biological shield wall, and (2) from the in-core instrument tunnel through the lower compartments. We express the dome transport fraction as

$$f_{\text{dome}} = f_{\text{gap}}(1 - f_{\text{noz/shld}}) + f_{\text{sub}}(1 - f_{\text{gap}}). \qquad (21)$$

Most of the debris dispersed through the annular gap reaches the upper dome, although a small amount is redirected by the nozzles through cutouts back into the lower compartments. On the other hand, much of the debris that is ejected into subcompartments will be trapped. These processes can be partially quantified based on insights and modeling of the Zion and Surry databases.

The annular gap around the RPV is an important debris-transport pathway to the upper dome. In the SNL IET-11 experiment [41], the melt-laden gas melted the insulation and swept it from the gap. The SNL HIPS-8C [148] experiment and the SNL IET-9 [41] experiment also simulated the gap but without the insulation. Pilch et al. (see [23, App. J]) analyzed these three experiments and concluded that the fraction of dispersed debris that goes through the gap is given by

$$f_{\text{gap}} = \frac{A_{\text{gap}}}{A_{\text{gap}} + A_{\text{cavity}}} \tag{22}$$

based on arguments of equal pressure drop through the gap and cavity dispersal paths. Here, A_{gap} represents the minimum flow area through the RPV annular gap between the RPV and biological shield wall; this flow area is typically taken at the nozzles and should be reduced by the projected area of the nozzles. The insulation, neutron shields, and refueling seal rings are not typically considered as obstacles to debris transport through the gap. The minimum flow area out of the cavity, A_{cavity}, typically occurs at the exit of the instrument tunnel. In the HIPS-10S experiment [126], the in-core instrument tubes and their supports were forcibly ejected from the cavity; consequently, the minimum tunnel flow area is evaluated without the presence of these structures. For many Westinghouse plants with large dry or subatmospheric containments, $f_{\text{gap}} \sim 17\%$ [24]. The partition of dispersed debris between the gap and the cavity in system-level codes is based on pressure drop considerations and generally follows the trend of Eq. (22) unless the user-specified loss coefficients are dramatically different in the two directions.

Credit for flow through the nozzle cutouts in the biological shield wall and nominal credit for knockdown by the missile shield is accounted for with $f_{\text{noz/shld}}$. Specifically, Pilch *et al.* [19] recommend $f_{\text{noz/shld}} \sim 0.10$. They based this estimate on water or Wood's metal experiments performed at Purdue University [58, 149], where nozzle cutouts and the missile shield were simulated. Their scaling analysis suggests that prototypic behavior is bounded by test observations with water or Wood's metal. In these experiments, ~ 17–20% of the melt going up the gap went through the nozzle cutouts and another $\sim 10\%$ was knocked down by the missile shield. However, these tests were performed at 6.8 MPa and their data indicate a decreasing trend with increasing pressure. Lacking a predictive model, Pilch *et al.* [19] recommended taking only nominal credit for flow out of the nozzle cutouts into the steam generator compartments.

Early scoping experiments using water [62, 150] and all subsequent integral effects tests have found that most of the debris dispersed into the subcompartments was trapped without ever entering the dome. There are two mechanisms by which debris will pass through the subcompartments to the dome: particles can be carried by gas through the subcompartments into the dome, and particles can follow inertial line-of-sight flow paths from the cavity exit to the dome, usually through the seal table room in most plants.

Simple analyses suggest that debris particles typical of DCH processes (~ 1 mm) do not easily follow gas streamlines. Cook [151] predicted that a fraction (f_c),

$$f_c = \frac{1}{\lambda W} \ln\left\{1 + \lambda W\left(\frac{W}{L} - \left(\frac{W}{L} - 1\right)f\right)\right\},$$

$$\lambda W = \frac{3}{4} C_d \frac{\rho_g}{\rho_d} \frac{W}{d}, \tag{23}$$

of dispersed particles would strike structures near the exit of the Zion cavity. The group W/L is preserved by geometric scaling of experiments, but the group λW is not because particle sizes do not scale with facility size. Nonetheless, Eq. (23) suggests that $\sim 96\%$ will strike the seal table overhang in a 1 : 10th scale experiment of Zion and $\sim 85\%$ at full scale [148]. Pilch et al. [148] also noted that debris was not permanently retained by such a structure because of splashing.

Bertodano and Sharon [149] and Wu [125] also recognized the importance of λW in scaling the "dispersion" or carryover of debris through the subcompartments to the dome. Performing scaled experiments with water or Wood's metal in Zion geometry, Wu [125] bounded the dome carryover fraction while investigating the effect of various parameters on carryover. Total carryover to the dome increased from 2.4/1.6% for (water/Wood's metal) to 3.4/5.3% for RCS pressures ranging from 6.9 to 14.2 MPa. Carryover through the seal table room was predominantly as films and independent of RCS pressure. Carryover to the dome also remained constant for a 275% increase in the melt simulant (i.e., water) mass. Wu concluded that carryover through the seal table room was a function of projected areas only while carryover through the subcompartment vents was dependent on the trajectories of small particles.

Wu [125] also noted that only the very smallest particles can follow gas through the subcompartment and less direct flow paths to the dome. In a more prototypic DCH experiment or reactor accident (i.e., hot melt), such small particles would not fully contribute to heating in the dome because they would have oxidized and transferred much of their thermal energy in the cavity or subcompartment before they ever reached the dome.

We note that Zion [40, 52] integral effects tests with hot melts typically showed $\sim 9\%$ (includes concrete contaminant from melt–structure interactions) carryover to the dome (except in those tests with potentially nonprototypic structural damage). Wu [125] measured 1.6/2.4% carryover in isothermal tests in Zion geometry under similar driving pressures. Wu attributed his lower values to distortion in the cavity chute, which was 2.7 times longer than prototypic, in the SNL/IET tests. This distortion favors the production of additional fine particles which are more likely to follow gas streamlines to the dome. Issue resolution efforts [18, 23] recommended 5% carryover as a conservative estimate for Zion-like and Surry-like plants.

Some two-loop Westinghouse plants (Ginna, Kewaunee, Point Beach 1 & 2, Prairie Island 1 & 2, and H. B. Robinson) have two floors between the cavity exit and the upper dome: the seal table room floor and the operating deck floor. However, these plants have significant direct line-of-sight debris-transport pathways from the instrument tunnel exit to the containment dome. For this reason, the subcompartment debris transport fraction (f_{sub}) which were determined from experiments and recommended for all Zion-like and Surry-like plants, cannot be used for these plants.

For plants with direct line-of-sight debris-transport pathways from the instrument tunnel exit to the containment dome, Pilch et al. [24] recommended that the subcompartment debris transport fraction (f_{sub}) could be calculated with a simple area ratio model that is tied to the Surry database, specifically SNL IET-10. In Surry, the area of the seal table room opening (A_{str}) over the area of the instrument tunnel exit ($A_{cav\,exit}$) was 0.28, and in IET-10 the fraction of the debris that was ejected from the cavity exit that entered the seal table room was about 0.28. Thus, debris ejected at an opening will go through that opening, and transport through openings can be calculated with simple area ratios. The exception to this rule is if the cavity exit is much larger than the seal table room opening and the seal table room opening is directly above the half of the cavity furthest from the RPV. DCH experiments have shown that almost all of the debris was entrained out of the instrument tunnel along the wall furthest away from the reactor pressure vessel. In some plants with very large cavity exit areas, virtually all of the debris would be ejected from the half of the opening furthest away from the RPV. Furthermore, if area ratios are greater than 1, the fraction should be limited to 1 since more than 100% of the debris cannot be ejected through an opening. The simple Pilch et al. [19] model for the subcompartment debris transport fraction (f_{sub}) is given by

$$f_{sub} = \min\left\{ \frac{A_{str}}{\frac{1}{2}A_{cav\,exit}} ; 1 \right\} \min\left\{ \frac{A_{op\,deck}}{A_{str}} ; 1 \right\} - 0.05, \qquad (24)$$

where A_{str} is the area of the seal table room opening, $A_{cav\,exit}$ is the area of the instrument tunnel (cavity) exit, and $A_{op\,deck}$ is the area of the opening in the operating deck that is directly above the seal table.

Modeling of dome carryover with CONTAIN has no provision for inertially dominated transport through line-of-flight flow paths (e.g., seal table room in Zion or Surry). Debris is assumed to be carried by gas with a user-specified slip ratio, $S = V_g/V_d$ [22, 75], which in application is taken as independent of particle size although it is within the code's capability to do otherwise. The intercell flow of debris is proportional to the amount of airborne debris in the donor cell, in which the amount of airborne debris is determined by a balance of inflow, outflow, and trapping rates. Wu's [125] observation that fractional carryover was insensitive to large variations of liquid mass would seem to agree with CONTAIN's assumption that carryover is proportional to airborne mass in the donor cell.

Particle velocities indicative of debris flow through the cavity have been measured in a few tests by placing breakwires at regular intervals above the cavity exit. Table XVIII summarizes the limited data. The cross-sectionally averaged gas velocities at the cavity exit are computed from the known molar blowdown rates by assuming the gas is heated to near equilibrium (~ 2500 K) with the debris. The data and analyses indicate that the slip ratio decreases with increasing hole size and may be a function of cavity geometry. There have been no attempts to correlate the data, so all the scaling implications have not been assessed. Using the HARDCORE model, Sienicki and Spencer [104] computed the coupled debris and gas velocities during dispersal and found that the slip ratio ranged from 5 to 3.7 at the beginning and end of the dispersal interval. CONTAIN calculations suggest that containment loads are not sensitive to uncertainties in debris velocities within the cavity.

Trapping in CONTAIN is modeled as a first-order rate equation

$$\frac{dM_d}{dt} = -\lambda M_d, \tag{25}$$

TABLE XVIII

MEASURED DEBRIS VELOCITIES IN DCH EXPERIMENTS

	SNL/TDS-5	SNL/TDS-6	SNL/TDS-8	SNL/LFP-8a	SNL/WC-1	SNL/WC-3
	← Allen et al. [28]			Allen et al. [29]	Allen et al. [31]	
Cavity	Surry	Surry	Surry	Surry	Zion	Zion
P^0 (MPa)	3.8	4.1	3.99	2.9	4.6	3.8
D_h (m)	0.067	0.066	0.067	0.042	0.041	0.10
$\dot{N}$ (mole/s)	209	267	250	100	197	652
$V_{g,\text{cav}}$ (m/s)	162	207	155	124	225	325
$V_{d,\text{meas}}$ (m/s)	36	34	35	16	17.5	54
$S = V_g/V_d$	4.5	6.1	4.4	7.8	12.8	6.0

where λ is an inverse time scale that can in some cases depend upon M_d; hence Eq. (25) is not always linear. The trapping parameter λ can be determined in a number of ways: it can be (1) user-specified, (2) based on the gravitational fall time of a particle, or (3) based on the time of flight between selected structures together with a Kutatuladze criterion for determining whether deentrainment occurs on any given structure impact.

In analyses of the Zion-geometry IET experiments, the CONTAIN model predicted the dome carryover fraction to within the scatter in the experimental data (which was substantial, about a factor of two). It predicted the Surry-geometry carryover fractions to within 20–40%. In the less prototypic SNL/LFP experimental geometries, however, the model overpredicted the carryover by up to an order of magnitude. Decoupling of debris trajectories from the gas flow was suggested by Williams *et al.* [22] as a possible explanation for the poor agreement in the SNL/LFP experiments.

VII. Chemical and Thermal Interactions of Debris with Gas

Energy in the debris can contribute to DCH loads only insofar as this energy is transferred to the containment atmosphere. Hence models for the processes controlling this energy transfer play a central role in DCH analysis. The two principal processes by which energy is transferred from the debris to the gas phase are debris–gas heat transfer and metal–steam reaction followed by combustion of the hydrogen produced. Direct metal–oxygen reaction can also occur, but in typical DCH scenarios the dominant oxidation process is reaction with steam. This is especially true for compartmentalized containment geometries, in which most debris is confined to cavity and subcompartment volumes that quickly become oxygen-starved.

In this section, we first discuss debris oxidation chemistry and the related chemical equilibria, and then move on to discuss the processes that control interaction rates for both heat transfer and chemical reaction. It is often helpful to provide numerical examples for some of the quantities of interest. For the sake of consistency, we define a set of "reference conditions" for numerical evaluation to be a 50:50 mixture of steam and hydrogen at 1500 K and 0.5 MPa pressure, with a debris temperature of 2500 K. Except where explicitly noted, numerical examples will be provided for the reference conditions. The conditions are typical of the cavity and subcompartment during DCH, and it is in the cavity and in the subcompartment volumes near the cavity that most of the debris–gas interactions occur for compartmentalized geometries. The reference conditions are less applicable to the main volume of the containment, for which temperatures are usually lower, hydrogen concentrations much lower, and other gases are present.

Substantial oxidation of the metallic components of core debris can occur in DCH events, and a relatively simple prescription in which the metals are oxidized in order of decreasing chemical reactivity appears to be adequate for DCH modeling. DCH modeling has traditionally emphasized the thermal and chemical interactions of gas and blowdown steam with debris that is airborne as dispersed particulate. Characteristic interaction time scales for the debris–gas interactions are estimated, and it is concluded that the debris–gas interactions can approach the equilibrium state in the cavity during the time of maximum debris dispersal rates; departure from equilibrium may be greater under some other conditions. Application of a heat–mass transfer analogy leads to the conclusion that the degrees to which thermal and chemical equilibrium are approached will be approximately equal, and comparison with experimental results provides support for the assumption that the violence of dispersal processes in the cavity, and possibly the subcompartments, keeps the debris well-mixed and minimizes any drop-side mass transport rate limitations on reaction rates as long as debris remains molten. Contributions to DCH from nonairborne debris interacting with blowdown steam are considered, and evidence is cited that these contributions might be significant in some cases. The remainder of this section discusses these subjects in more detail.

A. Debris Oxidation Chemistry

The oxidation of the melt has been well demonstrated in DCH experiments employing chemically reactive melt driven with steam (e.g., [28, 41, 52]). Substantial hydrogen production is observed in open-geometry experiments with inert containment atmospheres [28, 29, 31], indicating that much of the hydrogen production likely occurs in the cavity, since there would be little opportunity for additional hydrogen production once the melt leaves the cavity in these experiments. Tutu *et al.* [118, 152] were the first to recognize the potential for extensive hydrogen production in the cavity. Extensive hydrogen production in the cavity and/or the subcompartments was also a key prediction of early CONTAIN analyses of DCH [76]. All major models explicitly represent DCH-produced hydrogen.

Of the metals that can be present in core debris, the most important species that are subject to oxidation are zirconium, chromium, and iron. In addition, unreacted aluminum is potentially present in the thermite melts used in most of the DCH experiments. We consider only these four metals here, even though others (nickel and various control rod materials) may be present. Neglect of these other species is justified because they can be present only in relatively small amounts and most of them are chemically less reactive than the metals that are considered here.

The following oxidation processes are considered, at least implicitly, in most DCH models (including those discussed in Sect. III):

$$Zr + O_2 \rightarrow ZrO_2 \text{ and } Zr + 2H_2O \rightarrow ZrO_2 + 2H_2,$$

$$Al + \tfrac{3}{4}O_2 \rightarrow AlO_{1.5} \text{ and } Al + \tfrac{3}{2}H_2O \rightarrow AlO_{1.5} + \tfrac{3}{2}H_2,$$

$$Cr + \tfrac{3}{4}O_2 \rightarrow CrO_{1.5} \text{ and } Cr + \tfrac{3}{2}H_2O \rightarrow CrO_{1.5} + \tfrac{3}{2}H_2,$$

$$Fe + \tfrac{1}{2}O_2 \rightarrow FeO \text{ and } Fe + H_2O \rightarrow FeO + H_2,$$

$$H_2 + \tfrac{1}{2}O_2 \rightarrow H_2O. \tag{26}$$

Except for CONTAIN, the DCH models described in Sect. III impose a reaction hierarchy that favors oxygen over steam as long as any unreacted oxygen is available. In CONTAIN, it is assumed that steam and oxygen react in parallel at rates primarily controlled by the rate of diffusion to the surface. In this model, the default assumption is that hydrogen produced by metal–steam reactions immediately recombines with oxygen as long as any oxygen is present in the cell of interest.

The metals are listed in order of decreasing reactivity in Eq. (26) as evaluated from the standard Gibbs free energies of the reactions. The DCH models impose a reaction hierarchy such that the less reactive metals in any given parcel of debris do not begin to react until all the more reactive metals are completely consumed. In addition, the models assume that all reactions can go to completion except for the iron–steam reaction, for which chemical equilibrium is considered by most of the models.

The assumption that all reactions can go to completion introduces little error in the case of metal–oxygen reactions. The adequacy of these approximations for the metal–steam reactions can be judged from Table XIX, in which calculated hydrogen–steam ratios at chemical equilibrium are given for the four metal–steam reactions at temperatures of 2000 and

TABLE XIX

EQUILIBRIUM H_2 / H_2O RATIOS AND HEATS OF REACTION PER G-MOLE STEAM

| Metal | Equilibrium H_2/H_2O ratio | | ΔH_{rx} at 2500 K |
	2000 K	2500 K	kJ/g-mole steam
Zirconium	9.0×10^5	2.6×10^4	-299
Aluminum	3.0×10^5	7.2×10^3	-265
Chromium	59	13.0	-138
Iron	2.2	2.1	-2.0

2500 K. In each case it is assumed that metals and their oxides are immiscible and that both the metal and its oxide are pure and undiluted with other species. In addition, the table gives the enthalpy of reaction at 2500 K, expressed as kJ/g-mole of reacting steam.

The equilibrium steam–hydrogen ratios show that any error resulting from assuming complete reaction of zirconium and aluminum is totally negligible. Even for chromium the error is quite small. For iron the equilibrium effect is more important, and it is therefore taken into account in most DCH models. In CONTAIN, equilibrium is modeled assuming that metals and oxides are immiscible and that iron oxide forms an ideal solution with the other oxides present; iron metal is treated as being pure because iron cannot begin to react in this model until all other metals are consumed. With these assumptions, the hydrogen–steam ratio at equilibrium is given by

$$\frac{X_{H_2}}{X_{H_2O}} = K_{eq}/X_{FeO},\qquad (27)$$

where the X's are mole fractions and K_{eq} is the equilibrium constant. For a rather wide range of temperatures, $K_{eq} \sim 2$ for the iron–steam reaction. Hence, for a system including only iron and its oxide, the iron–steam reaction would proceed only until two-thirds of the steam is converted to hydrogen. In a typical DCH calculation, the FeO mole fraction in the oxide phase is considerably less than unity as a result of dilution by the other oxides present. Equation (27) then implies that the H_2–H_2O ratio at equilibrium is considerably greater than 2.

It also appears from Table XIX that the chemical reactivities (as represented by the equilibrium H_2–H_2O ratios) are sufficiently well separated that the reaction hierarchy assumption does not result in significant error. Zirconium and aluminum might appear to be a partial exception; however, these two species would not normally appear in the same calculation. In any case, the energies released by the steam reactions are sufficiently similar for zirconium and aluminum that it does not matter greatly which metal is assumed to react with a given quantity of steam.

B. Interactions of Gas with Airborne Debris

If we assume that chemical reactions are controlled by mass transport processes only (i.e., there are no intrinsic chemical kinetic limitations), heat transfer rates and chemical reaction rates can be written in the general form

$$\dot{Q} = h_{ht} A_{dg}(T_d - T_g) \quad \text{and} \quad \dot{N} = h_{mt} \rho_{flm}(X_g - X_e), \qquad (28)$$

where h_{ht} and h_{mt} are effective heat and mass transfer coefficients that account for both drop-side and gas-side limitations to transport processes,

$$\frac{1}{h_{ht}} = \frac{1}{\left(\dfrac{k}{L}\,\mathrm{Nu}\right)_d} + \frac{1}{\left(\dfrac{k}{L}\,\mathrm{Nu}\right)_g + \sigma\epsilon\left(T_d^2 + T_g^2\right)(T_d + T_g)},$$

$$\frac{1}{h_{mt}} = \frac{1}{\left(\dfrac{D}{L}\,\mathrm{Sh}\right)_d} + \frac{1}{\left(\dfrac{D}{L}\,\mathrm{Sh}\right)_g},$$

(29)

where k is the thermal conductivity and D is the diffusivity. The concentration X_g refers to steam concentration in the actual flow while the concentration X_e refers to steam concentration at equilibrium. For airborne debris, the characteristic length L is the particle diameter d.

DCH processes are sufficiently violent within the cavity that drop-side limits to heat and mass transfer are generally considered negligible; i.e., Nu_d and Sh_d are large owing to the violent mixing. In the dome, where particles begin to freeze, or when large quantities of water quench debris in the cavity and subcompartments, internal mixing of drops is less and drop-side limits to heat and mass transfer may become important. Baker [153] concluded that $\mathrm{Sh}_d \sim 4/\pi$ under such circumstances, and by analogy, we expect $\mathrm{Nu}_d \sim 4/\pi$ also. Baker found that the drop-side limits to mass transport could be important when the arguments for violent mixing do not apply, but that including the drop-side thermal diffusion limits had little effect upon the predictions of his model.

For airborne particles, the gas-phase convective heat and mass transfer contributions can be computed from the Ranz and Marshall [154] correlation

$$\mathrm{Nu} = 2 + 0.6\,\mathrm{Re}^{1/2}\,\mathrm{Pr}^{1/3} \quad \text{and} \quad \mathrm{Sh} = 2 + 0.6\,\mathrm{Re}^{1/2}\,\mathrm{Sc}^{1/3} \qquad (30)$$

for flow over a sphere. Here convective mass transfer coefficients are evaluated by analogy to heat transfer. Generally, gas phase convective heat transfer dominates the radiative contributions in the cavity. Neglecting drop-side limits and radiation, these correlations imply interaction rates that vary as $d^{-3/2}$, except at low Reynolds numbers.

Evaluating Eq. (30) requires knowledge of the velocity of the debris relative to the gas. In Sect. VI it was argued that the debris particles cannot follow the gas flow closely, and debris velocities may be only a fraction of the gas velocities. Hence it is adequate to assume that the relative debris–gas velocities are equal to the gas velocity. Mean gas velocities may be estimated from the blowdown rate and the cross-section for flow. For the Zion SNL/IET experiments the peak steam blowdown rates were ~ 350 g-mole/s and the cavity flow cross-section, A_x, was about 0.077 m². Assuming the ideal gas law and the DCH reference conditions gives a flow velocity of about 110 m/s in the cavity and about 170 m/s in the chute connecting the cavity to the containment ($A_x \sim 0.052$ m²). Since gas flow rates are approximately proportional to $P_{0,\mathrm{acc}} d_h^2 / A_x$, where d_h is the hole size in the melt generator (or RPV in NPP analysis), these velocities can vary considerably for different scenarios. It is also important to remember that velocity stratification effects can substantially enhance velocities adjacent to cavity surfaces, where debris entrainment is expected to occur. In the dome, however, bulk gas velocities are negligible and the relative debris–gas velocity is determined by the debris velocity, not the gas velocity.

1. DCH Time Scales

It is important to have some understanding of the relative rates of the various processes controlling DCH, and it is convenient to compare these rates in terms of the characteristic time scales of the processes involved. In what follows, we first consider compartmentalized geometries and are interested in processes in the cavity and the subcompartments; we consider the dome at the close of this subsection. We let the subscripts g, d, hx, and rx refer to gas, debris, heat transfer, and chemical reaction, respectively, and define the following quantities:

$\tau_{hx,d}$ = time scale for debris to transfer all its energy to the gas,

$\tau_{rx,d}$ = time scale for debris to completely react chemically with the gas,

$\tau_{c,g}$ = time scale for gas to flow out of the volume of interest,

$\tau_{c,d}$ = debris airborne residence time for the volume of interest,

ψ_{hx} = heat capacity ratio, $m_d c_{p,d} / m_g c_{p,g}$,

ψ_{rx} = molar equivalents ratio, $\Sigma \nu_i n_{i,d} / n_{g,ox}$,

$\tau_{\mathrm{eq},hx}$ = thermal equilibration time $\approx \tau_{hx,d}/(1 + \psi_{hx})$, and

$\tau_{\mathrm{eq},rx}$ = chemical equilibration time $\approx \tau_{rx,d}(1 + \psi_{rx})$.

The quantity ψ_{hx} is equivalent to the heat capacity ratio ψ that was defined in Sect. III except that here it is defined locally for a volume that is small compared with the total containment volume and for which the pressure can equilibrate with the total containment volume; hence we use c_p instead of c_v as in Sect. III.

In the definition of ψ_{rx}, n_i refers to the number of moles of the ith reactive metal in the debris, ν_i is the stoichiometry subscript of the corresponding oxide when written in the form MO_ν, and $n_{g,ox}$ is the number of oxygen equivalents of oxidizing gas. For chemical reaction, the quantity ψ_{rx} plays a role analogous to ψ_{hx} in limiting heat transfer; that is, if $\psi_{rx} > 1$, the extent of reaction is limited by the steam supply rather than the metal supply.

We estimate the first four time scales defined above from the following:

$$\tau_{hx,d} = \frac{c_{pd} m_d \, \Delta T_{dg}}{\dot{Q}} \quad \text{and} \quad \tau_{rx,d} \approx \frac{\Sigma_i \, \nu_i n_i}{\dot{n}_{rx}},$$

$$\tau_{c,g} = \frac{RT_g}{\dot{n}_g PV} \quad \text{and} \quad \tau_{c,d} \approx \frac{S_d \tau_{c,g}}{1 + \lambda_{tr} S_d \tau_{c,g}}. \tag{31}$$

Here, $\dot{n}_{rx}$ is the molar hydrogen generation rate, $\dot{n}_g$ is the molar gas flow rate through the volume V of interest, S_d is the debris–gas slip factor, and λ_{tr} is the debris trapping rate. During the period of debris entrainment and dispersal, any debris trapped in the cavity is likely reentrained and the net trapping rate in the cavity is small. In the subcompartments, where turbulence is less, trapping dominates and $\tau_{c,d} \approx 1/\lambda_{tr}$. Due to relatively rapid trapping, debris airborne residence times in the subcompartments are probably less than the time required to transport debris out of the cavity. If this is the case, cavity interactions dominate total debris–steam interactions, and in what follows we focus attention on the cavity, although a similar analysis could be developed for the subcompartments.

Values of $\tau_{hx,d}$ and $\tau_{rx,d}$ calculated using Eqs. (29)–(31) are plotted against particle diameter in Fig. 12. The DCH reference conditions defined previously and a debris–gas relative velocity of 100 m/s were assumed. Separate thermal time scales for the convective and radiative transfers are given, as well as the time scale for the combined effect. The convective heat transfer dominates radiative heat transfer (calculated assuming a debris emissivity, ϵ, of 0.8) except for the largest sizes, which normally would contribute little to the total debris–gas interaction. Furthermore, taking into account velocity stratification effects could increase the convective transfer rates by up to a factor of two while it would not affect radiative transfer. These results support the assumption cited previously that convection is the principal heat transfer mode.

Reaction time scales ($\tau_{rx,d}$) were calculated for two different melt compositions. The first consisted of 28.5 wt% iron, 12.9 wt% zirconium,

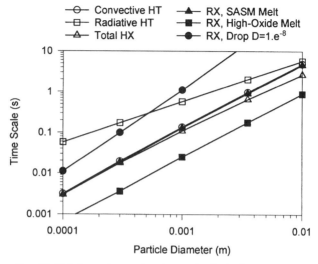

FIG. 12. Debris–gas interaction time scales for individual particles.

and 58.6 wt% UO_2. The second composition was 4.5 wt% iron, 2.25 wt% zirconium, and 93.2 wt% UO_2. The compositions are simplified representations of, respectively, the SASM melt [20] and the composition defined by Pilch *et al.* [18]. These compositions are labeled "SASM" and "high-oxide" in Fig. 12. For an isolated particle, the reaction time scale is proportional to the amount of metal in the particle when gas-phase mass transport is assumed to be the only factor limiting reaction rates. Hence, reaction time scales are longer for the SASM melt composition than for the highly oxidic composition.

If drop-side limits are controlling, the time scale for reaction is not necessarily less for the highly oxidic composition than it is for the SASM composition. The curve labeled "Drop $D = 10^{-8}$" is calculated assuming drop-side mass transport limitations control reaction rates with a diffusivity, D_d, of 10^{-8} m^2/s, which is estimated to be typical of a stagnant liquid drop composed of the materials of interest at temperatures of 2500 K [76, 153]. Drop-side reaction rate limitations this large would substantially slow reaction rates. For example, CONTAIN analyses of the SNL/IET-1R and SNL/IET-3 experiments performed with $D_d = 10^{-8}$ m^2/s underpredicted experimental hydrogen production by more than a factor of two and ΔP was underpredicted by 35 percent in SNL/IET-3, in which combustion of DCH-produced hydrogen was a major contributor to ΔP.

2. Validity of the Equilibrium Approximation

If the debris–gas equilibrium time scales are short compared with the debris and gas residence times, the equilibrium assumptions made by the simple models described in Sect. III may be expected to yield good results. For the SNL/IET Zion-geometry experimental conditions, a consideration of the cavity volume and the gas flow rates cited previously indicates that $\tau_{c,g}$ during the debris dispersal period is $\sim 0.01–0.015$ s, and about twice that if the volume of the chute connecting the cavity to the containment vessel is included.

For particle sizes of about 1 mm, Fig. 12 shows that $\tau_{hx,d} \sim 0.1$ s for debris to transfer all its stored energy. If we assume that the debris–gas slip factor is about 5, the debris residence times in the cavity and chute volumes will be about 5 times the gas residence times and hence of the same order of magnitude as $\tau_{hx,d}$. However, when ψ_{hx} is large, the debris needs to transfer only a small fraction of its energy before equilibrium with the gas is achieved; consequently, it is more appropriate to compare $\tau_{eq,hx}$ with the gas residence time, $\tau_{c,g}$, that was estimated above. Using the relationship $\tau_{eq,hx} \sim \tau_{hx,d}/(1 + \psi_{hx})$ and assuming $\psi_{hx} \sim 5$, as estimated for the Zion/IET experiments during the period of peak debris dispersal rates, we observe that $\tau_{eq,hx} \sim 0.017$ s, which is less than the debris residence time, $\tau_{c,d} \sim 0.06$ s, in the cavity. Consequently, debris–gas thermal equilibrium is favored.

When $\psi_{hx} \gg 1$, the preceding argument underestimates the plausibility of the equilibrium assumption because the particle size assumed, $d_{30} \sim 1$ mm, corresponds to the mass mean diameter typically observed in the experiments. Actually, if we consider the initial heating rate when hot debris is first introduced to the cool gas, the Sauter mean diameter, d_{32} is a more appropriate particle size for evaluating $\tau_{hx,d}$. For a lognormal size distribution with $d_{30} \sim 1$ mm and $\sigma_g \sim 4$ (as is typical of the experiments), $d_{32} \sim 0.38$ mm, in which case $\tau_{hx,d} \sim 0.03$ s and $\tau_{eq,hx} \sim 0.005$ s, assuming that $\psi_{hx} \sim 5$. This result more decisively favors the equilibrium approximation. As the debris–gas interaction continues, the smallest particles quickly cool and the mean size of those particles that remain hot increases, and basing heating rates on d_{32} then overestimates the rate of approach to equilibrium. However, when $\psi_{hx} \gg 1$, interaction of only the small end of the size distribution is sufficient to result in a reasonably close approach to equilibrium. For large values of ψ_{hx}, then, the assumption of debris–gas thermal equilibrium appears to be a reasonable approximation.

Considering the approach to chemical equilibrium, we may note that ψ_{rx} is also greater than unity in the Zion/IET experiments during the period of peak debris dispersal rates, and this is true in most other experiments performed to date. For this case, an application of the gas-phase heat–mass transfer analogy (discussed further in Sect. VII.C) leads to the conclusion $\tau_{eq,rx} \sim \tau_{eq,hx}$, implying chemical equilibrium is also a reasonable approximation. For more highly oxidic melts ψ_{rx} is typically less than unity, and a comparison of the individual particle reaction times, $\tau_{rx,d}$, with the airborne residence time $\tau_{d,c}$ provides a better measure of the approach to equilibrium. Since $\tau_{rx,d}$ is short for highly oxidic particles, we again can conclude that the equilibrium assumption is a reasonable approximation except, possibly, for the large end of the particle size distribution.

The fact that the time scales for the various competing processes are of the same order of magnitude might be taken to imply that there could be great sensitivity to the particle size and to various parameters affecting $\tau_{c,d}$ and $\tau_{c,g}$. Actually, CONTAIN analysis of the Zion SNL/IET experiments revealed only quite limited sensitivity to particle size. The broad size distribution tends to wash out the sensitivity to particle size that might otherwise be expected.

It obviously would be much more satisfying to have experimental evidence of the validity of the equilibrium assumption. However, convincing experimental evidence is not available. Equilibrium would imply that gas temperatures approaching 2500 K should exist at the cavity exit and adjacent subcompartment volumes. Experimental measurements of these temperatures [40, 41, 52] generally fall in the range 1000–2000 K, which would imply considerable departures from equilibrium exist. However, the experimental measurements are very difficult to perform and the thermal response times of the thermocouples used are of the same order as the duration of the peak temperatures expected. Hence, the failure to confirm temperatures as high as implied by the equilibrium assumption is not considered to be a very strong argument against this assumption.

To lowest order, energy densities and velocities do not depend strongly upon facility scale, and the residence times ($\tau_{c,g}$ and $\tau_{c,d}$) therefore tend to vary linearly with scale. The review of particle sizes in Sect. V.B indicated that particle sizes are approximately independent of scale, so one would expect full-scale events to favor even more strongly the equilibrium approximation.

3. Debris–Gas Interactions in the Dome

Unobstructed flight path lengths are much greater in the dome than in the cavity or subcompartments, and this might be thought to favor achievement of equilibrium. However, there are several factors that preclude drawing firm conclusions to this effect:

1. Multiple deentrainment and reentrainment processes in the cavity likely increase cavity interaction times, which may be greater than considerations based upon flight paths alone might suggest.
2. The presence of noncondensible gases other than hydrogen in the dome, as well as lower gas temperatures and somewhat lower pressures, reduce heat and mass transfer rates.
3. For the dome volume $\psi < 1$ (often $\ll 1$), which implies that even particles at the large end of the size distribution must interact efficiently in order for the equilibrium assumption to be accurate.

Of the equilibrium models described in Sect. III, those which consider equilibrium debris–gas interactions in the dome overpredict ΔP when applied to experiments in open geometries. There are two possible reasons for any failure to achieve equilibrium. The first is that the interaction time scales may not be short compared with the airborne residence time scales. This effect likely decreases with increasing scale as discussed previously, and equilibrium might therefore by more nearly approached at plant scale. However, a second reason for failure to reach equilibrium is that debris may be largely restricted to a jet or cloud occupying only a fraction of the dome volume, rather than being well-mixed throughout the volume. This "cloud effect" could restrict the amount of gas available to interact with the debris. In effect, transfer processes between the cloud and the dome atmosphere could become dominant and limiting. The cloud effect, as a first approximation, is not expected to decrease with increasing scale. Experimental data permitting these two effects to be distinguished are not available. We would also note that, if debris scatters off the dome and falls to the floor, debris–gas interaction times will again scale with the facility size, which could permit a closer approach to equilibrium at full scale than at experimental scale even if the cloud effect does limit the efficiency of the initial interaction.

C. HEAT AND MASS TRANSFER ANALOGY

There is a close relationship between heat transfer and mass transfer, especially in the gas phase. If gas-phase transport controls the extent of debris–gas heat transfer and mass transfer, one would expect to find a relationship between the efficiency of heat transfer and the efficiency of chemical reaction in DCH. Williams [80] applied the gas-phase heat–mass transfer analogy to obtain a quantitative relation between the amount of hydrogen produced and the amount of energy transferred during the debris–steam interactions, given the validity of certain assumptions summarized below. An important feature of the argument was that it did not depend upon any assumptions concerning the geometric configuration of the debris–gas interface or flow velocities. Thus, the argument applies whether the debris is present as discrete airborne particles, films on structures, or a churn-turbulent mixture in the cavity. The discussion that follows summarizes the principal results.

The assumptions required for the analysis are:

1. The atmosphere interacting with the debris consists only of hydrogen and steam.
2. Metal is in excess of available steam ($\psi_{rx} > 1$).
3. The local debris–gas heat capacity ratio, ψ_{hx}, is large and cooling of the debris may therefore by neglected.
4. Conduction–convection (not radiation) is the dominant debris–gas heat transfer mode.
5. The debris is well mixed, with the surface area available for chemical reaction therefore being equal to the surface area available for heat transfer.
6. Mass transfer processes within the debris do not limit chemical reaction rates.

The first assumption is valid for the cavity and subcompartment volumes close to the cavity, since these volumes are quickly purged of air, and it is in these volumes that most of the debris–gas interactions occur for compartmentalized geometries. The validity of the second and third assumptions depends upon the initial conditions. The fourth assumption is satisfied for particles (Sect. VII.B); is may be more questionable for debris films on structures. The last two assumptions are the key phenomenological assumptions of the model; Williams considered them plausible because the extreme turbulence of the DCH event was considered likely to keep the debris well mixed. An important goal of the analysis was to provide a test of these assumptions.

Based upon the gas-phase heat–mass transfer analogy, it was argued that Nu and Sh are approximately equal. With drop-side rate limitations and radiant heat transfer not considered, manipulation of Eqs. (28) and (29) led to the conclusion that the degree to which debris–gas thermal equilibrium would be reached is approximately equal to the degree to which chemical equilibrium is reached:

$$\frac{d\{\ln(\Delta T_{dg})\}/dt}{d\{\ln(\Delta C_{dg})\}/dt} = \frac{\text{LeNu}}{\text{Sh}} \approx \text{Le}^{2/3} \approx 1. \qquad (32)$$

The debris–gas heat transfer rate was predicted to be related to the hydrogen production rate by

$$\dot{Q} \approx \frac{k_{\text{flm}}}{D_{\text{flm}}} \frac{1}{\rho_{\text{flm}}} \frac{\Delta T_{dg}}{\Delta X_{dg}} \dot{n} = \frac{k_{\text{flm}}}{D_{\text{flm}}} \frac{R_u T_{\text{flm}}}{P} \frac{T_d - T_0}{X_0 - X_e} \dot{n}. \qquad (33)$$

Here the subscript 0 refers to steam conditions after exiting the accumulator but prior to debris interactions, and the subscript e refers to conditions at the debris surface, assuming local debris–gas chemical equilibrium exists there. The subscript flm refers to film properties, calculated assuming $T_{\text{flm}} = 0.5(T_g + T_d)$ and $X_{\text{flm}} = 0.5(X_g + X_e)$, as recommended by Bird et al. [108]. The dependencies of k_{flm} and D_{flm} upon temperature and pressure are such that the quantity $(k_{\text{flm}} T_{\text{flm}}/D_{\text{flm}} P)$ in Eq. (33) is only weakly dependent upon conditions as they evolve during the debris–gas interactions, and it is a reasonable approximation to treat this quantity as being a constant and evaluate it for the DCH reference conditions defined previously. This leads to the result that about 70 kJ of energy is transferred to the gas for each g-mole of hydrogen generated, assuming $X_0 = 1$ and $T_0 = 500$ K for the incoming blowdown steam.

Williams [80] considered the SNL/WC and SNL/LFP test series (Sect. II) as providing a good data set to use in testing the concepts outlined here because the test parameters were such as to satisfy the first three assumptions listed at the start of this discussion, and because the experiments were simple in design and most of the potentially important experimental parameters were held constant, simplifying the interpretation. The most important experimental parameters, for the present purpose, that were varied were the unobstructed flight path in the Surtsey containment vessel (varied from 0.95 to 7.7 m) and the amount of debris dispersed from the cavity (which varied from 20 to 85% as a result of variations in the driving steam pressure and the melt generator hole size).

Williams [80] concluded that hydrogen production did not correlate with the length of the flight path in the containment vessel, suggesting that most of the hydrogen generation takes place in the cavity. This result is reasonable since the containment atmosphere was inert (argon gas). However, the containment pressurization (ΔP) did correlate with both the flight path and the mass of debris dispersed from the cavity, M_{disp} [29]. In Fig. 13, $\Delta P/M_{\text{disp}}$ is plotted against flight path (open square symbol), and a least-squares fit is also shown. The correlation is only fair ($R^2 = 0.62$), and extrapolation back to zero flight path results in a large intercept on the ordinate axis. Williams [80] therefore concluded that there was an important component of total energy transfer to the containment atmo-

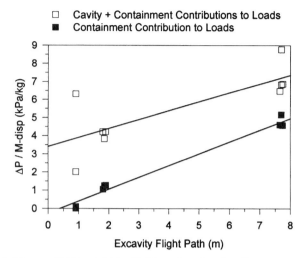

FIG. 13. $\Delta P/M_{\text{disp}}$ and $[\Delta P - \Delta P(Q_{\text{cav}})]/M_{\text{disp}}$ plotted against flight path in the SNL/LFP and SNL/WC experiments as a test of the heat–mass transfer analogy.

sphere that is not proportional to either the dispersed mass or the flight path length in the containment.

Debris–steam heat transfer occurring in the cavity is a plausible source of this additional energy, and the concepts outlined here were tested by using Eq. (33) and the experimental hydrogen production numbers to estimate the amount of pressurization, $\Delta P(Q_{\text{cav}})$, that resulted from heat transfer in the cavity:

$$\Delta P(Q_{\text{cav}}) = \frac{R_u}{VC_v} \frac{k_{\text{flm}}}{D_{\text{flm}}} \frac{R_u T_{\text{flm}}}{P_{\text{cav}}} \frac{T_d - T_g^0}{X_0 - X_e} N_{\text{H}_2}, \qquad (34)$$

where $N_{\text{H}2}$ is the moles of hydrogen produced in the experiment. The amount of pressurization, ΔP_{con}, that should be attributed to the flight of the debris through the containment was estimated by subtracting out $\Delta P(Q_{\text{cav}})$. A small contribution (10–20% of $\Delta P(Q_{\text{cav}})$) that results from the addition of blowdown steam to the containment vessel was also subtracted out. Resulting values of $\Delta P_{\text{con}}/M_{\text{disp}}$ are also plotted against flight path in Fig. 13 (filled squares), and the linear least-squares line is also given. The correlation is greatly improved ($R^2 = 0.984$), and the intercept on the ordinate axis is now close to the origin.

Williams [80] concluded from these results that assumptions 5 and 6 stated at the beginning of this section are probably valid, at least as a first approximation, because serious failure of these assumptions would reduce hydrogen production much more than heat transfer, and Eq. (34) would have underpredicted $\Delta P(Q_{cav})$. Assuming well-mixed debris and neglecting drop-side rate limitations has therefore been recommended in performing DCH calculations using the CONTAIN code [22].

Williams [80] also concluded that debris–gas heat transfer is unlikely to quench chemical reaction so long as the reactive metals zirconium, aluminum, or chromium are present, because the estimated heat transfer occurring in parallel with hydrogen generation (~ 70 kJ/g-mole) is less than the reaction energy (Table XIX). However, the iron–steam reaction does not generate sufficient energy to make up for this energy loss. Quenching of the iron–steam reaction due to debris cooling is therefore possible.

D. INTERACTIONS OF NONAIRBORNE DEBRIS

Traditional approaches to DCH analysis assume that interactions of debris with gas and blowdown steam may be ignored except for debris that is present as airborne particulate, in part because the surface–volume ratio is so much higher for airborne debris than for debris that is deposited on structures. However, this argument neglects the fact the airborne residence time is very short, while nonairborne debris may have considerably longer times in which it can interact. Although airborne debris interactions are expected to dominate whenever airborne debris is present in good supply, these interactions are largely limited to the supply of coherent steam in compartmentalized geometries, while nonairborne interactions may continue after the coherence interval. Hence they constitute a potential source of hydrogen and energy transfer in addition to the airborne interactions.

Interest in the nonairborne interactions was first highlighted by the observation [29] that a plot of hydrogen production versus mass of debris dispersed from the cavity in the LFP experiments did not come close to the origin when extrapolated back to zero debris dispersal. Allen et al. [29] therefore suggested that substantial hydrogen production could occur even if no debris is dispersed from the cavity. However, an alternative explanation is that hydrogen production is limited by the amount of steam coherent with the dispersal process, rather than by the melt mass. Although both explanations probably have some validity, there is currently no universal agreement as to the relative role played by each. One reason is that subjective differences in estimating the amount of coherent steam in these experiments can lead to conflicting conclusions. For example, Pilch et al. [23] concluded that the TCE model (which considers only coherent steam) does a reasonable job of estimating hydrogen production in the SNL/LFP and SNL/WC experiments, while Williams [80] concluded that coherent steam alone could not adequately explain the observed hydrogen production in the same experiments.

A more quantitative treatment of nonairborne interactions was given in Williams *et al.* (see [22, App. B]). They assumed the debris was present as films on structure surfaces and applied conventional heat transfer correlations for planar surfaces [108], together with the heat–mass transfer analogy, to examine the plausibility that nonairborne interactions can be significant. Quantitative results were given based upon a correlation of the form

$$Nu = 0.037Re_L^{0.8}Pr^{1/3} \quad \text{and} \quad Sh = 0.037Re_L^{0.8}Sc^{1/3}. \tag{35}$$

The subscript L refers to the characteristic length scale of nonairborne debris, which was taken to be that of cavity and subcompartment structures. Applying Eq. (35) to the Zion-geometry SNL/IET experiments, it was estimated that 20–40% of the hydrogen production and containment pressurization observed in these experiments could have been the result of the nonairborne debris interactions.

Williams *et al.* [22] therefore concluded that nonairborne interactions were potentially significant and should not be neglected. They also acknowledged large uncertainties in the analysis and that these uncertainties could either increase or decrease the importance of nonairborne interactions. For example, the Nusselt correlation in Eq. (35) is based upon flow parallel to a dry surface and may be too low for DCH conditions involving flows impinging on surfaces at various angles, and surfaces covered with wavy liquid films. On the other hand, debris–structure heat transfer could act to cool the films sufficiently rapidly to reduce the contribution of nonairborne debris. An approximate treatment of film cooling was given by Williams *et al.* [22], who concluded that this heat transfer could reduce the importance of nonairborne interactions, but it was unlikely to eliminate these interactions.

The simple equilibrium models do not generally include nonairborne interactions as contributions to containment loads. Including these interactions within the philosophy of the models would be difficult because a close approach to thermal and chemical equilibrium is not expected for the nonairborne interactions, and invoking an equilibrium assumption could be excessively conservative in many instances. MELCOR includes a representation of nonairborne interactions for both the debris–atmosphere interactions and the debris–structure heat transfer. However, the user must specify the controlling heat and mass transfer coefficients directly through the input, and little guidance is available as to the choice of these parameters.

The CONTAIN code models nonairborne debris by permitting the user to specify an effective particle size, d_t, for the nonairborne debris. Heat transfer and chemical reaction are then calculated using the same models as those applied to the airborne debris. An important limitation of the model is that heat transfer to the structures is not modeled. Empirically, d_t values of 0.01–0.02 m were found to give reasonable results for experiments conducted at 1/10-scale, and it was shown in Williams et al. [22, App. B] that this result is consistent with the treatment outlined in Eq. (35). A scaling rationale was provided for applying the model to DCH events at other scales, including application to NPP events. To a certain extent, the "nonairborne" model was viewed as a semiempirical means of representing any process that permits debris to interact with the noncoherent portion of the blowdown steam; the actual geometry of the debris–gas interface was considered to be uncertain. Uncertainties in the treatment were acknowledged to the important, and sensitivity studies exploring the potential impact of these uncertainties upon the results of interest were recommended.

In the CONTAIN analyses of the IET experiments in which hydrogen could burn, much of the contribution of nonairborne interactions to ΔP resulted from increased hydrogen production and the subsequent combustion of this hydrogen. Hence, the importance of the nonairborne interactions may be reduced in NPP applications, though not completely eliminated, if the melt composition is highly oxidic, as with Pilch et al. [18].

VIII. Hydrogen Combustion

Figure 14 illustrates the potential modes of hydrogen combustion during a DCH event. Hot hydrogen produced by metal steam reactions in the reactor cavity or subcompartment vents into the upper dome of the containment, where it can burn as a diffusion flame. The oxygen necessary to support the diffusion flame can only be supplied by entrainment of the preexisting atmosphere into the buoyant plume. This entrainment also carries H_2, present in the preexisting atmosphere, which can also burn within the plume.

The hot combustion products will rise to the some, possibly inducing mixing of the hot combustion products with the atmosphere. Again, there is the potential to burn preexisting hydrogen (and oxygen) as the atmosphere mixes with the hot combustion products.

An upward-propagating flame can be initiated if the preexisting atmosphere contains sufficient hydrogen and not too much steam. The

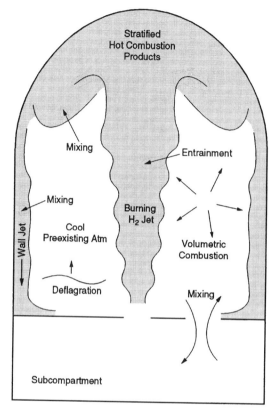

FIG. 14. Hydrogen combustion during DCH events.

preexisting atmosphere can be heated by radiation from the burning jet, by radiation from stratified combustion products in the dome, by particles of debris falling through the atmosphere, or by mixing of the containment atmosphere. With sufficient heating of the atmosphere, recombination (i.e., slow volumetric combustion) of H_2 and O_2 can occur. Because of the strong temperature dependence of the chemical reactions, a condition of more rapid volumetric combustion (autoignition) can occur if heating of the atmosphere exceeds a threshold temperature.

The hot jet vents into the upper dome on a time scale of the RCS blowdown, although most occurs during the debris dispersal period. The blowdown process may void the subcompartment of any oxygen. Some hydrogen, however, could remain in the subcompartment beyond the blowdown interval, but it can burn only if it mixes with oxygen-bearing regions in the dome. Owing to buoyancy forces, this hot hydrogen will eventually mix with the containment atmosphere and possibly burn. In addition, a cool boundary layer will sink along the wall of the containment causing mixing of hot combustion products in the dome with the cooler containment atmosphere. If the mixing region is sufficiently hot, then additional combustion of hydrogen initially in the atmosphere can occur.

Peak pressures in the containment occur on the time scale of the RCS blowdown. The key issue then is not how much hydrogen burns but how much hydrogen burns on the DCH time scale so it can contribute to peak containment pressure. Hydrogen combustion cannot contribute to peak containment pressure unless the energy release rate is greater than the heat transfer rate to structures.

A. JET COMBUSTION

Hot hydrogen can burn as a diffusion flame as it vents from the subcompartments to the dome. The hydrogen in the jet comes from two sources: hydrogen produced from reactions of blowdown steam with the metallic constituents of the melt, and hydrogen carried from the RCS with the blowdown steam.

The ignition of hydrogen jets requires both an energy source and sufficient oxygen in the atmosphere into which the jet vents. The energy source in a DCH event is both hot particles and the jet (gas) temperature itself. The jet temperature at which autoignition occurs is a function of the jet velocity and composition as well as of the containment atmosphere composition and temperature. Shepard [155] verified that pure hydrogen jets autoignited at temperatures above 903 to 1003 K. Zabetakis [156] also observed that pure hydrogen jets autoignited above ~ 943 K, and the autoignition temperature increased to ~ 1000 K when the steam concentration increased to $\sim 60\%$. A further increase to $\sim 70\%$ steam raised the jet autoignition temperature to ~ 1110 K. Zabetakis showed that the jet autoignition temperature was not a strong function of the velocity. However, the available data on dilute (low hydrogen concentrations) suggest that jet autoignition may be a strong function of composition.

There must also be sufficient oxygen in the atmosphere for a hydrogen jet to ignite. Observations from the Nevada Test Site (NTS) tests [157, 158] indicate that diffusion flames will not burn when the oxygen concentration drops below $\sim 5-8\%$. Although it has not been tested, the limiting oxygen concentration for diffusion flames is likely a function of temperature, so it is useful to examine the DCH database for additional insights. Jet inerting is sometimes observed in DCH experiments [159] when the oxygen concentration in the atmosphere is below $\sim 8\%$ and jet inerting was observed in the SNL/IET-5 test with the oxygen concentration below 5%. It should be noted that this oxygen limitation is a practical concern initially only when the containment contains ~0.18–0.39 MPa of steam. Should combustion by initiated, the NTS tests showed that burning would continue until the gas mixture became inert due to the combined effects of steam addition and oxygen depletion. Allen *et al.* [40] show that measured temperatures in the vent spaces in the Zion IET experiments exceed ~1000 K and the measurements were probably low because of slow thermocouple response times. Consequently, autoignition of the jet may occur for a sufficiently hydrogen-rich jet. Figure 15 shows that jet combustion was generally observed in the Zion SNL/IET tests when a reactive atmosphere was present.

Accident scenarios with large quantities of cavity water or large quantities of RPV water coejected into the containment with the melt might dilute and quench the hydrogen jet to the point that a diffusion flame cannot ignite or be sustained. The SNL/IET-8B experiment [40], conducted with a Zion cavity half full of water, indicated that significant quantities of hydrogen were both produced and burned. Gases venting to the dome never significantly exceeded saturation (~380 K) during the event, thus suggesting that hot particles could have acted as an ignition source for the jet. This observation, however, has not been reconciled with the need for particles to maintain temperatures above ~1000 K to be

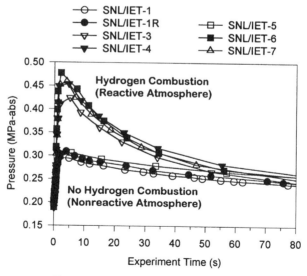

FIG. 15. Jet combustion in the SNL/IET tests.

effective as an ignition source. Note, however, that the jet composition in the IET-8B experiment cannot be determined.

The TCE and CLCH models bound uncertainties associated with ignition of hydrogen jets by assuming that hydrogen jets will always burn regardless of the source diameter, jet velocity, jet composition, and jet temperature. Ignition and burning of jets in the TCE model is subject only to the constraint that the preexisting atmosphere contain at least 5% oxygen. Some situations, such as large quantities of cavity or coejected water, could create conditions where jet combustion may not actually occur, but there is not clear indication of this in the existing database. The CONTAIN code has a provision for temperature-dependent and composition-dependent thresholds for jet combustion.

Details of the entrainment and jet combustion processes are difficult to predict. There can be multiple vent points of irregular shape (i.e., triangles, annuli) and size, and the source, strength, and composition can vary with the sequence. For these reasons, the TCE and CLCH models employ a stoichiometric argument in order to estimate the amount of hydrogen combustion in a reasonable fashion. CONTAIN [22] uses a similar model, diffusion flame burn (DFB), subject only to its more quantitative threshold criteria. In CONTAIN's standard prescription for DCH analyses, no temperature threshold is imposed as it is assumed that hot debris will provide ignition sources; it is also assumed that the molar ratio of diluent to hydrogen in the jet must be less than $9:1$ for combustion to occur.

The hot hydrogen–steam jet can burn only because oxygen is entrained into the jet, assuming, of course, that the atmosphere oxygen threshold criterion for jet combustion is met. Entrainment also carries some of the preexisting hydrogen into the hot jet, and this entrained hydrogen can also burn.

Assuming that entrainment into the jet carries sufficient oxygen to burn all the jet hydrogen plus any preexisting hydrogen carried along with the oxygen into the jet, the total number of hydrogen moles burned within the flame can be written as [23]

$$
N_{H_2}(\text{flame}) = N_{H_2}(\text{jet})\left(1 + \frac{X_{H_2}}{2X_{O_2} - X_{H_2}}\right), \tag{36}
$$

with the second term in brackets representing a correction for hydrogen entrained along with oxygen into the jet. The hydrogen must be bounded by $N_{H_2}(\text{jet}) + N_{H_2}(\text{ATM})$. Pilch [159] quantified the entrainment of preexisting atmosphere into a hydrogen jet from a purely hydrodynamic perspective and concluded, through comparison with some key experiments, that the stoichiometric criterion is a reasonable and potentially conservative estimate of the amount of preexisting hydrogen that is entrained and burned in the jet. The stoichiometric estimate and the hydrodynamic estimate are both independent of scale, but the latter is a function of the number and size of flow paths to the dome.

B. STRATIFICATION

The potential for jet combustion products to stratify at the top of the dome is a function of the normalized distance to the roof (H/d_j), the vessel aspect ratio (H/D_v), and the Froude number. Peterson [160], through scaling and data, concluded that stratification for a non-DCH application is expected when

$$\frac{H}{d_j} > \frac{\left(\dfrac{\rho_a V_j^2}{(\rho_a - \rho_j) g d_j} \right)^{1/3}}{\left(1 + 3.54 \dfrac{H}{d_j} \right)^{2/3}} . \tag{37}$$

For representative IET experiments, Pilch [159] concluded that there is a potential for the jet combustion products to stratify on the dome and displace the cooler preexisting atmosphere downward. Stratification was also observed in the NTS tests [157] when the source velocity of hydrogen was sufficiently low. The NTS tests, however, also showed that mixed conditions could be produced when containment sprays are actuated.

The experimental evidence for stratification is straightforward. Temperatures in the SNL/IET-11 experiment (Surry geometry) ranged from ~ 1500 K on the dome to ~ 600 K on the operating deck. Strong temperature stratification persisted for ~ 20–40 s, and this represents a lower bound to material mixing. Gas samples taken in the SNL/IET-11 experiment indicate that a well-mixed state was not reached in the dome until ~ 2 min. Thermal stratification in the dome on DCH time scales has also been reported by Allen *et al.* [36, 37] for Zion geometry in SNL/IET-7. Thermocouples did not show evidence of stratification in SNL/IET-4,5,6, but it remains true that thermocouples indicate that the dome was not well mixed on DCH time scales.

In the SNL/CE tests in which most debris was sourced directly to the some, Blanchat [57] noted that gas concentrations differed dramatically in the dome and subcompartments for up to 30 min, and well mixed conditions were only achieved by turning mixing fans on. The Kiva code, which has particle tracking capabilities superimposed on a finite difference gas flow calculation, has been applied so some SNL/DCH tests by Marx [66,

161] in open geometry. Consistent with temperature measurements (e.g., [26]), these calculations also showed thermal stratification on DCH time scales. In addition, Kiva also predicted strong compositional variations of the atmosphere on DCH time scales.

The SNL/IET-11 experiment is close to Peterson's threshold for mixing, and the potential for mixing could increase with increasing vessel aspect ratio (H/D_v). However, recent work by Hihara and Peterson [162] suggests that stable stratification is independent of aspect ratio for Froude numbers characteristic of the IET experiments. The operating conditions of the experiments were chosen so as to yield prototypic velocities and densities. The experiments were also geometrically scaled; consequently, the Froude number will be smaller at plant scale and the potential for stratification will increase. Stratification may not be a major effect in every containment geometry or accident scenario, but failure to achieve well-mixed conditions on DCH time scales needs to be carefully assessed when control volume codes are applied to DCH.

C. Deflagrations

The DCH event itself, with hot particles and autoigniting plumes, is sufficient to ignite an upward-propagating flame if the atmosphere contains sufficient H_2 and not too much steam. Such conditions are typically determined from a hydrogen–air–steam flammability chart. Flammability charts such as Kumar's [163] are applicable when the atmosphere is ~ 375 K. For a fixed concentration of steam, the hydrogen concentration must exceed the lower flammability limit (LFL) for a flame to propagate through the mixture. There also is a critical steam concentration of ~ 50–65% beyond which all concentrations of hydrogen are inerted. The potential temperature dependence of deflagrations, however, must be recognized in DCH events where heating of the atmosphere could induce a deflagration for conditions that would not normally be considered flammable.

The amount of hydrogen burned will approach zero at the LFL. Kumar [163], Wong [164], and Carcassi and Fineschi [165] all show that (in the absence of steam) the combustion completeness is a linear function of the hydrogen concentration, with the combustion completeness going to 0 and 1 at the LFL for upward- and downward-propagating flames, respectively. The combustion completeness is known to decrease with the addition of steam, and the LFL for upward and downward propagating flames increases slightly with steam concentration and decreases with temperature. With these in mind, Pilch [159] writes the hydrogen combustion completeness as

$$\eta_c = \frac{X_{H_2} - X_{H_2}(\text{UP}; X_{\text{STM}}; T)}{X_{H_2}(\text{DWN}; X_{\text{STM}}, T_{a,m}) - X_{H_2}(\text{UP}; X_{\text{STM}}; T)}, \qquad (38)$$

where

$$X_{H_2}(\text{UP}, X_{\text{STM}}, T) = 0.037 + 0.02381 X_{\text{SMT}} - 5 \times 10^{-5}(T - 373)$$

$$X_{H_2}(\text{DWN}, X_{\text{STM}}, T) = 0.075 + 0.02381 X_{\text{SMT}} - 1.0135 \times 10^{-4}(T - 373).$$
$$(39)$$

The temperature dependence of the upward and downward flammability limits were taken from a review by Stamps and Berman [166]. The data ranged from ~ 375 to 900 K. That the upward and downward flammability limits converge at ~1100 K implies that complete combustion of hydrogen will occur for higher temperatures.

Experimental data [164, 165] indicate that the combustion completeness is slightly higher in very turbulent environments (~ 1.06 times higher than combustion completeness in a quiescent environment); however, Pilch [159] showed that the DCH environment is more closely analogous to the "quiescent" database. Pilch [159] validated the combustion completeness model against data with different steam concentrations at ~ 373 K. Data do not exist for combined validation with both composition and temperature varying simultaneously.

Flammability charts show that there is a steam concentration beyond which hydrogen combustion is always inerted, regardless of hydrogen concentration. Consistent with Kumar's [163] data, the inerting limit recommended by Pilch [159] is given by

$$X_{\text{STM}}(\text{inert}) = 0.63 + 3 \times 10^{-4}(T - 373), \qquad (40)$$

where the temperature derivative, $dX_{\text{STM}}/dT \sim 3$ vol%/100 K was measured by Kumar over the 373–473 K temperature range.

The appropriate conditions at which to evaluate deflagrations is ambiguous for DCH applications. At the start of the event, the containment is at its initial temperature and composition. For initial conditions typical of some DCH scenarios (i.e., $X_{\text{STM}} \sim 0.45$), deflagrations are only marginally possible at the upper end of the distribution of hydrogen concentration in the containment if fan coolers or sprays are not operational. During the DCH event, Pilch [159] expected the containment to favor a stratified state with blowdown and hot combustion products (of the jet) on the dome and the cooler preexisting atmosphere displaced downward if sprays are not operational. This preexisting atmosphere will have undergone some heating (radiation from the jet, compression heating from blowdown, and possibly some mixing with the blowdown and jet combustion products). Since the completeness of reaction and deflagration rate both increase with temperature, the TCE model attempts to bound the treatment of deflagrations by assuming a well-mixed atmosphere, although credit is taken for the additional inerting potential of the blowdown steam. The assumption of well-mixed conditions is inherent in a control volume code like CONTAIN.

The characteristic time over which deflagrations release their energy is controlled by the building height (H) and the flame speed (V_f):

$$\tau_{H_2} = \frac{H}{V_f}.$$ (41)

Flame speeds are often experimentally determined by dividing the facility scale by the pressure rise time. Wong [164] has correlated the flame velocity in quiescent environments from a broad database. Pilch [159] modified Wong's correlation to account for temperature effects:

$$V_f = 23.7 H^{1/3} \left(\frac{X_{H_2}}{1 - X_{STM}} - X_{H_2}(V_f, T) \right) \left(\frac{T}{373} \right)^{1/3} \exp(-A),$$ (42)

where

$$X_{H_2}(V_f, T) = 0.036 - 5 \times 10^{-5}(T - 373),$$

$$f(V_f, T) = \left(\frac{T}{373} \right)^{1/3},$$

$$A = 4.877 X_{STM}(1 + 0.61677 X_{STM}).$$ (43)

Pilch [159] concluded that flame propagation is difficult to achieve in a steam-laden containment atmosphere characteristic of station blackout accidents with no containment heat removal. The burning process, when it does occur, is too slow and inefficient to contribute to peak loads except possibly in certain unlikely situations. Deflagrations, however, can significantly enhance DCH loads with nearly complete combustion in scenarios where active containment cooling is operational prior to the DCH event. Such TMI-like scenarios have essentially no steam in the atmosphere. However, Pilch et al. [19] showed that enhanced DCH loads when the atmosphere steam concentration is low are essentially offset by the lower initial containment pressure; consequently, the total pressure in the containment was nearly identical in the extremes of ~0 and ~45% steam in the atmosphere.

System-level codes such as CONTAIN all employ deflagration models of the sort presented here, only the temperature dependence of the processes are not generally taken into account. In the recommended usage of the deflagration model [22], the flame speed is set to a high value (≥ 10 m/s) in order to take into account the expectation that hot debris particles flying through the containment will provide multiple ignition sources, substantially shortening burn times. No adjustment is made to the flammability limits or the burn-completeness correlations to take into account elevated temperatures (see, however, Sect. VIII.D). In all the experiments, bulk hydrogen concentrations were below the CONTAIN model deflagration flammability limits. Hence, there are no experimental tests of this treatment.

Figure 15 shows that jet combustion was generally observed in the Zion SNL/IET tests when a reactive atmosphere (containing oxygen) was present. The SNL/IET-3 and SNL/IET-4 experiments had reactive atmospheres with no preexisting hydrogen, so jet combustion was the only mode possible. The SNL/IET-6 and SNL/IET-7 experiments with reactive atmospheres contained ~ 2.6 and $\sim 4.0\%$ H_2, respectively. Given that the observed loads were comparable with and without preexisting hydrogen, it is likely that the preexisting hydrogen did not burn; or if it did burn to some extent in these tests, then it burned too slowly to contribute to containment pressurization. Bulk average dome temperatures in the SNL/IET-6,7 experiments were ~ 700 K, and Eq. (39) predicts that the upward flammability limit would be $\sim 3.3\%$ hydrogen. Consequently, little or no combustion of preexisting hydrogen is predicted. The presence of multiple ignition sources from dispersed debris apparently had no effect (beyond heating the atmosphere) on inducing combustion or enhancing burn completeness.

The bulk average dome temperatures were much higher in the Surry SNL/IET tests. For instance, the bulk averaged dome temperature in the SNL/IET-11 test was ~ 1100 K; at this temperature Eq. (38) predicts that all the preexisting hydrogen (2.4%) would burn. Adiabatic combustion of this preexisting hydrogen could contribute 0.14 MPa to the DCH loads (total pressure rise in the test was 0.43 MPa). Posttest gas analyses suggests that $\sim 75\%$ of the preexisting hydrogen burned [41], but analyses by Pilch [159] suggest that most of the burning occurred on a time scale too long to contribute to peak pressure. CONTAIN analyses [22] were not conclusive as to whether this hydrogen contributed to loads in any significant way. The SNL/IET-11 test simulated the annular gap around the RPV, and ~ 22 kg of hot thermite was sprayed into the dome so there were many potential ignition sources.

D. Volumetric Combustion

Hydrogen reacts with oxygen at all temperatures, although the reaction is strongly temperature-dependent. These types of reactions occur uniformly through a constant temperature mixture; thus, we refer to this as volumetric combustion. The atmosphere is not likely to be well-mixed on DCH time scales. As noted above, the preexisting atmosphere may be displaced downward and remain cooler than the bulk average temperature. Nonetheless, the displaced atmosphere will be heated by radiation from the jets and the portion of the debris that is dispersed to the dome. Pilch [159] concluded that any volumetric combustion in these situations would be too slow to contribute to DCH loads except possibly in the SNL/IET-11 experiment where autoignition would be predicted.

Autoignition can occur if a uniformly heated gas exceeds a critical temperature. Autoignition is defined as a sudden volumetric combustion of preexisting hydrogen. Autoignition, like the slow volumetric combustion discussed above, does not need a spark or pilot flame for combustion to occur. In DCH events, the difficulty in achieving well-mixed conditions on DCH time scales would seem to preclude autoignition events. Examination of autoignition criteria for well-mixed uniformly heated mixtures lends more credibility to the exclusion of autoignition events. Stamps and Berman [166] have summarized autoignition data for hydrogen–air–steam mixtures. Stamps and Berman show that the autoignition temperature becomes asymptotically large at the hydrogen concentrations ($<6\%$) of interest to DCH and that the addition of steam makes autoignition even more difficult.

Autoignition data have generally been taken near stoichiometric conditions; however, Conti and Hertzberg [167] observed autoignition at 873 K for a dry mixture with 6% hydrogen. The data of Zabetakis [156] indicate that the addition of 30% steam increases the autoignition temperature by ~ 30 K for conditions near stoichiometric. For lean mixtures (5–20% H_2), Tamm et al. [168, 169] has shown that the glow plug temperature necessary to ignite a dry mixture increases ~ 80 K with the addition of 30% steam. To ensure a conservative treatment of hydrogen combustion, Pilch [159] recommended evaluation of autoignition using the bulk average dome temperature and an autoignition temperature of ~ 950 K.

As noted previously, the CONTAIN deflagration model does not consider widened flammability limits at evaluated temperatures. Partly for this reason, a bulk spontaneous recombination (BSR) model is provided to permit volumetric oxidation at sufficiently elevated temperatures [22]. The threshold temperature is user-specifiable, with a relatively conservative value, 773 K, being chosen as default for NPP analysis, in part to take into account the expectation that the effective threshold declines with increasing scale. The reaction rate is also user-specified, with the standard value being $5/V_{\text{cell}}^{1/3}$ s^{-1}, based upon the rate at which the highly heated regions in the dome expanded in the IET experiments, as inferred from the temperature measurements. This value yields reaction times comparable to DCH time scales but not so rapid that mitigation by atmosphere–structure heat transfer is completely ineffective. BSR was not predicted to occur in the Zion IET experiments, in agreement with experimental results. BSR was predicted in the Surry IET experiments but the contribution of BSR to the calculated ΔP, ≤ 0.05 MPa, was considered too small to permit a test of the model by comparing with the experimental ΔP results, and Williams *et al.* [22] believed that no conclusions could be drawn as to whether preexisting hydrogen contributed to ΔP in these experiments. Williams *et al.* [22] also noted that the BSR treatment is likely to be overly conservative if stratification effects are important and suggested sensitivity calculations to evaluate uncertainties if BSR contributes substantially to calculated containment loads.

E. MIXING LIMITED COMBUSTION

Combustion of hydrogen can occur in a mixing zone if oxygen is present and the mixing zone is sufficiently hot. The hydrogen and oxygen can be brought into the mixing zone by either the hot or cold stream, or both. If the combustion is to be mixing limited, then the temperature of the mixing zone must be sufficiently hot that the reaction time scale is short compared with the mixing time scale; as a practical matter, this means that the mixture temperature must be ~900–1000 K.

There are three potential mixing processes that must be considered: forced mixing of the containment atmosphere by the blowdown gases, buoyant mixing between the subcompartment and the dome, and mixing due to a cool wall "jet" sinking along the vessel walls. Pilch [159] analyzed the potential for hydrogen combustion during these mixing processes and concluded that any potential combustion would be too slow to contribute to peak DCH loads.

IX. Heat Transfer to Structures

Heat transfer to structures is a mitigator of DCH loads because energy absorbed by structures is either not deposited in the gas or it is drained from the gas before peak loads occur. Direct energy exchange with structures can occur by one of two processes: impingement of debris on structures or by radiation heat transfer between airborne particles and structures. Although generally considered a second-order effect, impingement heat transfer has never been assessed explicitly in the context of DCH; however, freezing of melt on cavity surfaces is sometimes credited as a mechanism to reduce the melt mass that participates in DCH.

Condiff *et al.* [170] noted that the mean free path of thermal radiation in steam was comparable to characteristic containment dimensions, particularly in the subcompartments. Thus, they argued that debris-to-structure radiation heat transfer was a first-order mitigator of DCH loads. We note, however, that radiation accounts for $< 50\%$ of the particle heat loss. More importantly, DCH experiments have always been accompanied by large aerosol clouds [25, 29, 67]. For measured aerosol loadings, the mean free path of thermal radiation is only ~ 0.06 m (using Mie theory) for open geometry tests. The aerosol density (particles/m^3) is expected to be higher in the subcompartments when structures are simulated, and the mean free path will be correspondingly shorter. Consequently, direct debris-to-structure heat transfer is generally considered to be insignificant. We note, however, that CONTAIN can parametrically simulate this process; but as a default, direct debris-to-structure radiation is set to zero.

Structures can indirectly mitigate DCH loads by draining energy from gas already heated by airborne debris. At issue is whether gas–structure heat transfer is significant on the DCH time scale that is comparable to or less than the blowdown interval.

First, we note that on a global basis gas–structure heat transfer in the dome is a slow process relative to the DCH event. In the SNL/IET-3, 4, 6, 7 experiments [38], the DCH event produced maximum pressures in ~ 2–3 s, while the initial cooldown rate of the atmosphere would take ~ 25 s to reduce the pressure back to its initial value; consequently, $\dot{Q}_{\mathrm{LOSS}}/\dot{E}_{\mathrm{DCH}} \sim 2.5/25 = 0.1$. All DCH experiments exhibit similar trends. The energy loss rate in the dome on the DCH time scale can be represented as

$$
\frac{\dot{Q}_{\mathrm{LOSS}}}{\dot{E}_{\mathrm{DCH}}} = \frac{\int h_{\mathrm{eff}}(T_g - T_w)\,dA_{\mathrm{dome}}}{\dfrac{\Delta PV}{\gamma - 1}\dfrac{1}{\tau_b}}. \tag{44}
$$

We emphasize the surface integral because temperatures are expected to be nonuniform in the dome on DCH time scales, although such nonuniformities do not alter the conclusion that heat transfer in the dome is insignificant on DCH time scales. Equation (44) also illustrates that heat losses relative to the DCH event are to first order independent of physical scale because the blowdown time, τ_b, is proportional to the facility size.

Heated blowdown gases have a greater potential to lose energy as they flow through the subcompartment before venting to the dome. The fraction of energy lost due to heat transfer to subcompartment structures,

$$
\frac{\dot{Q}_{\text{Loss}}}{\dot{E}_{\text{flow}}} = \frac{\displaystyle\int \frac{1}{\dfrac{1}{\epsilon_g} + \dfrac{1}{\epsilon_w} - 1}\, \sigma\left(T_{g,\text{cav}}^4 - T_w^4\right) dA_{\text{sub}}}{\dot{m}_g C_p T_{g,\text{cav}}}, \tag{45}
$$

is also, to first order, independent of physical scale because the gas flow rate, $\dot{m}_g$, is proportional to the flow area in the RPV.

This result for the subcompartment, in conjunction with a similar result for the dome Eq. (44), is significant because structure heat transfer processes are not distorted in small-scale experiments. Thermal resistances in the wall are negligible on DCH time scales, and radiative heat transfer is typically much larger than convective losses in the subcompartments [22]. A bound, $\dot{Q}_{\text{LOSS}}/\dot{E}_{\text{flow}} \leq 1 - 1000/2500 \sim 0.6$, can be estimated by using measured gas temperatures (~ 1000 K) as they vent to the dome [40, 52] and an assumed temperature (~ 2500 K) of gases exiting the cavity. Such an estimate is bounding because thermocouples in the vent spaces are not capable of tracking the transient (comparable time constants), and because some of the apparent gas cooling is due to mixing of hot coherent blowdown gases with cooler subcompartment gases. Williams *et al.* [22], using the CONTAIN code, concluded that heat transfer to structures is an essential element of experiment and plant analyses.

Both SNL/IET experiments [40] and ANL/IET experiments [52] show that there can be ~ 600–700 K differences in measured gas temperatures between subcompartment vents (in the roof of the subcompartment above the reactor coolant pumps (RCPs) located close to the cavity) and positions below the RCPs located farthest from the cavity exit. Thus, there is a front-to-back and top-to-bottom gradient in gas temperatures that reflects highly nonuniform mixing and flow of gas through the subcompartment. The SNL/IET experiments also indicate that nonuniform mixing can evidence itself with substantially higher gas temperatures on one side of the subcompartment compared to the other. Nonuniform flow and strong temperature gradients have been observed with pyrometers and graphite calorimeters near the cavity exit [32, 33, 34, 35, 36, 37]. CONTAIN calculations indicate that $\sim 75\%$ of the blowdown vented to the dome through openings closest to the cavity.

Assessments of subcompartment heat transfer are also potentially sensitive to the values used for wall and gas emissivities. Wall emissivities range from 0.6 to 0.9 for surfaces of interest to experiment and plant analyses, and CONTAIN's standard input prescription recommends $\epsilon_w = 0.8$. Williams *et al.* [22] recommend an effective gas emissivity of $\epsilon_g = 0.8$, although the basis is more speculative because the mixture is composed of an emitting gas (steam) and aerosols, and a validated mixture law does not exist for the conditions of interest. For $T_g \sim 1800{-}2500$ K, CONTAIN's modified Modak correlation gives $\epsilon_g \sim 0.6{-}0.7$ for steam alone in the cavity and subcompartments. Menguc and Viskanta [171] performed a detailed calculation of radiative transport with scattering as a key feature through an optically inert gas laden with Fe_2O_3 aerosol with a number density typical of DCH experiments in open geometry. The effective gas–wall emissivity was only ~ 0.4 for a wall emissivity of 0.5. Correcting to a wall emissivity of 0.8, the gas emissivity is ~ 0.66, so that the effective gas–wall emissivity is ~ 0.57. The effective gas–wall emissivity could be substantially lower in the subcompartments where the aerosol density is much greater. Allen *et al.* [32, 33, 34, 35, 36, 37] note that pyrometers focused on the cavity exit read only ~ 1500 K, while similar measurements in open geometry experiments gave readings consistent with the debris temperature (~ 2500 K). Allen *et al.* concluded that the intense aerosol cloud (so dense that a video camera could not see debris exiting the cavity) was significantly reducing radiation to the pyrometer.

Assessments of subcompartment heat transfer are sensitive to nonuniform flow through the subcompartments [23]. There are some potentially important features of flow through the subcompartment that cannot be captured by a control volume code. Given the difficulties in accurately predicting subcompartment heat transfer, the CLCH model [72] treats DCH as an adiabatic process, and the TCE model is essentially adiabatic, although some minimal credit is given to the impact of structure heat transfer (in the dome) on deflagrations.

Atmosphere–structure heat transfer is an important mitigating effect in the CONTAIN analyses of both the Zion and Surry geometry IET experiments [22]. One reason for this result was that oxygen starvation in the subcompartments delayed combustion of DCH-produced hydrogen until it reached the dome. This delay increased the time available for atmosphere–structure heat transfer to operate. Eliminating the effects of heat transfer and hydrogen hold-up in the subcompartments increased the calculated ΔP by 60 to 75%. Williams *et al.* [22] performed sensitivity studies investigating the uncertainties resulting from uneven temperature and gas distributions in the subcompartments, atmospheric emissivity uncertainties, and uncertainty in the effective heat transfer area. They concluded that the resulting uncertainties in the loads were $\sim 10{-}15\%$, which is small compared with the total mitigation effect calculated; hence, it was concluded that DCH calculations could take considerable credit for this mitigation. However, it was also noted that the extent of mitigation calculated could depend upon the scenario. High steam flow rates resulting from rapid blowdown or debris–water interactions could reduce the effectiveness of atmosphere–structure heat transfer by accelerating transport of thermal energy and hydrogen to the dome.

X. Impact of Water on DCH

Water can play a role in DCH in various contexts: water on the basement floor can interact with debris dispersed from the cavity; water initially in the cavity can be codispersed with the debris; and RPV water overlying the debris prior to vessel breach can be coejected with the debris. Water is expected to be present on the basement floor in almost any accident scenario, and Pilch *et al.* [18] have concluded that at least some water will be coejected with the debris in any DCH event. The presence of water in the reactor cavity is plant-specific and often depends upon small details. A review of the industry's Individual Plant Evaluations (IPEs) for Westinghouse plants by Pilch *et al.* [24] shows that many U.S. reactor cavities will be deeply or partially flooded if the refueling water storage tank (RWST) has discharged; otherwise, most cavities will be almost dry or only partially flooded.

Both systems code calculations, and simple arguments based upon thermodynamic limits to the extent of debris–gas–water interactions indicate that water has the potential to either mitigate or augment DCH loads, depending upon the scenario. Independent reviews of DCH phenomenology [20, 96] have concurred that the effect can be in either direction and that its potential importance to DCH loads is high. Controlling factors in model predictions include the debris–water mass ratio, the containment geometry, and whether debris–gas interactions would be steam-limited in the absence of additional steam generated by vaporizing water. Major uncertainties include the amount of water that actually interacts and the fate of that water which does not interact initially.

Experimentally, water on the basement floor appears to have only limited impact while the situation with respect to cavity water is unclear. Results indicate that cavity water does increase the amount of thermal energy extracted from the debris, but part of the energy typically goes into generating steam rather than heating the atmosphere. Measured impacts upon containment temperatures have ranged from eliminating any temperature rise to enhanced temperature rises and rates of rise. There are no examples in which water clearly had a large effect (in either direction) upon containment pressurization, and there are no clear tests of the prediction that water can result in either substantial augmentation or mitigation of DCH loads under the appropriate conditions. Where dry-cavity comparison cases are available, cavity water has increased hydrogen production to some extent. Melt–water interactions in the cavity often result in increased cavity pressurization, possibly to the point that some cavity designs might sustain structural damage. Issues concerning cavity loads and their coupling with cavity structural response have generally been considered to be outside the scope of the DCH programs.

A. Potential Effects of Water

Simple arguments show that water has the potential to either mitigate or augment DCH loads. Possible mitigation effects include quenching of debris, suppression of hydrogen combustion by steam inerting, and quenching of hydrogen combustion energy by aerosolized water. Possible augmentation effects include increasing the supply of coherent steam available for thermal and chemical interactions with the debris, accelerating the transport of energy and hydrogen to the dome, and reducing subcompartment temperatures for the same amount of sensible heat transfer to the gas. The accelerated transport and reduced temperatures can reduce the mitigating effect of atmosphere–structure heat transfer [76].

Except for the effects of reduced atmosphere–structure heat transfer rates, the potential effects of water noted above involve thermodynamic arguments that can be illustrated with simple hand calculations without resort to elaborate computer codes. In compartmentalized containment geometries, reasoning similar in concept to the modeling used in TCE and CLCH imply that small or moderate amounts of water can augment DCH in scenarios for which debris–gas interactions would otherwise be steam-starved, while large amounts of water can have a mitigating effect.

We can illustrate these potentials using the 1/10-scale SNL/IET Zion-geometry experiments as examples. All the experiments other than SNL/IET-8B had 3.48 kg (193 g-moles) of water in the cavity. Vaporization of all the water to produce saturated steam would extract about 9 MJ of energy. Adding 193 g-moles of saturated steam (T ~ 400 K) to the containment atmosphere (volume = 89.8 m^3) would contribute only ~0.0071 MPa to the vessel pressure. Adding 9 MJ of thermal energy would contribute ~0.034 MPa. Energy used to vaporize water is only 20–25% as efficient in pressurizing the containment as is atmospheric heating.

The Zion geometry, however, is highly compartmentalized, with most of the debris failing to reach the dome. In the cavity and subcompartments, $\psi_{hx} > 5$ in these experiments (see Sect. VII.B). If we assume that only a fraction $1/(1 + \psi_{hx})$ of the debris energy is available for transfer to the blowdown steam, vaporizing the water actually reduces the available energy by less than 2 MJ. In addition, the 193 g-mole of steam produced is not enough to prevent the 250–300 g-moles of hydrogen that were generated in these experiments from burning. Thus, the potential for mitigation is probably minor in this case.

On the other hand, if the 193 g-moles of steam equilibrate thermally with the debris, about 17 MJ of additional energy is transferred, sufficient to pressurize the containment by about 0.06 MPa. Furthermore, $\psi_{rx} \sim 2$ in these experiments if only the coherent blowdown steam is available to react with the metal; if the steam generated by vaporizing the water equilibrates chemically (according to Eq. (27)) as well as thermally with the debris and if the resulting hydrogen is burned, the additional thermal and chemical energy transferred is sufficient to pressurize the containment vessel by about 0.19 MPa. Evidently, the potential for augmentation is considerably greater than the potential for mitigation in this instance.

In contrast, the thermal energy of the debris would be sufficient to vaporize $\sim 89\%$ of the 62 kg of water present in the SNL/IET-8B experiment, with no thermal energy left to heat the steam or the containment atmosphere. Furthermore, the 3060 g-moles of steam that would be produced could be sufficient to inert the combustion of the ~ 300 g-moles of hydrogen that were produced. Unvaporized water might remain airborne long enough to provide an atmospheric heat sink, reducing pressurization. For large amounts of water, therefore, the potential for strong mitigation clearly exists.

In open containment geometries it is less clear that there is a potential for substantial augmentation, whatever the amounts of water. Both ψ_{hx} and ψ_{rx} will usually be considerably less than unity, and the steam provided by vaporizing water does not significantly increase the thermodynamic limit for either debris–gas heat transfer or for chemical reaction. On the other hand, the potential quenching effects of the water on the debris and on hydrogen combustion can still arise. It is likely, therefore, that the balance between augmentation and mitigation is shifted in favor of mitigation, relative to compartmentalized-geometry containments. Note, however, that these arguments are based upon thermodynamic limits; the possibility exists that the water could affect particle size and other parameters affecting rates of thermal and chemical interactions.

Implicit in these scoping calculations is an assumption that debris interacts with all the water that is available. Key questions then become:

(1) to what extent does the available water actually interact with the melt, and (2) what effect does water that does not interact directly with the melt have on DCH loads?

B. EXPERIMENTAL EVIDENCE

Debris–water interactions are very complex and dynamic processes, and it is not to be expected that reality will show effects that are either as large or as simple as those given by the thermodynamic treatment. Available experimental evidence is insufficient to demonstrate conclusively the degree to which the trends outlined previously may or may not be realized in reality, but the data do provide sufficient insights to be worth reviewing.

In tests with low driving pressures, cavity water appeared to induce or enhance debris dispersal from the cavity, as observed in the ANL/CWTI-5 test [42] and the SNL/IET-8A test [40]. At higher driving pressures, dispersal is largely complete with a dry cavity, and water did not result in much additional debris dispersal in SNL/IET-8B [40].

Visual records (films, video, and/or flash X-rays) indicate that significant amounts of water can be ejected from the cavity prior to ejection of debris [30, 40, 42, 172]. In some cases, water may have been ejected as a slug rather than as dispersed droplets [42, 172]. In SNL/IET-8B, appearance of grayish clouds preceded appearance of highly luminous clouds of debris, aerosols, and combusting hydrogen. Comparable grayish clouds were not observed in other Zion-geometry experiments, and it is likely that water droplets were an important component of the clouds. It is noteworthy that neither water nor steam suppressed hydrogen combustion in the SNL/IET-8B experiment, but airborne water droplets may have reduced the contribution of combustion to containment pressurization. Quantitative conclusions as to how much water was driven out of the cavity without interacting with the debris cannot be drawn from the visual records, but the fraction could have been substantial in some cases.

An attempt to make some quantitative comparisons of experiments with and without cavity water is summarized in Table XX. The debris and water masses given reflect initial conditions only, without implications as to how much water actually interacted with how much melt. The experiment pairs FAI/DCH-2 versus FAI/DCH-3 and WC-1 versus WC-2 are approximate counterparts permitting direct wet/dry comparisons. Other experiments within a series are related but either are not close counterparts or do not provide direct wet/dry comparisons. The following observations may be offered.

TABLE XX

EXPERIMENTS INVOLVING CAVITY WATER

Experiment	Reference	Scale	Driving gas	Water: debris (kg)	Geometry	ΔP (MPa)	H₂ prod. (g-mole or vol%)
CWTI-5[a]	Spencer et al. [42]	1:30	Ar	5.6:3.94	Compart.	0.38	15.6[d]
CWTI-6[a]		1:30	Ar	0.0:3.75	Compart.	0.13	9.7
CWTI-12	Spencer et al. [43]	1:30	Ar	4.6/2.69	Open	0.27	11
	Blomquist et al. [44]						
	Binder and Spencer [45]						
FAI/DCH-1[a]	Henry and Hammersley [173]	1:20	N₂	0.7:20	Compart.	0.064 (0.25)[b]	2.4 v/o[c]
DCH-2[a]		1:20	N₂	3.3:20	Compart.	0.052 (0.22)[b]	3.4 v/o[c]
DCH-3[a]		1:20	N₂	0.0:20	Compart.	0.052 (0.11)[b]	0.8 v/o[c]
DCH-4[a]		1:20	Steam	0.7:20	Compart.	0.070 (0.17)[b]	5.4 v/o[c]
WC-1	Allen et al. [30]	1:10	Steam	0.0:50	Open	0.272	145
WC-2		1:10	Steam	11.76:50	Open	0.286	179
IET-7[a]	Allen et al. [40]	1:10	Steam	3.48:43	Compart.	0.271	274
IET-8B[a]		1:10	Steam	62:43	Compart.	0.244	299

[a] Water was also present on the basement floor in these experiments.
[b] Values in parentheses are ΔP in the subcompartment.
[c] Hydrogen results were reported in terms of volume percent (vol%) rather than g-moles.
[d] This measurement was considered to be uncertain.

1. *CWTI Experiments*

The containment was compartmentalized in a Zion-like manner, water was present on the basement floor, and the atmosphere was inert in both experiments; the cavity was flooded in CWTI-5 and dry in CWTI-6. These two experiments appear to show the largest effects of cavity water on containment pressures; however, in CWTI-6 the apparatus delivering the high-pressure driving gas did not function properly, cutting short the blowdown. About twice as much debris was dispersed from the cavity in CWTI-5, but it is unclear as to what extent this resulted from the water versus the truncated blowdown in CWTI-6. These differences explain, at least in part, the differences in observed containment pressures. The containment atmosphere was preheated to ~410–420 K in these experiments. Atmosphere temperatures actually decreased slightly in CWTI-5 and increased only slightly in CWTI-6. Evidently, vaporization of water and addition of blowdown gas, not containment heating, were the principal causes of the pressure increase in both experiments. Hydrogen production was apparently increased in the case with cavity water, although this measurement was considered uncertain. Cavity pressurization was increased by a factor of four when the water was present.

The CWTI-12 test was conducted in open geometry, water was not present on the basement floor, the cavity was deeply flooded, and the air atmosphere was heavily inerted with steam (X_{STM} ~ 0.85). Atmosphere temperatures essentially followed saturation, and analyses indicate that ~97% of the debris energy went into vaporizing water. Consequently, the resulting pressurization was dominated by blowdown and vaporization, not heating of the atmosphere. Enhanced cavity pressurization was also observed in this test.

2. *FAI/DCH Experiments*

The FAI/DCH series was conducted in Zion-like compartmentalized geometries. All tests had water on the basement floor, with FAI/DCH-1 having ten times the amount of the others. The atmosphere was partially inerted with nitrogen in all the tests to simulate sequences with steam in the atmosphere. A very nonprototypic feature was that the subcompartment volume and the dome volume were actually separate vessels connected by a length of 6-inch pipe. This pipe had sufficient flow resistance that there was substantial transient pressurization of the subcompartment with respect to the dome, something that never happened in the more prototypic IET experimental geometries. This feature may have tended to suppress pressurization of the dome, and hence the extent of pressurization of the subcompartment is also given in Table XX as a possible measure of DCH energy release.

Pressurization in the dome vessel on the ~1.5 s DCH time scale was only somewhat greater in the cases with cavity water than in the case without cavity water, while pressure rises in the subcompartment vessel were about a factor of two greater in the cases with cavity water. Temperature rises and rates of rise were more rapid in the cases with cavity water in both the subcompartment and the dome volumes, although the rise was minimal (<55 K) in the dome volume in all cases. Henry and Hammersley [173] concluded that, in all four experiments, most of the observed rise in containment temperature and pressure resulted from vaporization of water, not direct heating of the containment atmosphere.

Comparison of hydrogen produced in FAI/DCH-3 with the other experiments suggests water on the subcompartment floor could have given a ~15–33% contribution to hydrogen production relative to cases with cavity water or water plus blowdown steam. Henry and Hammersley [173] give no evidence that hydrogen combustion occurred in these experiments, even though the initial oxygen concentration was ~10%. Henry and Hammersley suggested that rapid vaporization of basement water could have inerted combustion of DCH-produced hydrogen.

3. WC-1 and WC-2 Experiments

These tests were conducted with a Zion-geometry cavity and an open containment geometry, without any representation of subcompartment structures. The atmosphere was inerted. Cavity water had very little effect upon ΔP, while hydrogen production increased by ~23%. The additional hydrogen production was small compared with what could have been produced by the water and the available metal, suggesting an inefficient interaction. Measured atmosphere temperatures were comparable in the two tests.

4. SNL/IET (Zion) Experiments

Like the FAI/DCH experiments, five of these tests had atmospheres with an oxygen content of ~10%; unlike the FAI/DCH tests, DCH-produced hydrogen burned and contributed substantially to containment pressurization in all five cases. These tests included two pairs of counterpart tests with and without water on the basement floor. The basement water did nothing to suppress hydrogen combustion and comparisons of containment pressure response with the temperature response indicated that the basement water was not vaporized on DCH time scales to any large extent [40]. Measured ΔP values and hydrogen production were the same to within 10 and 20%, which is probably within experiment replication uncertainties. It has been suggested [23] that water on the basement floor can be a source of long-term hydrogen production, but experimental evidence supporting this hypothesis is not available.

SNL/IET-7 and SNL/IET-8B are the only counterpart pair in the series in which the amount of cavity water was varied. These tests yielded similar results for ΔP, hydrogen production, and hydrogen combustion despite the large differences in the amount of cavity water. Peak atmosphere temperatures in IET-8B were considerably lower than in IET-7, about 462 K versus ~695 K, respectively. Although considerable amounts of hydrogen were produced and burned in IET-8B, some of the energy from the hydrogen combustion apparently went into vaporizing water. This result is consistent with the hypothesis that water will either have little effect or else mitigate DCH [19]. However, no dry cavity comparison case is available, and even the water in IET-7 could potentially have had a substantial effect, as was noted in Sect. X.A.

From a comparison of the temperature and pressure rises, it may be inferred that about 75% of the cavity water in SNL/IET-8B was vaporized on DCH time scales, and addition of steam rather than containment heating was responsible for a significant fraction (~44%) of the experimental pressure increase. These results imply an efficient thermal interaction with the cavity water in SNL/IET-8B, in contrast with the WC-1/WC-2 results cited above. Continued debris interactions with dispersed water in the cavity was suggested as a possible reason for this difference [40]. Williams *et al.* [22] note that the total net energy addition to the atmosphere (latent plus sensible heat) was considerably greater in SNL/IET-8B than in SNL/IET-7: about 146 MJ versus 80 MJ, respectively. Both estimates neglect energy losses. Despite the enhanced thermal efficiency in SNL/IET-8B, oxidation efficiencies were comparable in the two tests. Gas analyses suggest that most of the DCH-produced hydrogen burned in both experiments, indicating that the larger amount of water had little effect upon hydrogen combustion in SNL/IET-8B.

Allen *et al.* [40] note that transport of debris to the dome in IET-8B was approximately three times greater than all other SNL/IET tests, which only had condensate levels of water in the cavity. The increased carryover to the dome did not result in higher DCH loads because of the large quantities of cavity water.

C. WATER AND DCH MODELS

Nourbakash *et al.* [63] developed a single-cell thermodynamic equilibrium model that includes cold water from sprays in its formulation. The model predicts enhanced loads for some combinations of debris and water.

However, the model always predicts that water mitigates DCH loads for conditions representative of large dry or subatmospheric containments. In these parametric calculations, the quenching effect of water on the atmosphere temperature always dominated the effect on steam addition.

The cavity dispersal model of Sienicki and Spencer [104] includes water interactions; however, no predictions are reported on the impact of water on containment loads. The TCE and CLCH models do not explicitly model the impact of water on DCH. The MELCOR code does include some parametric capability to represent the impact of debris–water interactions on DCH. The MAAP code has also been used to assess the impact of cavity water and coejected RCS water on DCH loads [86]. MAAP3B predicted that 43 mt of coejected water (which flashes on blowdown) would enhance DCH loads while a deeply flooded cavity (227 mt of water) would mitigate DCH loads (Fig. 16).

Boyack *et al.* [96] performed an NRC-sponsored peer review of the CONTAIN code. With regard to the modeling of possible debris–water interactions in CONTAIN, Boyack *et al.* concluded that energetic FCIs are

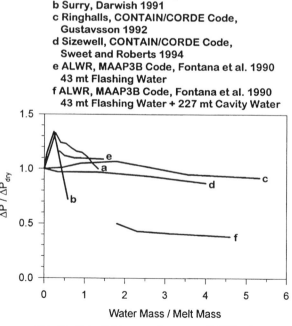

a Surry, CONTAIN Code, Williams et al. 1987
b Surry, Darwish 1991
c Ringhalls, CONTAIN/CORDE Code,
   Gustavsson 1992
d Sizewell, CONTAIN/CORDE Code,
   Sweet and Roberts 1994
e ALWR, MAAP3B Code, Fontana et al. 1990
   43 mt Flashing Water
f ALWR, MAAP3B Code, Fontana et al. 1990
   43 mt Flashing Water + 227 mt Cavity Water

FIG. 16. Potential impact of water on DCH loads.

poorly understood and are not modeled. Less-energetic interactions are better understood, but there are no features in the code to predict processes such as water slug ejection from the cavity. Heat transfer and oxidation of the debris can be represented parametrically but without capturing the essential features of the physical processes. The recommended procedure [22] is therefore to perform sensitivity studies, varying the fraction of the available water that is assumed to actually interact. Results follow trends expected from the thermodynamic arguments but are less extreme; nonetheless, substantial enhancements have been predicted when debris–gas thermal and/or chemical interactions would be heavily steam-limited without the water. Similar results have been predicted by Darwish [71], as shown in Fig. 16. Using a different version of CONTAIN that includes the CORDE module, Sweet and Roberts [92] and Gustavsson [91] have reported calculations in which cavity water had relatively little effect (Fig. 16). In the CORDE code, simple parametric models for water ejection as a slug are modeled, and the melt therefore interacts with only a fraction of the available water.

Analysis of the WC-2 experiment with CONTAIN correctly predicted that the water has little effect upon ΔP but overpredicted hydrogen production unless it was assumed that only a small fraction ($\sim 15\%$) of the water interacts. Analysis of the IET experiments with small amounts of water [22] yielded better agreement between calculated and measured pressure–time histories when it was assumed the water did participate; interaction with dispersed water in the subcompartment was the suggested reason for the difference with respect to WC-2. Analysis of the SNL/IET-8B experiment overpredicted the mitigating effect of the water. The inability of the code to model enhanced heat transfer rates associated with FCIs was the suggested reason. Systematic experiments have not been performed to test code predictions that water can augment or mitigate DCH loads under appropriate conditions.

Results of a number of model calculations involving debris–water interactions are presented in Fig. 16, in which the ratio ΔP (wet)$/\Delta P$ (dry) is plotted against the water–debris mass ratio. The differences in the trends reflect both model differences and the fact that the models predict that there are other parameters that are important in addition to the water–debris mass ratio. In particular, the CORDE code has a parametric model for slug ejection so melt interacts with only a fraction of the water while the other calculations assumed complete mixing of melt with all the prescribed water. Other parameters that could lead to differences include the containment geometry and the degree to which debris–gas interactions would be steam-limited in the absence of additional steam provided by vaporizing water.

D. COEJECTED RPV WATER

Water coejected from the RPV with the melt raises some of the same issues as codispersed cavity water, but there are also important differences. Coejected water accompanies and/or follows the melt, and therefore cannot be dispersed from the cavity as a slug in advance of the melt. Instead, much of the water may follow the dispersal event. RPV water will partially flash to steam upon depressurization, and the remaining water is likely to be highly fragmented. Flashing tends to add to containment pressurization, while aerosolized water can act as a heat sink. Pilch *et al.* [19] have concluded that the contribution to pressurization from flashing is secondary (~ 0.07 MPa), assuming that the RCS is partially filled and the water is subcooled by ~ 100 K. These scoping estimates by Pilch *et al.* assumed that any pressurization from flashing was additive to possible DCH loads.

MAAP3B predicted that coejected water would enhance DCH loads (Fig. 16). CONTAIN calculations for an ice condenser plant with the ice depleted prior to vessel breach predicted substantial (factor-of-two) enhancement could result from small amounts (~ 10,000 kg) of coejected water in a scenario in which thermal interactions of steam and debris would have been heavily steam-limited ($\psi_{hx} \approx 4$) in the absence of water. In a different scenario with more blowdown steam available, sensitivity to either 10,000 or 75,000 kg of coejected water was less. In this particular case, compartmentalization prevented enough aerosolized water from reaching the dome to result in substantial mitigation; aerosolized water was calculated to be a significant mitigating effect in other analyses, however.

Blanchat [57] has performed some scoping experiments to investigate the effects of RPV water in Calvert Cliffs geometry with a 1 : 10 linear scale factor. This containment geometry is categorized as open geometry, because the dominant dispersal pathway is to the dome. The thermite (33 kg) was located in the cavity and ignited prior to the blowdown event; hence, these experiments do not constitute true simulations of coejected melt and water. The SNL/CES-2 experiment was steam-driven, while the SNL/CES-3 experiment was driven with 100 kg of saturated water. The pressure rise and the hydrogen production were comparable in the two tests, while the temperature rise was less in the water-driven test and the temperature dropped very quickly following the initial atmosphere heating and pressurization event. Evidently, the aerosolized water and other mitigation effects approximately compensated for the additional steam provided by the flashing water. A third test, SNL/CES-1, driven by nitrogen and 100 kg of cold water, resulted in loads and hydrogen production that were about 25% lower than in the other tests. No model or code analyses are yet available for any of these experiments.

E. Cavity Pressurization

Substantial pressurization of the cavity has been observed during codispersal of melt and water from the cavity. Table XXI shows that peak cavity pressures up to ~6 MPa have been reported in experiments where the cavities are at least half full. The time intervals indicate that the melt–water interactions are sometimes explosive. Pressures of this magnitude may raise structural concerns for certain "free-standing" cavities, and these pressures were certainly responsible for the destruction of some early cavity mockups [172]. We note, however, that free-standing cavities are much less likely to contain large quantities of water at vessel breach [24], and there was no attempt to scale the structural strength of experiment cavities to their NPP counterpart. Concerns may also exist with respect to displacement of the RPV. Issues concerning cavity loads and their coupling with cavity structural response have generally been considered to be outside the scope of the DCH programs. The experimental observations of

TABLE XXI

Cavity Pressurization During HPME and Codispersal of
Melt and Water from a Reactor Cavity

Experiment	Cavity geometry	Scale	Melt	Water depth	Peak cavity pressure (MPa)	Time interval (s)
ANL/CWTI-5 Spencer et al. [42]	Zion-like	1:30	UO_2–steel thermite	Full	0.9	0.40
CWTI-12 Blomquist et al. [44]	Zion-like	1:30	UO_2–steel thermite	Full	1.4	0.40
SNL/HIPS-4W Tarbell et al. [172]	Zion	1:10	$Fe–Al_2O_3$ thermite	Full	4.5	0.01
SNL/HIPS-6W Tarbell et al. [172]	Zion	1:10	$Fe–Al_2O_3$ thermite	Full	6.0	0.01
SNL/HIPS-9W Tarbell et al. [172]	Zion	1:10	$Fe–Al_2O_3$ thermite	Full	3.0	0.01
SNL/IET-8B Allen et al. [40]	Zion	1:10	$Fe–Al_2O_3$ thermite	Half full	2.50	0.3[a]
FAI/ASCo Hammersley et al. [56]	ASCo	1:25	$Fe–Al_2O_3$ thermite	Half full	3.0	3.0

[a] Multiple energetic events occurring over 0.3 s.

potentially explosive interactions are only indicative of a concern because reduced physical scale and the use of melt simulants preclude any statement regarding prototypicality.

XI. DCH in Pressure-Suppression Containments

A. DCH in Boiling-Water Reactors

There are three basic containment designs for boiling-water reactors (BWRs) with each using a pressure suppression pool as a means of mitigating the threats of containment overpressurization and fission product releases. The basic system in each case involves the use of a large pool of water (contained within a "wetwell") through which hot steam and gaseous fission products would be directed in case of a major accident. The important effects of the pool are to condense the steam (mitigating overpressurization) and to scrub (i.e., remove) fission product gases and aerosols. The Mark I and Mark II containment types have inerted drywell atmospheres, while the Mark III design uses hydrogen ignitors to control the accumulation of hydrogen in the containment prior to vessel failure. Differences in plant geometry and safety systems can be found elsewhere [174].

DCH phenomena in BWRs have received only limited attention because of the perception that DCH is unlikely due to automatic depressurization systems (ADS). Depressurization of the RPV prior to vessel breach reduces the in-vessel hydrogen production, delays the source term releases from the RPV, and enhances the likelihood of using the available low-pressure systems for water injection into the RPV. A brief review by Schmidt and Pilch [175] confirms that the core damage frequency is lower in BWRs because of reliable ADS systems; however, the probability of vessel failure at high pressure (~ 7 MPa) can be relatively high (~ 20–80%) given that core damage does occur. Here we review only the phenomenological modeling approaches that have been applied to BWRs. Unlike the situation for PWRs, this modeling has developed without the advantage of experimental data on containment loads.

DCH in BWRs was first examined in the QUEST study [176, App. B], which recognized the importance of vent clearing in mitigating DCH loads. DCH phenomena in BWRs was also addressed in NUREG-1150 [16], where pressure loading on the drywell came from two sources that could not be treated separately: explosive melt–water interactions (steam explosions) in the pedestal region under the RPV and DCH interactions with drywell atmosphere. NUREG-1150 treated steam explosions and DCH simultaneously by eliciting experts to supply probability distributions for the combined loads. To date, the more phenomenological approaches have not addressed the steam explosion contribution, although Murata and Louie [82] do consider nonexplosive melt–water interactions in the pedestal region.

Table XXII provides a brief overview of DCH modeling in BWRs. Luangdilok's [178] model postulates very restrictive kinetic limitations to debris entrainment (rate and quantity) and debris–gas energy exchange (far from thermal equilibrium). Specifically, entrainment is assumed to occur in the pedestal region only within a solid angle leading through the pedestal door and during a period of ~0.87 s required to displace melt films out of the pedestal region. Thus, entrainment and recirculation of melt within the pedestal region is ignored as a mechanism to heat blowdown gases. Debris–gas interactions are limited by the time required for a debris particle to traverse across the pedestal, through the door, and impact on the liner. This time of flight is very short and significantly limits the amount of energy liberated by the debris before deentrainment.

The models of Khalil [177], Murata and Louie [82], Mast *et al.* [176], and ABWR [179] all assume or predict thermal equilibrium between debris and drywell gases. One of two processes limits peak drywell pressures: vent clearing, or gas flow from the drywell to the wetwell through the vents. The latter is limiting when the pressurization rate due to DCH exceeds the depressurization rate associated with gas flow from the drywell to the wetwell. Murata and Louie [82] noted that flow from the pedestal region to the rest of the drywell may be limiting (rather than flow to the wetwell), thus leading to overpressurization and possible failure of the pedestal. Well after the DCH event, Murata and Louie noted that flooding of the pedestal region could lead to fission product bypass of the suppression pool.

The model of Kajimoto and Muramatsu [180] differs from the others in that very slow pressurization of the containment is predicted (~40 s). Slow pressurization such as this is apparently traceable to small blowdown and melt entrainment rates resulting from a small (0.1 m) assumed failure in the RPV.

Scaling insights can be derived from the modeling of Khalil, Murata and Louie, and ABWR. Assuming first that vent clearing limits drywell pressure, we have

$$P \approx P^0 + \dot{P}_{\text{DCH}} t_{\text{VENT}} \approx P^0 + 1.6 \left(\rho_w L_{\text{VENT}}^2 \left(\frac{\Delta P_{\text{DCH}}}{t_e} \right)^2 \right)^{1/3}, \quad (46)$$

where the vent clearing time is taken from Moody [105] for a Mark I

TABLE XXII

DCH Modeling for Boiling-Water Reactors

Reference/plant type	Vent clearing and gas flow	Entrainment rate (time) and participating mass	Particle/gas heat transfer/oxidation kinetics, metal oxidation
Khalil [177] Mark I	Instantaneous vent clearing with calculated gas flow to wetwell, peak pressure limited by gas flow through vents.	Parametric entrainment time 0.2–2 s, 7–70 mt participating mass as parametric input.	Thermal equilibrium, 50% of Zr melt oxidizes.
Luangdilok [178] Mark I	Vent clearing and gas flow to wetwell not modeled.	Entrainment rate (~10 mt/s) calculated with Ricou–Spalding correlation, entrainment (~9 mt) limited to flow towards pedestal door. The entrainment interval is ~ 0.87 s.	Time of flight limitation (~0.27 s) to particle–gas kinetics for heat transfer and metal oxidation.
ABWR [179]	Vent clearing and gas flow to wetwell calculated, peak pressure limited by gas flow through vents.	Entrainment time (~2 s) guided by ANL/CWTI-13 test, participating debris mass limited by impaction, flow splits, and failure to fragment.	Thermal equilibrium, 50% of Zr melt oxidizes.
Kajimoto and Muramatsu [180] Mark II	Vent clearing in ~3 s with calculated gas flow to wetwell.	Source rate equated as the product of melt ejection rate (through 0.1-m diameter hole in RPV) and a parametric fraction dispersed into atmosphere.	Kinetics of debris–gas heat transfer and metal oxidation modeled. Equilibrium predicted.
Mast et al. [176]	Vent clearing in ~0.4 s limits loads.	DCH time scale (~6 s) taken as the sum of the melt ejection time and particle settling time.	Kinetics of individual particles during settling time. Equilibrium often predicted.
Murata and Louie [82] Mark III	Vent clearing (~1 s) followed by gas flow to wetwell limits pressure rise in drywell but not the pedestal pressure.	Parametric debris mass (~140 mt) source over a parametric interval (~ 5–8 s).	Kinetics of debris–gas heat transfer and metal oxidation modeled, equilibrium predicted, nonexplosive interactions with water in the pedestal region modeled.

containment subject to a constant pressurization rate. Furthermore, the pressurization rate is approximated by $\Delta P_{DCH}/t_e$, where ΔP_{DCH} could be taken from the one-cell or two-cell equilibrium model and t_e is an effective pressurization interval. If vent clearing does not limit drywell pressurization, then choked isentropic flow of gas from the drywell to the wetwell will achieve equilibrium with the pressurization rate when the drywell pressure becomes

$$P \approx \frac{\dfrac{\Delta P_{DCH}}{t_e}}{\gamma C_d \, \dfrac{A_h}{V_{drywell}} \left(\dfrac{R_u T}{MW_g} \gamma \left(\dfrac{2}{\gamma+1} \right)^{(\gamma+1)/(\gamma-1)} \right)^{1/2}}. \qquad (47)$$

The peak pressure, in general, will be the maximum of the two limits. Note that there can be no contribution from hydrogen combustion in Mark I or Mark II containments because they are nitrogen-inerted.

The controlling unknown in both expressions is the effective entrainment interval t_e. Table XXII shows that most modeling efforts treat t_e parametrically. Luangdilok [178] calculated the entrainment rate using the Ricou–Spalding correlation. This may overestimate the entrainment time as debris can be entrained throughout the pedestal region, heating blowdown gases. Another possible estimate of the entrainment time might be obtained from $t_e = R_\tau \tau_b$, where R_τ is the coherence ratio of the form of Eq. (19) or (20), and τ_b is the blowdown time. The current state of knowledge relies on phenomenological intuition in the absence of definitive experiments.

The dynamics of steam quenching in the suppression pool are well known for design-basis accidents (DBAs). DCH events differ from DBAs in that the steam will be superheated to perhaps 2500 K. Failure of the suppression pool to fully quench the superheated steam could result in enhanced radiological release and enhanced pressurization of the wetwell. Luangdilok [178] proposed that the jet quenching length could be calculated from

$$\frac{Z}{D} \approx 3.1 \left\{ \frac{h_{fg,j} + C_{p,j}(T_j - T_{sat})}{C_{p,w}(T_{sat} - T_w^0)} \right\} \left(\frac{\rho_j}{\rho_w} \right)^{1/2}. \qquad (48)$$

For a Mark I containment subject to a DCH event, Luangdilok concluded that the steam jet would completely quench in the suppression pool. Mast et al. [176] noted that the suppression pool could be saturated in some DCH-relevant sequences. Steam as well as noncondensibles could then fluidize the suppression pool, compromising its ability to scrub aerosols from the gas.

B. DCH in Ice-Condenser Containments

Ice-condenser containments are a class of PWR containments in which most of the containment volume is divided between upper and lower compartments configured such that almost all flow from the lower to the upper compartment must pass through an intermediate compartment filled with baskets containing approximately 10^6 kg of ice. Over 50% of the total containment volume is in the upper compartment, while all reactor coolant systems are in the lower compartment. If there is a break anywhere in the parts of the reactor cooling systems that are within containment, the steam is released to the lower compartment and then flows to the ice chest, where it is condensed.

This system has been demonstrated to be very effective in suppressing the pressure rise resulting from DBAs. Hence, ice-condenser containments have been designed with volumes that are about half as great as typical PWR large dry containments, and the design pressures are also lower. However, the ice condenser is not effective in suppressing some other types of loads that can arise in degraded-core and severe accidents, notably hydrogen combustion loads. Indeed, the small containment volume combined with condensation of steam by the ice (thereby preventing inerting) makes hydrogen combustion a special concern. Hence, ice condensers are equipped with igniters intended to burn off hydrogen before dangerous concentrations develop, and air return fans (ARFs) are provided to prevent high local concentrations of hydrogen from developing. However, these systems may not be available in all accident scenarios of potential interest to DCH.

Debris transport to the upper compartment is not believed to be an issue in ice-condenser containments. No line-of-sight paths exist, and the potential transport path through the ice condenser is sufficiently tortuous that transport of substantial debris with the gas flow is even less likely than in compartmentalized large dry containments. Tutu *et al.* [181] have reported experimental simulations of transport through the ice chest, with results indicating that only ~20% of the debris reaching the ice chest will pass through it, and CONTAIN analyses of DCH in ice condenser plants indicate that most of the debris will not even get as far as the ice chest entrance.

Hydrogen production and combustion are often the overriding issues for DCH loads in ice condenser plants. If hydrogen released to the containment has not been burned off prior to vessel breach, combustion of this hydrogen alone can result in dangerous loads [16, 78], and DCH need do little more than provide the ignition source. Williams and Gregory [78] performed a series of CONTAIN analyses indicating that containment survival was problematic even if preexisting hydrogen was burned off prior to vessel breach. The melt compositions in their calculations were based upon results given by Gieseke *et al.* [182]. These melts had a very high metallic content that generated large amounts of hydrogen during the DCH event, and combustion of this hydrogen could produce threatening loads even in the absence of substantial preexisting hydrogen.

CONTAIN calculations have been performed recently for ice-condenser plants in which melt compositions were assumed to be those given by Pilch *et al.* [18], which have a much lower metallic content than was assumed by Williams and Gregory [78]. As in the earlier calculations, the question of whether ignition sources were available prior to vessel breach was found to be crucial. Given that the preexisting hydrogen has been burned off, the low metal contents of the melts assumed in the more recent analyses resulted in much less threatening loads than those calculated in the previous work provided the ice condenser remains effective at the time of vessel breach. Without a large contribution from DCH-produced hydrogen, DCH threats result primarily from thermal energy transfer to gas, steam, and water, and the ice condenser was calculated to be effective in suppressing this threat. However, it must be noted that there is currently no validation base for the CONTAIN ice condenser model under these conditions. Furthermore, the ice condenser may not be effective in some accident scenarios of potential interest: for example, owing to depletion of the ice, highly uneven melting of the ice prior to vessel breach, and/or damage to the ice condenser from hydrogen detonations.

In ice-condenser containments, the seal table room is located adjacent to the containment shell that forms the pressure boundary. If melt impacting the seal table causes it to fail, it has been suggested that sufficient melt could accumulate against the shell to bring about a thermal failure [147]. This mechanism has been addressed in NUREG-1150 [16] and the plant IPEs.

XII. Closure

An extensive integral effects database closely tied to scaling analyses and plant accident sequences relevant to DCH has been developed over the last 10 years. The three most important conclusions derived from these experiments and their analysis are: (1) there is no evidence of a strong effect of physical scale on DCH loads in plants with compartmentalized geometry, (2) DCH-produced hydrogen is a significant contributor to DCH loads, and (3) containment compartmentalization significantly mitigates DCH loads. Ongoing experiments are addressing DCH phenomena in plants where most debris is dispersed to the dome. No integral experiments exist for BWR geometries.

There have also been numerous separate effects research programs to investigate details of specific phenomena relevant to DCH, especially debris dispersal and entrainment from the cavity and particle fragmentation during the entrainment process. Key insights include: (1) no existing reactor cavity has been found to be retentive at the RCS pressures of interest to DCH, and (2) some melt dispersal may be in the form of films, but the remainder is sufficiently fragmented that thermal and chemical equilibrium are reasonably approached in the cavity during dispersal for many DCH scenarios.

Results from these integral and separate effects tests have allowed the development of simple physics models and DCH-specific models for system-level codes for containment analyses. All of the models have been validated against the integral effects database and reproduce the major trends of containment loads reasonably well. The various modeling tools are not always fully consistent in their details; consequently, extrapolation outside the database should be approached cautiously. Modeling the impact of codispersed and coejected water on DCH is considered inadequate, which adds to the uncertainties in predicting DCH loads for scenarios involving water.

PRA analyses, recent assessments of spontaneous depressurization mechanisms, and recently implemented accident management procedures to depressurize have led to an increasing perception that RPV failure at high pressure is highly unlikely. In addition, analysis of the in-vessel core melt progression have led to the perception that melts with a high metallic content are unlikely in any accident sequence in which the RPV does remain highly pressurized. Given a melt consisting principally of oxides, current modeling tools generally predict that compartmentalized containments of the PWR large dry or subatmospheric type will not be threatened by DCH loads even if vessel breach does occur at high pressure. Consequently, the perceived DCH threat has been greatly diminished as a result of 10 years of research.

Acronymns and Abbbreviations

	ARF	air-return fan
ADS	automatic depressurization system	BSR	bulk spontaneous reaction
		BWR	boiling-water reactor
ANL	Argonne National Laboratory	CLCH	Convection Limited Containment Heating
ANL/IET	Argonne National Laboratory–Integral Effects Test	CLWG	Containment Loads Working Group

CTTF	Containment Technology Test Facility	RCS	reactor coolant system
		RPV	reactor pressure vessel
DBA	design-basic accident	RWST	refueling water storage tank
DCH	direct containment heating	SASM	Severe Accident Scaling Methodology
DFB	diffusion-flame burn		
EPRI	Electric Power Research Institute	SASM-TPG	Severe Accident Scaling Methodology–Technical Program Group
FAI	Fauske and Associates, Inc.		
FCI	fuel–coolant interaction	SCE	Single-Cell Equilibrium
HPME	high-pressure melt ejection	SNL	Sandia National Laboratories
IPEs	individual plant examinations	SNL/IET	Sandia National Laboratories–Integral Effects Test
LFL	lower flammability limit		
LFP	limited flight path		
MAAP	Modular Accident Analysis Program	SNL/LFP	Sandia National Laboratories–Limited Flight Path
mt	metric ton, mass		
NPP	nuclear power plant	SNL/TDS	Sandia National Laboratories–Technology Development Series
NRC	Nuclear Regulatory Commission		
NTS	Nevada Test Site of the DOE	SNL/WC	Sandia National Laboratories–Wet Cavity
PRA	probabilistic risk assessment		
PWR	pressurized-water reactor	TCE	Two-Cell Equilibrium
RCB	reactor containment building	ZPSS	Zion Probabilistic Safety Study
RCP	reactor coolant pump		

Nomenclature

a_0	sonic velocity		atmosphere
A_{cavity}	minimum flow area out of the cavity	d_{30}	mass mean diameter of melt particles
$A_{\text{cav exit}}$	the area of the instrument tunnel (cavity) exit	d_{32}	Sauter mean diameter of melt particles
A_{dg}	debris–gas interface area	d_h	hole size in the melt generator or RPV
A_{dome}	area of dome surface		
A_{gap}	minimum flow area through the annular gap between the RPV and biological shield wall	d_j	diameter of hydrogen jet
		d_t	effective particle size for nonairborne debris with the CONTAIN code
$A_{\text{op deck}}$	the area of the opening in the operating deck that is directly above the seal table	D	diffusivity
		D_h	RPV hole diameter
		$D_{h,f}$	final RPV hole diameter
		D_h^0	initial RPV hole diameter
A_{str}	the area of the seal table room opening	$\dot{D}_{\text{ref}}$	reference hole diameter ablation rate
A_{sub}	area of subcompartment surface	D_v	RPV diameter
		e_{bias}	relative bias of measured to predicted data sets
A_x	cavity flow cross-section		
C_v	constant-volume molar heat capacity of the containment	$\dot{E}_{\text{DCH}}$	DCH energy generation rate in containment dome

$\dot{E}_{\text{flow}}$ energy carried into sub-compartment by flow

Eu_c Euler number

f fraction of melt remaining in RPV at start of gas blowthrough

f_c fraction of dispersed particles predicted to strike structures near the exit of the Zion cavity

f_{coh} fraction of blowdown gas that is coherent with debris dispersal

f_{disp} fraction of melt dispersed from cavity

f_{dome} fraction of debris transported from reactor to containment dome

f_{gap} fraction of debris entering the annular gap for dispersal to the containment dome

f_m fraction of melt that can leave the cavity under its own momentum during HPME before the onset of gas blowdown into the cavity

$f_{\text{noz/shld}}$ fraction of debris entering the annular gap that is redirected by nozzles through cutouts back into the lower compartments instead of the containment dome

f_{sub} fraction of debris transported to dome from subcompart-ments

Fr_D Froude number

g acceleration of gravity

h_c critical depth of remaining melt in RPV at blowthrough

$h_{d,w}$ hole ablation heat transfer coefficient

h_{eff} effective heat transfer coefficient

h_{ht} heat transfer coefficient

h_H height of side hole in RPV at blowthrough

h_{mt} mass transfer coefficient

H containment dome height

k thermal conductivity

K_{eq} equilibrium constant

Le   Lewis number

$\dot{m}_g$ gas flow rate

M_d mass of debris

M_d^0 initial debris mass

M_g^0 initial gas mass in RCS

M_d^0 initial melt mass available to flow through RPV holes

$\dot{n}_g$ molar gas flow rate through the volume of interest

$n_{g,\text{ox}}$ number of oxygen equivalents of oxidizing gas

$\dot{n}_{rx}$ molar hydrogen generation rate

N number of data points

$\dot{N}$ chemical reaction rate for the debris–gas mixture

N_{H2O} moles of steam in the accumulator in the experiment

N_i moles of the subscript material

Nu   Nusselt number

P^0 initial containment pressure

Pr   Prandtl number

$\dot{Q}$ heat transfer rate for the debris–gas mixture

$\dot{Q}_{\text{LOSS}}$ energy loss rate on DCH time scale

Re   Reynolds number

Re_f Reynolds number, fluid

Re_g Reynolds number, gas

R_τ coherence ratio, the ratio of characteristic dispersal time to characteristic blowdown time

R_u universal gas constant

S_d debris–gas slip factor

Sc   Schmidt number

Sh   Sherwood number

t_1 time to heat and begin melting the RPV substrate

t_{VENT} vent clearing time or time for gas flow from the drywell to the wetwell of a BWR

T_d debris temperature

T_g gas temperature

T_g temperature of gas in the containment dome

$T_{g,\text{cav}}$ temperature of gas in the cavity

T_w	temperature of the wall
U^0	initial containment atmosphere internal energy
V	containment volume
V_{CAV}	cavity volume
V_d	debris velocity in coupled gas/debris flow
V_f	flame speed
V_g	gas velocity in coupled gas/debris flow
V_j	velocity of the hydrogen/steam jet
We	Weber number
X_i	concentration of subscript material

GREEK

α_c	thermal diffusivity of the frozen crust
α_g	transient gas void fraction in the orifice
ΔD_h	change in RPV hole diameter
ΔE_i	energy term components for SCE model, defined in Table III
ΔP	containment pressure change
ΔP_{DCH}	containment pressure change due to DCH
ΔP_{meas}	measured pressure change
ΔP_{pred}	predicted pressure change
ΔP_{SCE}	SCE model predicted pressure change
ΔT_r	reference temperature difference between melt and RPV substrate
ΔU	change in atmosphere internal energy
ϵ_g	gas emissivity
ϵ_w	wall emissivity
η	efficiency of debris/gas interactions ($\eta = \Delta P_{meas}/\Delta P_{SCE}$)
η_c	combustion completeness of hydrogen
η_1	efficiency of DCH processes in the subcompartment

η_2	efficiency of DCH processes in the dome
σ_g'	geometric standard deviation of melt particle size
σ_{rms}	relative standard deviation of measured to predicted data sets
ρ_a	density of air in the containment dome
ρ_d	debris density
ρ_{flm}	film density
ρ_g	blowthrough gas density
ρ_j	density of the hydrogen–steam jet
τ_b	characteristic RCS blowdown time
τ_c	effective pressurization interval
$\tau_{c,d}$	debris airborne residence time for the volume of interest
$\tau_{c,g}$	time scale for gas to flow out of the volume of interest
τ_d	time to eject all melt through available RPV holes
$\tau_{d,c}$	airborne residence time
τ_D	characteristic time to double the RPV hole size by ablation
$\tau_{eq,hx}$	thermal equilibration time $\approx \tau_{hx,d}/(1 + \psi_{hx})$
$\tau_{eq,rx}$	chemical equilibration time $\approx \tau_{rx,d}/(1 + \psi_{rx})$
$\tau_{hx,d}$	time scale for a debris particle to transfer its energy to the gas
τ_{H2}	characteristic hydrogen deflagration duration
τ_M	characteristic time to eject all melt from RPV in absence of ablation
τ_g	single-phase gas discharge time for independent flow
$\tau_{rx,d}$	time scale for a debris particle to react chemically with the gas
τ_{tran}	characteristic time for melt to reach the cavity exit
ψ	heat capacity ratio

ψ_{hx}	heat capacity ratio, $m_d c_{p,d}/m_g c_{p,g}$	λ_c	debris trapping parameter in CONTAIN
ψ_{rx}	molar equivalents ratio, $\sum \nu_i n_{i,d}/n_{g,ox}$	λ_{tr}	debris trapping rate
λ	growth rate constant for freezing of a superheated melt on concrete	γ	ratio of gas specific heats

Acknowledgments

The authors wish to express their appreciation to a number of people who have reviewed and commented on the entire article or on specific sections where they have special expertise. Drs. T. Y. Chu and R. O. Griffith, the internal Sandia reviewers, provided comments and suggestions that greatly improved the quality of this article. The authors are grateful to several DCH experts outside of Sandia who have reviewed and commented on this article: Professor T. G. Theofanous of the University of California at Santa Barbara, Dr. N. K. Tutu of Brookhaven National Laboratory, Dr. J. L. Binder of Argonne National Laboratory, Dr. Y. F. Khalil of Northeast Utilities and Yale University, and Dr. Q. Wu of Purdue University. Dr. D. W. Stamps of Evansville University also reviewed the section on hydrogen behavior.

This work was sponsored by the Accident Evaluation Branch of the Office of Research, United States Nuclear Regulatory Commission. The authors wish to express their appreciation to Drs. C. G. Tinkler and R. Y. Lee of the NRC for their review and comments on this article.

References

1. Rempe, J. L., Chavez, S. A., Thinnes, G. L., Allison, C. M., Korth, G. E., Witt, R. J., Sienicki, J. J., Wang, S. K., Stickler, L. A., Heath, C. H., and Snow, S. D. (1993). *Light-Water Reactor Lower Head Failure Analysis*. NUREG/CR-5642, EGG-6218, Idaho National Engineering Laboratory, ID.

2. ZPSS (1981). *Zion Probabilistic Safety Study*. Commonwealth Edison Co., Chicago.

3. Spencer, B. W., Bengis, M., Baronowsky, S., and Sienicki, J. J. (1982). *Sweepout Thresholds in Reactor Cavity Interactions*. ANL/LWR/SAF 82-1, Argonne National Laboratory, Argonne, IL.

4. Spencer, B. W., Kilsdonk, D., Sienicki, J. J., and Thomas, G. R. (1983). Phenomenological investigations of cavity interactions following postulated vessel melt-through, NUREG/CP-0027, In *Proc. Int'l. Meeting on Thermal Nuclear Reactor Safety*, Chicago, Vol. 2.

5. Tarbell, W. W., Brockmann, J. E., and Pilch, M. (1984). *High-Pressure Melt Streaming (HIPS) Program Plan*. NUREG/CR-3025, SAND82-2477, Sandia National Laboratories, Albuquerque, NM.

6. Powers, D. A., Brockmann, J. E., Bradley, D. R., and Tarbell, W. W. (1983). The role of ex-vessel interactions in determining the severe reactor accident source term for fission products. In *Proc. Int. Mtg. on Light Water Reactor Severe Accident Evaluation*, Cambridge, MA.

7. Brockmann, J. E., and Tarbell, W. W. (1984). Aerosol source term in high-pressure melt ejection. *Nucl. Sci. Eng.* **88**, 342–356.

8. Brockmann, J. E., and Tarbell, W. W. (1984). Aerosol generation by pressurized melt

ejection. In *Aerosols: Science, Technology, and Industrial Application of Airborne Particles* (Liu, Pui, and Fissan, eds.). Elsevier Science Publishing Co., Inc., New York, NY.

9. Brockmann, J. E., Appendix C in Lipinski, R. J., Bradley, D. R., Brockmann, J. E., Griesmeyer, J. M., Leigh, C. D., Murata, K. K., Powers, D. A., Rivard, J. B., Taig, A. R., Tills, J., and Williams, D. C. (1985). *Uncertainty in Radionuclide Release Under Specific LWR Accident Conditions*. SAND84-0410, Vol. 2, Sandia National Laboratories, Albuquerque, NM.

10. Williams, D. C., and Griesmeyer, J. M. (1986). Risk-significance of chemical issues affecting the source term. Presented at *ACS Severe Accident Chemistry Symp.*, Anaheim, CA.

11. Tarbell, W. W., Pilch, M., Brockmann, J. E., Ross, J. W., and Gilbert, D. W., (1986). *Pressurized Melt Ejection into Scaled Reactor Cavities*. NUREG/CR-4512, SAND86-0153, Sandia National Laboratories, Albuquerque, NM.

12. CLWG (1985). *Estimates of Early Containment Loads from Core Melt Accidents*. NUREG-1079, U.S. Nuclear Regulatory Commission, Washington, DC.

13. Bergeron, K. D., and Williams, D. C. (1985). CONTAIN calculations of containment loading of dry PWRs. *Nucl. Eng. Des.* **90**, 153−159.

14. Pilch, M., and Tarbell, W. W. (1986). *Preliminary Calculations on Direct Heating of a Containment Atmosphere by Airborne Core Debris*. NUREG/CR-4455, SAND85-2439, Sandia National Laboratories, Albuquerque, NM.

15. NRC (U.S. Nuclear Regulatory Commission) (1987). *Reactor Risk Reference Document*. NUREG-1150, Vols. 1−3, Washington, DC (draft for comment).

16. NRC (U.S. Nuclear Regulatory Commission) (1990). *Severe Accident Risks: An Assessment for Five U.S. Nuclear Power Plants*. NUREG-1150, Vol. 1, Final Summary Report, Washington, DC.

17. Knudson, D., and Dobbe, C. (1993). *Assessment of the Potential for High Pressure Melt Ejection Resulting from a Surry Station Blackout Transient*. NUREG/CR-5949, EGG-2689, EG & G Idaho, Inc., Idaho Falls, ID.

18. Pilch, M. M., Allen, M. D., Knudson, D. L., Stamps, D. W., and Tadios, E. L., (1994). *The Probability of Containment Failure by Direct Containment Heating in Zion*, NUREG/CR-6075, Suppl. 1, Sandia National Laboratories, Albuquerque, NM.

19. Pilch, M. M., Allen, M. D., Spencer, B. W., Bergeron, K. D., Quick, K. S., Knudson, D. L., Tadios, E. L., and Stamps, D. W., (1995). *The Probability of Containment Failure by Direct Containment Heating in Surry*. NUREG/CR-6109, SAND93-2078, Sandia National Laboratories, Albuquerque, NM.

20. Zuber, N. (1991). *An Integrated Structure and Scaling Methodology for Severe Accident Technical Issue Resolution*. NUREG/CR-5809, EGG-2659 (draft for comment).

21. Kmetyk, L. N. (1993). *MELCOR 1.8.2 Assessment: IET Direct Containment Heating Tests*. SAND93-1475, Sandia National Laboratories, Albuquerque, NM.

22. Williams, D. C., Griffith, R. O., Tadios, E. L., and Washington, K. E., (1996). *Assessment of the CONTAIN Direct Containment Heating (DCH) Model*. SAND94-1174, Sandia National Laboratories, Albuquerque, NM.

23. Pilch, M. M., Yan, H., and Theofanous, T. G. (1994). *The Probability of Containment Failure by Direct Containment Heating in Zion*. NUREG/CR-6075, SAND93-1535, Sandia National Laboratories, Albuquerque, NM.

24. Pilch, M. M., Allen, M. D., and Klamerus, E. W. (1996). *Resolution of the Direct Containment Heating Issue for all Westinghouse Plants with Large Dry Containments or Subatmospheric Containments*. NUREG/CR-6338, Sandia National Laboratories, Albuquerque, NM.

25. Tarbell, W. W., Brockmann, J. E., Pilch, M., Ross, J. W., Oliver, M. S., Lucero, D. A.,

Kerley, T. E., Arellano, F. E., and Gomez, R. D. (1987). *Results from the DCH-1 Experiment*. NUREG/CR-4871, SAND86-2483, Sandia National Laboratories, Albuquerque, NM.

26. Tarbell, W. W., Nichols, R. T., Brockmann, J. E., Ross, J. W., Oliver, M. S., and Lucero, D. A. (1988). *DCH-2: Results from the Second Experiment Performed in the Surtsey Direct Heating Test Facility*. NUREG/CR-4917, SAND87-0976, Sandia National Laboratories, Albuquerque, NM.

27. Allen, M. D., Pilch, M., Nichols, R. T., Brockmann, J. E., Sweet, D. W., and Tarbell, W. W. (1991). *Experimental Results of Direct Containment Heating by High-Pressure Melt Ejection into the Surtsey Vessel: The DCH-3 and DCH-4 Tests*. SAND90-2138, Sandia National Laboratories, Albuquerque, NM.

28. Allen, M. D., Blanchat, T. K., and Pilch, M. M. (1994a). *Test Results on Direct Containment Heating by High-Pressure Melt Ejection into the Surtsey Vessel: The TDS Test Series*. SAND91-1208, Sandia National Laboratories, Albuquerque, NM.

29. Allen, M. D., Pilch, M., Nichols, R. T., and Griffith, R. O. (1991b). *Experiments to Investigate the Effect of Flight Path on Direct Containment Heating (DCH) in the Surtsey Test Facility: The Limited Flight Path (LFP) Tests*. NUREG/CR-5728, SAND91-1105, Sandia National Laboratories, Albuquerque, NM.

30. Allen, M. D., Pilch, M., Griffith, R. O., and Nichols, R. T. (1992a). *Experiments to Investigate the Effect of Water in the Cavity on Direct Containment Heating (DCH) in the Surtsey Test Facility: The WC-1 and WC-2 Tests*. SAND91-1173, Sandia National Laboratories, Albuquerque, NM.

31. Allen, M. D., Pilch, M., Griffith, R. O., and Nichols, R. T. (1992b). *Experimental Results of Tests to Investigate the Effects of Hole Diameter Resulting from Bottom Head Failure on Direct Containment Heating (DCH) in the Surtsey Test Facility: The WC-1 and WC-3 Tests*. SAND91-2153, Sandia National Laboratories, Albuquerque, NM.

32. Allen, M. D., Pilch, M., Griffith, R. O., Nichols, R. T., and Blanchat, T. K. (1992c). *Experiments to Investigate the Effects of 1:10 Scale Zion Structures on Direct Containment Heating (DCH) in the Surtsey Test Facility: The IET-1 and IET-1R Tests*. SAND92-0255, Sandia National Laboratories, Albuquerque, NM.

33. Allen, M. D., Pilch, M., Griffith, R. O., Williams, D. C., and Nichols, R. T. (1992d). *The Third Integral Effects Test (IET-3) in the Surtsey Test Facility*. SAND92-0166, Sandia National Laboratories, Albuquerque, NM.

34. Allen, M. D., Blanchat, T. K., Pilch, M., and Nichols, R. T. (1992e). *Results of an Experiment in a Zion-Like Geometry to Investigate the Effect of Water on the Containment Basement Floor on Direct Containment Heating (DCH) in the Surtsey Test Facility: The IET-4 Test*. SAND92-1241, Sandia National Laboratories, Albuquerque, NM.

35. Allen, M. D., Blanchat, T. K., Pilch, M., and Nichols, R. T. (1992f). *Experimental Results of an Integral Effects Test in a Zion-Like Geometry to Investigate the Effects of a Classically Inert Atmosphere on Direct Containment Heating: The IET-5 Experiment*. SAND92-1623, Sandia National Laboratories, Albuquerque, NM.

36. Allen, M. D., Blanchat, T. K., Pilch, M., and Nichols, R. T. (1992g). *An Integral Effects Test in a Zion-Like Geometry to Investigate the Effects of Preexisting Hydrogen on Direct Containment Heating in the Surtsey Test Facility: The IET-6 Experiment*. SAND92-1802, Sandia National Laboratories, Albuquerque, NM.

37. Allen, M. D., Blanchat, T. K., Pilch, M., and Nichols, R. T. (1992h). *An Integral Effects Test to Investigate the Effects of Condensate Levels of Water and Preexisting Hydrogen on Direct Containment Heating in the Surtsey Test Facility: The IET-7 Experiment*. SAND92-2021, Sandia National Laboratories, Albuquerque, NM.

38. Allen, M. D., Blanchat, T. K., Pilch, M., and Nichols, R. T. (1993a). *An Integral Effects Test to Investigate the Effects of Condensate Levels of Water and Preexisting Hydrogen on*

Direct Containment Heating in the Surtsey Test Facility: The IET-7 Experiment. SAND92-2021, Sandia National Laboratories, Albuquerque, NM.

39. Allen, M. D., Blanchat, T. K., Pilch, M., and Nichols, R. T. (1993b). *Experiments to Investigate the Effects of Fuel/Coolant Interactions on Direct Containment Heating: The IET-8A and IET-8B Experiments.* SAND92-2849, Sandia National Laboratories, Albuquerque, NM.

40. Allen, M. D., Pilch, M. M., Blanchat, T. K., Griffith, R. O., and Nichols, R. T. (1994b). *Experiments to Investigate Direct Containment Heating Phenomena With Scaled Models of The Zion Nuclear Power Plant in the Surtsey Test Facility.* NUREG/CR-6044, SAND93-1049, Sandia National Laboratories, Albuquerque, NM.

41. Blanchat, T. K., Allen, M. D., Pilch, M. M., and Nichols, R. T. (1994). *Experiments to Investigate Direct Containment Heating Phenomena With Scaled Models of the Surry Nuclear Power Plant.* NUREG/CR-6152, SAND93-2519, Sandia National Laboratories, Albuquerque, NM.

42. Spencer, B. W., Sienicki, J. J., and McUmber, L. M. (1987). *Hydrodynamics and Heat Transfer Aspects of Corium–Water Interactions.* EPRI NP-5127, Argonne National Laboratory, Argonne, IL.

43. Spencer, B. W., Sienicki, J. J., Sehgal, B. R., and Merilo, M. (1988). Results of EPRI/ANL DCH investigations and model development. Presented at *ANS/ENS Conf. on Thermal Reactor Safety,* 2–7 October, Avignon, France.

44. Blomquist, C. A. (1989). *High-Pressure Corium Injection Into a Simulated RPV Cavity: Results of Direct Containment Heating Tests CWTI-11 to 15.* EPRI-RP 1931-2, Electric Power Research Institute, Palo Alto, CA.

45. Binder, J. L., and Spencer, B. W. (1996). Investigations into the physical phenomena and mechanisms that affect direct containment heating loads. *Nucl. Eng. Des.* (accepted for publication).

46. Binder, J. L., McUmber, L. M., and Spencer, B. W. (1992a). *Quick Look Data Report on the Integration Effects Test 1R in the COREXIT Facility at Argonne National Laboratory.* Argonne National Laboratory, Argonne, IL.

47. Binder, J. L., McUmber, L. M., and Spencer, B. W. (1992b). *Quick Look Data Report on the Integral Effects Test 1RR in the COREXIT Facility at Argonne National Laboratory.* Argonne National Laboratory, Argonne IL.

48. Binder, J. L., McUmber, L. M., and Spencer, B. W. (1992c). *Quick Look Data Report on the Integral Effects Test-3 in the COREXIT Facility at Argonne National Laboratory.* Argonne National Laboratory, Argonne, IL.

49. Binder, J. L., McUmber, L. M., and Spencer, B. W. (1992d). *Quick Look Data Report on the Integral Effects Test-6 in the COREXIT Facility at Argonne National Laboratory.* Argonne National Laboratory, Argonne, IL.

50. Binder, J. L., McUmber, L. M., and Spencer, B. W. (1992e). *Quick Look Data Report on the Integral Effects Test-7 in the COREXIT Facility at Argonne National Laboratory.* Argonne National Laboratory, Argonne, IL.

51. Binder, J. L., McUmber, L. M., and Spencer, B. W. (1992f). *Quick Look Data Report on the Integral Effects Test-8 in the COREXIT Facility at Argonne National Laboratory.* Argonne National Laboratory, Argonne, IL (draft for review).

52. Binder, J. L., McUmber, L. M., and Spencer, B. W. (1994). *Direct Containment Heating Integral Effects Tests at 1/40 Scale in Zion Nuclear Power Plant Geometry.* NUREG/CR-6168, ANL-94/18, Argonne National Laboratory, Argonne, IL.

53. Binder, J. L., McUmber, L. M., and Spencer, B. W. (1993a). *Direct Containment Heating Experiments in Zion Nuclear Power Plant Geometry Using Prototypic Core Materials: The U1A and U1B Tests.* ANL/RE/LWR 93-3, Argonne National Laboratory, Argonne, IL.

54. Binder, J. L., McUmber, L. M., and Spencer, B. W. (1993b). *Direct Containment Heating*

Experiments in Zion Nuclear Power Plant Geometry Using Prototypic Core Materials: The U2 Test. ANL/RE/LWR 93-6, Argonne National Laboratory, Argonne, IL.

55. Henry, R. E., and Hammersley, R. J. (1991). Direct containment heating experiments in a Zion-like geometry. In *ANS Proc. 26th Natl. Heat Transfer Conf.*

56. Hammersley, R. J., Cirauqui, C., Faig, J., and Henry, R. E. (1995). Direct containment heating experiments for Vandellos and Asco nuclear power plants. In *ANS Proc. Natl. Heat Transfer Conf.* 5–9 August, Portland, OR, Vol. 8, pp. 384–392.

57. Blanchat, T. K. (1996). DCH tests using CE geometry with coejected water. Presented at *ASME-AIChE/ANS Natl. Heat Transfer Conf.*, 3–6 August, Houston.

58. Bertodano, M. L. (1993). Direct containment heating DCH source term experiment for annular reactor cavity geometry. In *Proc. Ninth Nucl. Thermal Hydraulics ANS Winter Mtg.*, San Francisco, pp. 111–120.

59. Williams, D. C. (1991). *DHEAT2: A Computer Code for Adiabatic Analyses of Hydrogen Burns, Direct Containment Heating, and Metal Oxidation Events.* Internal Memorandum, Sandia National Laboratories, Albuquerque, NM.

60. Pilch, M. (1992). Adiabatic equilibrium models for direct containment heating. In *Proc. 19th Water Reactor Safety Information Mtg.*, NUREG/CP-0119.

61. Pilch, M. (1992). Kinetic limitations to adiabatic equilibrium models for direct containment heating. In *ANS Proc. Natl. Heat Transfer Conf.*, San Diego.

62. Ginsberg, T., and Tutu, N. K. (1988). Progress in understanding direct containment heating phenomena in pressurized light water reactors. Presented at *Third Int. Topical Mtg. on Nuclear Power Plant Thermal Hydraulics and Operations*, Seoul, Korea.

63. Nourbakash, H. P., Perez, S. E., and Lehner, J. R. (1993). *Effectiveness of Containment Sprays in Containment Management.* NUREG/CR-5982, BNL-NUREG-52354, Brookhaven National Laboratory, Upton, NY.

64. Schneider, R. E., and Sherry, R. R. (1993). Application of a two-cell adiabatic model for direct containment heating to the ABB CE System 80+ ALWR. In *Trans. Am. Nucl. Soc.* (14–18 November, San Francisco), Vol. 69, pp. 509–510.

65. Sienicki, J. J., and Spencer, B. W. (1987). The PARSEC computer code for analysis of direct containment heating by dispersed debris. In *Heat Transfer, Pittsburgh 1987* (R. W. Lyczkowski, ed.). *AIChE Symp. Ser.* **83**, 355.

66. Marx, K. D. (1989). A computer model for the transport and chemical reaction of debris in direct containment heating experiments. *Nucl. Sci. Eng.* **102**, 391–407.

67. Tarbell, W. W., Brockmann, J. E., Washington, K. E., Pilch, M., and Marx, K. D. (1988). Direct containment heating and aerosol generation during high pressure melt ejection experiments, SAND88-1504C. In *Proc. ANS/ENS Winter Mtg.* (T. M. Overson, ed.). ANS, Washington, DC.

68. Sweet, D. W., Washington, K. E., and Pilch, M. (1991). *Further Development of the KIVA-DCH Code for the Analyses of the Transport and Chemical Reaction of Molten Debris on Direct Containment Heating Experiments.* SAND90-2535, Sandia National Laboratories, Albuquerque, NM.

69. Corradini, M. L., Huhtiniemi, I., Kim, M. H., Moses, G. A., and Pong, L. C. (1986). *Application of a Direct Heating Model to Sandia Surtsey Tests.* UWRSR-50, University of Wisconsin, Madison, WI.

70. Ginsberg, T., and Tutu, N. K. (1987). DCHCVIM: A Direct heating containment vessel interactions module. *AIChE Symp. Ser.* **83**, 347–354.

71. Darwish, S. (1991). Containment response to direct containment heating in pressurized water reactors. Ph.D. Dissertation, University of Michigan, Ann Arbor.

72. Yan, H., and Theofanous, T. G. (1993). The prediction of direct containment heating. *ANS Proc. Natl. Heat Transfer Conf.*, Atlanta, GA.

73. Bergeron, K. D., Carroll, D. E., Tills, J. L., Washington, K. E., and Williams, D. C. (1986). Development and applications at the interim direct containment heating model

for the CONTAIN computer code. *Trans. 14th Water Reactor Safety Information Mtg.*, Gaithersburg, MD.

74. Gido, R. G. (1992). *Burn Modeling Improvements for the CONTAIN Code*. Change Document 96, Update C110T, Dept. 6429, Sandia National Laboratories, Albuquerque, NM.

75. Washington, K. E., and Williams, D. C. (1995). *Direct Containment Heating Models in the CONTAIN Code*. SAND94-1073, Sandia National Laboratories, Albuquerque, NM.

76. Williams, D. C., Bergeron, K. D., Carroll, D. E., Gasser, R. D., Tills, J. L., and Washington, K. E. (1987). *Containment Loads Due to Direct Containment Heating and Associated Hydrogen Behavior: Analysis and Calculations with the CONTAIN Code*. NUREG/CR-4896, SAND87-0633, Sandia National Laboratories, Albuquerque, NM.

77. Williams, D. C., and Louie, D. L. Y. (1988). CONTAIN analyses of direct containment heating events in the Surry plant. In *Proc. ANS/ENS Winter Meeting* (T. M. Overson, ed.). ANS Thermal Hydraulics Division, Washington, DC.

78. Williams, D. C., and Gregory, J. J. (1990). *Mitigation of Direct Containment Heating and Hydrogen Combustion Events in Ice Condenser Plants*. NUREG/CR-5586, SAND90-1102, Sandia National Laboratories, Albuquerque, NM.

79. Gido, R. G., Williams, D. C., and Gregory, J. J. (1991). *PWR Dry Containment Parametric Studies*, NUREG/CR-5630, SAND90-2339, Sandia National Laboratories, Albuquerque, NM.

80. Williams, D. C. (1992). *An Interpretation of the Results of Some Recent Direct Containment Heating (DCH) Experiments in the Surtsey Facility*. SAND92-0442C, Proceedings of the NURETH-5 Conf. Salt Lake City, UT.

81. Tutu, N. K., Park, C. K., Grimshaw, C. A., and Ginsberg, T. (1990). *Estimation of Containment Pressure Loading Due to Direct Containment Heating for the Zion Plant*. NUREG/CR-5282, BNL/NUREG-52181, Brookhaven National Laboratory, Upton, NY.

82. Murata, K. K., and Louie, D. L. Y. (1988). Parametric CONTAIN calculations of the containment response of the Grand Gulf Plant due to reactor pressure vessel failure at high pressure. In *Proc. Fourth Workshop on Containment Integrity*, NUREG/CP-0095, SAND88-1836. 14–17 June, Washington, DC, pp. 127–146.

83. EPRI (1990). *MAAP-3.0B-Modular Accident Analyses Program for LWR Power Plants*, Vols. 1 and 2. EPRI NP-7071-CCML, Electric Power Research Institute, Palo Alto, CA.

84. EPRI (1994). *MAAP-Modular Accident Analysis Program User's Manual*, Vols. 1–4. EPRI report prepared by Faukse & Associates, Inc., Burr Ridge, IL, for Electric Power Research Institute, Palo Alto, CA.

85. Carter, J. C., Fontana, M. H., Additon, S. L., Summitt, R. L., Henry, R. E., and Hammersley, R. J. (1990). *Prevention of Early Containment Failure Due to High-Pressure Melt Ejection and Direct Containment Heating for Advanced Light Water Reactors*. DOE/ID-10271, MICS-90032.

86. Fontana, M. H., Carter, J. C. and Additon, S. L. (1990). Bounding analysis of containment of high pressure melt ejection in advanced light water reactors. In *ANS Proc. 6th Nucl. Thermal Hydraulics Division*. 11–16 November.

87. Morris, B. W., and MacBeth, R. V. (1988). Experimental modeling and computer simulation of core debris dispersal, IAEA-SM-296/70. Presented at *Int. Symp. on Severe Accidents in Nuclear Power Plants*, Sorrento, Italy.

88. Morris, B. W., and Roberts, G. J. (1991). *User's Manual for CORDE and the CONTAIN/CORDE Interface*. AEA-TR-5033, Winfrith Technology Center, UK.

89. Morris, B. W. (1990). *DCH Calculations with CORDE Pre-Release Version 2*. AEEW-R2634, PWR/SAWG/P(90)593, Winfrith Technology Center, UK.

90. Lowenhielm, G., Espefalt, R., and Soderman, E. (1992). Follow-on activities of the

Swedish severe accident mitigation program. Presented at *8th Pacific Basin Nuclear Conf.*, 12–16 April, Taipei, Taiwan.

91. Gustavsson, V. (1992). Direct containment heating calculations for Ringhals 3. Presented at *CSARP Mtg.*, 4–8 May, Bethesda, MD.

92. Sweet, D. W., and Roberts, G. J. (1994). Analysis of conditions in a large dry PWR containment during a TMLB' accident sequence. In *Third Int. Conf. on Cont. Design and Operation*, Toronto, Ontario, Vol. 1.

93. Fruttuoso, G. A., Manfredini, A., Oriolo, F., Sandrelli, G., and Valisi, M. (1992). A model for the analysis of containment behavior during high-pressure melt ejection scenarios. In *Proc. ANS/ENS Int. Conf.*, Chicago.

94. Summers, R. M., Cole, Jr., R. K., Smith, R. C., Stuart, D. S., Thompson, S. L., Hodge, S. A., Hyman, C. R., and Sanders, R. L. (1995) "MELCOR Computer Code Manuals: Vol. 1, Primer & Users Guides, Version 1.8.3, Vol. 2, Reference Manuals, Version 1.8.3," NUREG/CR-6119, SAND93-2185, Sandia National Laboratories, Albuquerque, NM.

95. Kmetyk, L. N. (1994). *MELCOR 1.8.2 Assessment: Surry PWR TMLB' (with a DCH study).* SAND93-1899, Sandia National Laboratories, Albuquerque, NM.

96. Boyack, B. E., Corradini, M. L., Denning, R. S., Khatib-Rahbar, M., Loyalka, S. K., and Smith, P. N. (1995). *CONTAIN Independent Peer Review.* LA-12866, Los Alamos National Laboratory, Los Alamos, NM.

97. IDCOR (1983). MAAP: *Modular Accident Analysis Program User's Manual*, Vols. 1 and 2. IDCOR Technical Report 16.2. Fauske & Associates, Inc., Burr Ridge, IL.

98. Henry, R. E. (1989). Evaluation of fission product release rates during debris dispersal, PSA 1989. In *Proc. ANS/ENS Int. Topical Mtg. on Probability Reliability and Safety Assessment*, pp. 375–383.

99. Sienicki, J. J., and Spencer, B. W. (1986). Superheat effects on localized vessel breach enlargement during corium ejection. *Trans. ANS*, Tansad 52 1-658, pp. 522–524.

100. Pilch, M. M., and Tarbell, W. W. (1985). *High-Pressure Ejection of Melt from a Reactor Pressure Vessel.* NUREG/CR-4383, SAND85-0012, Sandia National Laboratories, Albuquerque, NM.

101. Sehgal, B. R., Andersson, J., Dinh, N., and Okkonen, T. (1994). Scoping experiments on vessel hole ablation during severe accidents. Presented at *Severe Accident Workshop*, SARJ-4, Japan.

102. Dinh, T. N., Bui, V. A., Nourgaliev, R. R., Okkonen, T., and Sehgal B. R. (1995). Modeling of heat and mass transfer processes during core melt discharge from a reactor pressure vessel. Presented at *Seventh Int. Mtg. on Nucl. Reactor Thermal Hydraulics*, NURETH-7, Saratoga Springs, NY.

103. Epstein, M. (1976). The growth and decay of a frozen layer in forced flow. *Int. J. Heat Mass Transfer* **19**, 1281–1288.

104. Sienicki, J. J., and Spencer, B. W. (1986). A multifluid multiphase flow and heat transfer model for the prediction of sweepout from a reactor cavity. In *Proc. 4th Miami Int. Symp. on Multi-Phase Transport and Particulate Phenomena*, Miami Beach, FL.

105. Moody, F. (1990). *Introduction to Unsteady Thermofluid Mechanics.* Wiley, New York.

106. Riemann, J., and Khan, M. (1983). Flow through a small break at the bottom of a large pipe with stratified flow. Presented at *2nd Int. Topical Mtg. on Nucl. Reactor Thermal Hydraulics*, 11–14 January, Santa Barbara, CA.

107. Ardon, K. H., and Bryce, W. M. (1990). Assessment of horizontal stratification entrainment model in RELAP5/MOD2 by comparison with separate effects experiment. *Nucl. Eng. Des.* **122**, 263–271.

108. Bird, R. B., Stewart, W. E., and Lightfoot, E. N. (1960). *Transport Phenomena.* Wiley, New York.

109. Pilch, M., and Griffith, R. O. (1992). *Gas Blowthrough and Flow Quality Correlations for*

Use in the Analysis of High-Pressure Melt Ejection (HPME) Events. SAND91-2322, Sandia National Laboratories, Albuquerque, NM.

110. Ishii, M., and Grolmes, M. (1975). Inception criteria for droplet entrainment in two-phase concurrent film flow. *AIChE J.* **21**, 308–318.

111. MacBeth, R. V., and Trenberth, R. (1987). *Experimental Modeling of Core Debris Dispersion from the Vault Under a PWR Pressure Vessel, Part 1: Preliminary Experimental Results.* AEEW-R1888, Winfrith, UK.

112. Rose, P. W. (1987). *Experimental Modeling of Core Debris Dispersion from the Vault Under a PWR Pressure Vessel, Part 2: Results of Including the Instrument Tubes Support Structure in the Experiment.* AEEW-R2143, Winfrith, UK.

113. MacBeth, R. W., Rose, P. W., and Mogford, D. J. (1988). *Experimental Modeling of Core Debris Dispersion from The Vault Under a PWR Pressure Vessel, Part 3: Results of Varying the Size Scaling Factor of the Model Used.* AEEW-R2426, PWR/SAWG/P(88)489, Winfrith, UK.

114. Hall, J. A. (1989). *High-Pressure Melt Ejection Experiments: Results and Carryover Measurements at 1/25th Scale.* TD/B/FPR/040/M89, PWR/SAWG/P(89)529, Berkeley Nuclear Laboratories, UK.

115. MacBeth, R. V., and Rose, P. W. (1987). *Report on Simulated Debris Dispersion Experiments Using Scale Models of the Vault Region under Ringhals 2 and Ringhals 3 and 4 PWR.* Swedish State Power Board, Vattenfall, Vallingby, Sweden.

116. Nichols, R. T., and Tarbell, W. W. (1988). Low-pressure debris dispersal from scaled reactor cavities. In *Proc. ANS/ENS Int. Mtg.* ANS Thermalhydraulics Division, Washington, DC.

117. Pilch, M., Nichols, R., and Oliver, M. (1989). Core debris dispersal and the low-pressure cutoff to direct containment heating. Presented at *Severe Accident Research Program Partners Mtg.*, U.S. Nuclear Regulatory Commission, Washington, DC.

118. Tutu, N. K., Ginsberg, T., Finfrock, C., Klages, J., and Schwarz, C. E. (1988a). *Debris Dispersal from Reactor Cavities during High-Pressure Melt Ejection Accident Scenarios.* NUREG/CR-5146, BNL-NUREG-52147, Brookhaven National Laboratory, Upton, NY.

119. Tutu, N. K., Ginsberg, T., and Finfrock, C. (1988). Low-pressure cutoff for melt dispersal from reactor cavities. In *Proc. Fourth Nucl. Thermal Hydraulics Division*, 30 October-4 November, Washington, DC, pp. 29–37.

120. Tutu, N. K., and Ginsberg, T. (1990). *A Letter Report on the Results of Melt Dispersal Experiments With the Surry and Zion Cavity Models.* Submitted to U.S. Nuclear Regulatory Commission, Brookhaven National Laboratory, Upton, NY.

121. Tutu, N. K., Ginsberg, T., Finfrock, C., Klages, J., and Schwarz, C. E. (1990a). *Melt Dispersal Characteristics of the Watts Bar Cavity.* Technical Report A-3024, Brookhaven National Laboratory, Upton, NY.

122. Chun, M. H., So, D. S., and Lee, C. S. (1991). A parametric study of high-pressure melt ejection from two different scale reactor cavity models. *Int. Comm. Heat Mass Transfer* **18**, 619–628.

123. Kim, M., Chung, C., Kim, H., and Kim, S. (1992). Experimental study on direct containment heating phenomena. Presented at *ANS Winter Mtg.*, Chicago.

124. Ishii, M., Revankar, S. T., Zhang, G., Wu, Q., and O'Brien, P. (1993). *Separate Effects Experiments on Phenomena of Direct Containment Heating: Air–Water Simulation Experiments in Zion Geometry.* PU NE-93/1, Purdue University, West Lafayette, IN.

125. Wu. Q. (1995). Transient two-phase flow and application to severe nuclear reactor accident. Ph.D. Dissertation, Purdue University, Lafayette, IN.

126. Allen, M. D., Nichols, R. T., and Pilch, M. (1990). *A Demonstration Experiment of Steam-Driven, High-Pressure Melt Ejection: The HIPS-10S Test.* NUREG/CR-5373, SAND89-1135, Sandia National Laboratories, Albuquerque, NM.

127. Kim, S. B., Yoo, K. J., Lee, H. Y., and Kim, M. H. (1994). Improvement of the reactor cavity design for mitigation of direct containment heating. Presented at *SARJ-94 Workshop on Severe Accident Research in Japan*, 31 October–1 November, Tokyo.

128. Levy, S. (1991). Debris dispersal from reactor cavity during low temperature simulant tests of direct containment heating. Part I: Tests with constant gas flow rates, Part II: Tests with blowdown gas conditions. In *ANS Proc. Natl. Heat Transfer Conf.*, Minneapolis, MN.

129. Kataoka, I., and Ishii, M. (1982). *Mechanism and Correlation of Droplet Entrainment and Deposition in Annular Two-Phase Flow*. NUREG/CR-2885, ANL-82-44, Argonne National Laboratories, Argonne, IL.

130. Ricou, F. B., and Spalding, D. B. (1961). Measurements of entrainment by axisymmetric turbulent jets. *J. Fluid Mech.* **11**, 21–32.

131. Morris, B. W., Smith, P. N., and Forbes, S. W. (1989). *A Study of Debris Dispersal and Related Phenomena at Relatively Low Vessel Pressures*. PWR/SAWG/P(89)508, Winfrith, UK.

132. Williams, D. C., and Griffith, R. O. (1996). *Assessment of Cavity Dispersal Correlations for Possible Implementation in the CONTAIN Code*. SAND94-0015, Sandia National Laboratories, Albuquerque, NM.

133. Whalley, P. B., and Hewitt, G. F. (1978). *The Correlation of Liquid Entrainment Fraction and Entrainment Rate in Annular Two-Phase Flow*. AERE-R9187, HTFS RS237, AERE Harwell, England.

134. Brockmann, J. E., Beck, D. F., Berman, M., Bradley, D. R., Copus, E. R., Nelson, L. S., Pilch, M., Power, D. A., Stamps, D. W., and Tarbell, W. W. (1989). *Conceptual Reactor Cavity Design for Accident Mitigation*. SAND88-1192, Sandia National Laboratories, Albuquerque, NM.

135. Tutu, N. K., Ginsberg, T., and Klages, J. R. (1991). Nuclear Reactor Melt Retention Structure to Mitigate Direct Containment Heating. United States Patent Number 5,049,352.

136. Frid, W. (1988). *Behavior of a Corium Jet in High Pressure Melt Ejection from a Reactor Pressure Vessel*. NUREG/CR-4508, SAND85-1726, Sandia National Laboratories, Albuquerque, NM.

137. Baker, L., Jr., Pilch, M., and Tarbell, W. W. (1988). Droplet structure interactions in direct containment heating. In *Proc. ANS/ENS Int. Mtg.*, ANS Thermalhydraulics Division, Washington, DC.

138. Wu. Q., and Ishii, M. (1994). Investigation of liquid film flow in a reactor cavity for DCH separate effect experiment. In *Tenth Proc. ANS Winter Meeting on Nuclear Thermal Hydraulics*, Washington, DC.

139. Ishii, M., and Mishima, K. (1989). Droplet entrainment correlation in annular two-phase flow. *Int. J. Heat Mass Transfer* **32**(10), 1835–1846.

140. Zhang, G. J., Wu, Q., Ishii, M., Revankar, S. T., and Lee, R. Y. (1994). Simulation experiment on corium dispersion in direct containment heating using air–water and air–woods metal. In *Tenth Proc. ANS Winter Mtg. on Nuclear Thermal Hydraulics*, Washington, DC.

141. Blomquist, C. A., Spencer, B. W., McUmber, L. M., Gregorash, D. A., Aeschlimann, R. W., and Banez, B. T. (1985). *Data Report for Corium/Water Thermal Interaction Test CWTI-13*. Argonne National Laboratory, Argonne, IL.

142. Blomquist, C. A., Spencer, B. W., McUmber, L. M., Gregorash, D. A., Aeschlimann, R. W., and Banez, B. T. (1986). *Data Report for Corium/Water Thermal Interaction Test CWTI-14*. Argonne National Laboratory, Argonne, IL.

143. Sienicki, J. J., and Spencer, B. W. (1986). Corium droplet size in direct containment heating. *Trans. ANS*, Tanso 53, pp. 557–558.

144. Zhang, G. J., and Ishii, M. (1995). Entrance effect on droplet entrainment in DCH. *ANS Proc. Natl. Heat Transfer Conf.*, 5–9 August, Portland, OR.

145. Pilch, M., Erdman, C. A., and Reynolds, A. B. (1981). *Acceleration-Induced Breakup of Liquid Drops*. NUREG/CR-2247, University of Virginia, Charlottesville, VA.

146. Tutu, N. K., and Ginsberg, T. (1987). *Direct Heating of Containment Atmosphere by Core Debris*. NUREG/CR-2331, BNL-NUREG-51454.

147. Pilch, M., Tarbell, W. W., Carroll, D. E., and Tills, J. L. (1986). High-pressure melt ejection and direct containment heating in ice condenser containments. In *Proc. ANS/ENS Int. Topical Meeting on Operability of Nuclear Power System in Normal and Adverse Environments*, 29 September–3 October, Albuquerque, NM.

148. Pilch, M., Tarbell, W. W., and Brockmann, J. E. (1988). *The Influence of Selected Containment Structures on Debris Dispersal and Transport Following High-Pressure Melt Ejection from the Reactor Vessel*. NUREG/CR-4914, SAND87-0940, Sandia National Laboratories, Albuquerque, NM.

149. Bertodano, M. L., and Sharon, A. (1996). DCH dispersal and entrainment experiment in scaled annular cavity. *Nucl. Eng. Des.* (accepted for publication).

150. FAI (1991). *Zion IPE Position Paper on Direct Containment Heating*. FAI/91-18, submitted to Commonwealth Edison Co., Chicago.

151. Cook, I. (1987). Contribution to *Reactor Safety Research Semiannual Report*. NUREG/CR-5039 (2 of 2), SAND87-2411 (2 of 2), Vol. 38 (J. V. Walker, ed.). Sandia National Laboratories, Albuquerque, NM, 1987.

152. Tutu, N. K., and Ginsberg, T. (1986). *Direct Heating of Containment Atmosphere by Core Debris*. NUREG/CR-2331, BNL-NUREG-51454.

153. Baker, L., Jr. (1986). Droplet heat transfer and chemical reactions during direct containment heating. In *Proc. ANS/ENS Int. Topical Mtg. on Thermal Reactor Safety*, 2–6 February, San Diego.

154. Ranz, W. E., and Marshall, W. R. (1952). Evaporation from drops. *Chem. Eng. Prog.* **48**(3), 141–146, 173–180.

155. Shepard, J. (1985). *Hydrogen-Steam Jet Flame Facility and Experiments*. NUREG/CR-3638, SAND84-0060, Sandia National Laboratories, Albuquerque, NM.

156. Zabetakis, M. G. (1956). *Research on the Combustion and Explosion Hazards of Hydrogen–Water Vapor–Air Mixtures*. AECU-3327, U.S. Atomic Energy Commission.

157. Thompson, R. T., Torok, R. C., Randall, D. S., Sullivan, J. S., Thompson, L. B., and Haugh, J. J. (1988). *Large-Scale Hydrogen Combustion Experiments, Vol. 1: Methodology and Results*. EPRI NP-3878, Electric Power Research Institute, Palo Alto, CA.

158. Shepard, J. E. (1987). *Analysis of Diffusion Flame Tests*. NUREG/CR-4534, SAND86-0419, Sandia National Laboratories, Albuquerque, NM.

159. Pilch, M. M. (1995). Hydrogen Combustion during DCH events. In *ANS Proc. Natl. Heat Transfer Conf.*, 5–9 August, Portland, OR.

160. Peterson, P. F. (1994). Scaling and analysis of mixing in large stratified volumes. *Int. J. Heat Mass Transfer* **37** (Suppl. 1), 97–106.

161. Marx, K. D. (1988). *A Model for the Transport and Chemical Reaction of Molten Debris in Direct Containment Heating Experiments*. NUREG/CR-5120, SAND88-8213, Sandia National Laboratories, Albuquerque, NM.

162. Hihara, E., and Peterson, P. F. (1995). Mixing in thermally stratified fluid volumes by buoyant jets. In *ASME/JSME Thermal Engineering Conf.*, Maui, HI, Vol. 1, pp. 155–162.

163. Kumar, R. K. (1985). Flammability limits of hydrogen–oxygen–diluent mixtures, AECL-8890. *J. Fire Sci.* **3**, 245–262.

164. Wong, C. C. (1988). *HECTR Analyses of the Nevada Test Site (NTS) Premixed Combustion Experiments*. NUREG/CR-4916, SAND87-0956, Sandia National Laboratories, Albuquerque, NM.

165. Carcassi, M., and Fineschi, F. (1987). Air–hydrogen deflagration tests at the University of Pisa. *Nucl. Eng. Des.* **104**, pp. 244–247.
166. Stamps, D. W., and Berman, M. (1991). High-temperature hydrogen combustion in reactor safety applications. *NSE* **109**, 39–48.
167. Conti, R. S., and Hertzberg, M. (1988). Thermal autoignition temperatures for hydrogen–air and methane–air mixtures. *J. Fire Sci.* **6**, 348–355.
168. Tamm. H., McFarlane, R., and Lui, D. D. S. (1985). *Effectiveness of Thermal Ignition Devices in Lean Hydrogen–Air–Steam Mixtures.* EPRI NP-2956, Electric Power Research Institute, Palo Alto, CA.
169. Tamm, H., Ungurian, M., and Kumar, R. K. (1987). *Effectiveness of Thermal Ignition Devices in Rich Hydrogen–Air–Steam Mixtures.* EPRI NP-5254, Electric Power Research Institute, Palo Alto, CA.
170. Condiff, D. W., Cho, D. H., and Chan, S. H. (1986). Heat radiation through steam in direct containment heating. *Trans. Am. Nucl. Soc.* **53**, 561–562.
171. Menguc, M. P., and Viskanta, R. (1986). Radiative transfer in a cylindrical vessel containing high temperature corium aerosols. *Nucl. Sci. Eng.* **92**, 570–583.
172. Tarbell, W. W., Pilch, M., Ross, J. W., Oliver, M. S., Gilbert, D. W., and Nichols, R. T. (1991). *Pressurized Melt Ejection into Water Pools.* NUREG/CR-3916, SAND84-1531, Sandia National Laboratories, Albuquerque, NM.
173. Henry, R. E., and Hammersley, R. J. (1991). Direct containment heating experiments in Zion-like geometry, *AIChE Symp. Ser.* **87**, 86–98.
174. Lobner, P., ed. (1989). *Nuclear Power Plant System Sourcebook.* SAIC-89/1541, Science Applications International Corp., San Diego.
175. Schmidt, R. C., and Pilch, M. M. (1994). *Assessment of the Importance of High Pressure Melt Ejection Events in BWR Plants.* Letter Report to U.S. Nuclear Regulatory Commission, Accident Evaluation Branch, Washington, DC.
176. Mast, P. K., Bradley, D. R., Brockmann, J. E., Lipinski, R. J., and Powers, D. A. (1985). *Uncertainty in Radionuclide Release Under Specific LWR Accident Conditions.* SAND84-0410, Vol. 4, App. B, Sandia National Laboratories, Albuquerque, NM.
177. Khalil, Y. F. (1991). *Hydrodynamic Modeling of Direct Containment Heating by Molten Debris.* Probabilistic Risk Assessment Safety Analysis Branch, Northeast Utilities, Hartford, CT.
178. Luangdilok, W. (1994). Scoping calculations of direct containment heating in BWR Mark I containments. In *Third Int. Conf. on Containment Design and Operation*, Vol. 1, Toronto, Ontario.
179. ABWR, General Electric Company (1994). *Advanced Boiling Water Reactor Standard Safety Analysis Report.* 23A6100 Revs. 1–4, Direct Containment Heating, Amendments 31–34.
180. Kajimoto, M., and Muramatsu, K. (1992). Analysis of direct containment heating in a BWR Mark II containment. In *ANS Proc. Natl. Heat Transfer Conf.*, San Diego, Vol. 6, pp. 386–400.
181. Tutu, N. K., Ginsberg, T., Finfrock, C., Klages, J. and Schwarz, C. E. (1986). *Trapping of Melt Droplets in Ice Condenser Channels during High Pressure Melt Ejection Accident: Preliminary Scoping Experiments and Analysis.* Tech. Report A-3024 9-16-86, Brookhaven National Laboratory, Upton, NY.
182. Gieseke, J. A., Cybulskis, P., Denning, R. S., Kuhlman, M. R., and Lee, K. W. (1983). *Radionuclide Release Under Specific LWR Accident Conditions: PWR Ice Condenser Containment Design.* Vol. IV, BMI-2104, Battelle Columbus Laboratories, Columbus, OH.

INDEX

Sources for Locating and Obtaining Documents

Due to the nature of the subject matter covered in this volume and consequently the unique type of materials in the reference sections, a guide is included for locating and obtaining documents. It is important to know that there are numerous ways of locating and obtaining the technical documents listed in the reference sections contained in this volume. This guide includes some of the most reliable sources for most of the technical documents. Although there are no guarantees that a particular document will be available from the sources listed, they should offer a good starting point for document location and retrieval efforts.

The National Technical Information Services (NTIS) is generally the best source for formal technical reports and for select conference proceedings with report number designations. Technical reports with the following designations can be ordered from NTIS: NUREG, NUREG/CR, NUREG/CP, ANL, BNL, EGG, LANL, NASA, NASA-SP, NASA-TR, ORNL, PNL, and SAND. In addition, NTIS is a good source for other technical reports, such as WCAP, CEGB, BMFT, AEEW, among others. In general, for most technical reports, start with NTIS; their holdings include very recent publications as well as extensive historical and archival materials.

Another source for these technical documents, specifically those related to the nuclear industry, is the U.S. Nuclear Regulatory Commission Public Document Room. This source can be very helpful especially with some of the U.S. national laboratory informal publications, nuclear electric utility and owners-group related materials (i.e., Westinghouse and General Electric), and older NUREG documents that were not available from NTIS. The NRC PDR is not a source for conference and journal material.

National Technical Information Service
Technology Administration
U.S. Department of Commerce
Springfield, VA 22161
(800-553-6847; 703-487-4650)

U.S. Nuclear Regulatory Commission
Public Document Room
2120 L Street NW
Washington, DC 20555
(800-397-4209; 202-634-3387)

Other sources for formal and informal technical reports and documentation are the issuing organizations themselves, such as the national laboratories, universities, companies, or publishers. Listed below are selected national laboratories and company contacts for assistance in locating information. For a complete listing of Department of Energy national laboratories, consult the following World Wide Web location: (http://www.doe.gov). The Electric Power Research Institute is the source for all EPRI documents and research project information (e.g., EPRI-RP). A number of their documents are proprietary and/or require licensing agreements, and generally are expensive to obtain. For personal communications, internal memorandum, or other obscure documents listed in the reference sections, contact the author(s) directly.

National Laboratories:

Argonne National Laboratory (ANL)
Technical Information Services Department
9700 South Cass Avenue
Argonne, IL 60439-4832
(630-252-5610)

Brookhaven National Laboratory (BNL)
Information Services Division
Building 477A
Upton, NY 11973-5000
(516-344-7860; 516-344-3484)

Idaho National Engineering Laboratory/EGG (INEL/EGG)
Technical Library
P.O. Box 1625
Idaho Falls, ID 83415
(208-526-1191)

Lawrence Berkeley Laboratory (LBL)
Information Resources Department
1 Cyclotron Road, Building 50B-4206
Berkeley, CA 94720
(510-486-6504)

Lawrence Livermore National Laboratory (LLNL)
LLNL Technical Information Department
P.O. Box 808
Livermore, CA 94550
(510-422-4636)

Los Alamos National Laboratory (LANL)
Library Services-Reports Collection
P.O. Box 1663
Los Alamos, NM 87545
(505-667-4446)

Oak Ridge National Laboratory (ORNL)
ORNL Central Research Library
P.O. Box 2008
Oak Ridge, TN 37831-0117

Pacific Northwest National Laboratory (PNNL)
Hanford Technical Library & Information Services
P.O. Box 999, K1-06
Richland, WA 99352
(509-375-2583)

Sandia National Laboratories (SAND)
Sandia Technical Library-Technical Reports
P.O. Box 5800
Albuquerque, NM 87185
(505-845-8187)

Corporate Sources:

American Nuclear Society
555 North Kensington Avenue
La Grange Park, IL 60525
(708-579-8210; fax 708-352-0499)

The Electric Power Research Institute
3412 Hillview Avenue
Palo Alto, CA 94304-1395
(415-855-2000)

General Atomics
Document Center
3550 General Atomics Court
San Diego, CA 92121-1194
(619-455-2502)

General Electric Company
Nuclear Division
175 Curtner Avenue
San Jose, CA 95125
(408-925-1000)

Westinghouse Electric Corporation
Technical Library
P.O. Box 355
Pittsburgh, PA 15230
(412-374-4200)

If unsuccessful in locating materials from the sources noted above, another way to obtain conference papers and other conference materials, journal articles, international publications, and older/esoteric documents is to contact a commercial document delivery vendor, document supply company, or private information consultant. Institutional and public libraries can identify local document vendors that specialize in technical and scientific materials.

This information has been supplied by Helen K. Todosow, Information Manager, Energy and Technology Information Resources, Brookhaven National Laboratory.

ISBN 0-12-020029-5